Lecture Notes in Physics

The Lecture Notes in Physics

The series Lecture Notes in Physics (LNP), founded in 1969, reports new developments in physics research and teaching – quickly and informally, but with a high quality and the explicit aim to summarize and communicate current knowledge in an accessible way. Books published in this series are conceived as bridging material between advanced graduate textbooks and the forefront of research and to serve three purposes:

- to be a compact and modern up-to-date source of reference on a well-defined topic

- to serve as an accessible introduction to the field to postgraduate students and nonspecialist researchers from related areas

- to be a source of advanced teaching material for specialized seminars, courses and schools

Both monographs and multi-author volumes will be considered for publication. Edited volumes should, however, consist of a very limited number of contributions only. Proceedings will not be considered for LNP.

Volumes published in LNP are disseminated both in print and in electronic formats, the electronic archive being available at springerlink.com. The series content is indexed, abstracted and referenced by many abstracting and information services, bibliographic networks, subscription agencies, library networks, and consortia.

Proposals should be sent to a member of the Editorial Board, or directly to the managing editor at Springer:

Christian Caron
Springer Heidelberg
Physics Editorial Department I
Tiergartenstrasse 17
69121 Heidelberg / Germany
christian.caron@springer.com

Francesca Bacciotti
Emma Whelan
Leonardo Testi (Eds.)

Jets from Young Stars II

Clues from High Angular Resolution Observations

 Springer

Francesca Bacciotti
Leonardo Testi
Osservatorio Astrofisico di
Arcentri
Largo E. Fermi, 5
50125 Firenze, Italy
fran@arcetri.astro.it
lt@arcetri.astro.it

Emma Whelan
Dublin Institute for Advanced
Studies
31 Fitzwilliam place
Dublin 2, Ireland
ewhelan@cp.dias.ie

F. Bacciotti, E. Whelan and L. Testi (Eds.), *Jets from Young Stars II: Clues from High Angular Resolution Observations*, Lect. Notes Phys. 742 (Springer, Berlin Heidelberg 2008), DOI 10.1007/ 978-3-540-68032-1

Marie Curie Research Training Network Contract, JETSET: Jet Simulations, Experiments and Theory
(Contract No. MRTN-CT-2004-005592)

ISBN: 978-3-540-68031-4   e-ISBN: 978-3-540-68032-1

Lecture Notes in Physics ISSN: 0075-8450
Library of Congress Control Number: 2007936167

Cover design: eStudio Calamar S.L.

Printed on acid-free paper

9 8 7 6 5 4 3 2 1

springer.com

Preface

Atomic jets and molecular outflows from Young Stellar Objects (YSOs) are among the most spectacular of astrophysical phenomena. Approaching this topic, one gets immediately trapped by the breathtaking beauty of the optical images taken from Earth or Space with modern telescopes, only to discover soon after that the physics that regulates their properties is even more intriguing. Outflows are believed to play a fundamental role in the formation process of a new star and its planets, as they could be the principal agent for the removal of the excess angular momentum from the star-disk system. Their generation mechanism, still poorly understood, involves a complex interplay of gravity, turbulence, and magnetic forces, while their propagation into the surrounding medium affects through the action of shocks and ionizing fronts the way the parent cloud evolves, and hence the future generations of young stars. A deep understanding of stellar jets is therefore fundamental to any theory of Star Formation.

The motivation of the four-year JETSET (Jet Simulations, Experiments and Theory) Marie Curie Research Training Network is to build an interdisciplinary European research community focused on the study of jets from young stars. The network's scientific goals focus on understanding (1) the driving mechanisms of jets around young stars; (2) the cooling-heating processes, instabilities, and shock structures in stellar and laboratory jets; and (3) the impact of jets on energy balance and star formation in the galactic medium. Central to these overall goals are the series of JETSET schools dedicated to the training of our young researchers in key jet topics.

The second JETSET school "Jets from Young Stars: High Angular Resoution Observations", was held at Marciana Marina, Elba Island, Italy, in September 2006. The school focused on the current techniques for observing jets, outflows, and circumstellar disks at High Angular Resolution and on the procedures used in making an initial analysis of the data. This book collects the lectures presented there. Although YSO jets can span several parsecs, many of their properties, related to important and as yet unclear aspects of their nature, are defined on distances of tens of AU or less. This is the case for example, of the cooling regions behind the internal shocks, where their emission is generated, or the zone close to the star and the accretion disk where jets are accelerated and collimated. Thus, to test models of their generation and propagation, even for the nearest star formation region, jets must be observed on sub-arcsecond scales. In the recent past new High Angular Resolution tools, such as, for example, observations using Adaptive Optics have allowed astronomers to begin to probe the smallest scales. Moreover, through the application of dedicated diagnostic techniques, the physical conditions in key regions of the jet could be determined for the first time. This has helped enormously to constrain models. The aim of this school was, therefore, to provide the attendees with a background in modern High Angular Resolution observational astrophysics, illustrating how top-quality results are obtained for the environment of YSO jets.

The structure of the book sees first a general introduction to stellar jets and their emission properties, followed by a detailed description of various observational instruments and techniques that allow the investgation of the jet phenomenon at high angular resolution. The first contribution consists of a comprehensive description of general jet properties and of the role of jets in star formation by Suzan Edwards. Pat Hartigan then summarizes the physics needed to interpret the spectra of stellar jets and gives a historical review of the techniques of spectral analysis. Simone Esposito and Enrico Pinna then describe the basic principles of Adaptive Optics (AO) and illustrate its use in existing telescopes. The realm of Infra-red observations and diagnostic analyses of stellar jets is then the topic of the contribution by Brunella Nisini, who also mentions a few observations obtained with AO. The latter are then discussed in the contribution by Catherine Dougados, who presents the full potentiality of ground observations with AO, in both the infrared and Optical wavelength ranges. Next, one finds a description of the technique of spectro-astrometry by Emma Whelan and Paulo Garcia. This novel technique allows one to obtain very accurate positional information, down to a few milliarcseconds, from seeing-limited spectral data. The domain of applicability of spectroastrometry, its limits, and a few recent results are described in this contribution. The topic of observations from Space that in the past few years have led to impressive advancements in the description of jet phenomena is covered in the contribution by Francesca Bacciotti, Mark McCaughrean and Tom Ray. They describe the current facilities and the most recent results obtained, as well as the opportunities that will be offered by the next generation of space observatories.

Salvatore Orlando and Fabio Favata then illustrate the realm of X-ray observations of stellar jets. The new space intruments working in this wavelength range provide a much higher angular resolution than previous attempts, thus allowing us to compare the X-ray emission with optical features and to identify the source of high energy radiation. The next four lectures deal with Interferometry, that, to date, sets the ultimate frontier for our abililty to reach small scales in observations of jets and disks. First, Neal Jackson reviews the basic principles of this complex technique, making it simple and plain with plenty of practical examples and explanations. Second, Henrik Beuther describes the opportunity offered by interferometry in the millimeter wavelength range and illustrates the results obtained for wide molecular outflows and collimated molecular jets. The two contributions by Fabien Malbet, Eric Tatulli, and G. Duvert then deal with the new instrumentation offered to conduct interferometric studies in the Infra-red, and in particular illustrate the Very Large Telescope Interferometer facility, together with first results and software packages. Finally, in the last lecture of this book, Leonardo Testi shows how the application of several different high angular resolution techniques to the study of protoplanetary disks allow one to constrain their structure and to derive the physical properties of their dust component.

The editors would like to thank all the lecturers for their excellent presentations, including Antonella Natta, as well as Felipe Pires and Malcolm Walmsley, whose lectures covered aspects of skill development in scientific research that are not treated in this book. We are also thankful to all school participants who created a very enjoyable atmosphere on the Island. We would like to especially acknowledge the huge amount of work done by Emanuela Masini, who coordinated the efforts of the local organising committe, and brilliantly solved all the organizational difficulties. We also acknowledge Fabrizio Massi, Andrea Fornari, and Deirdre Coffey, who helped us on different practical aspects of this school. Many thanks also to Eileen Flood, who helped Emanuela at the registration desk all over the week. Finally, our hearthful thanks go to the Heavens, for having arranged a spectacular Moon eclipse right during the conference dinner.

<table>
<tr><td>INAF-Osservatorio Astrofisico di Arcetri</td><td align="right">Francesca Bacciotti</td></tr>
<tr><td>Dublin Institute for Advanced Studies</td><td align="right">Emma Whelan</td></tr>
<tr><td>NAF-Osservatorio Astrofisico di Arcetri</td><td align="right">Leonardo Testi</td></tr>
<tr><td></td><td align="right">February, 2007</td></tr>
</table>

Contents

Part III Interferometry: Technique and Applications

Principles of Interferometry

(Sub)mm Interferometry Applications in Star Formation Research

Observing Young Stellar Objects with Very Large Telescope Interferometer

Presentation of AMBER/VLTI Data Reduction

High Angular Resolution Observations of Disks

Part I

Jet Emission

Stellar Jets: Clues to the Process of Star and Planet Formation

S. Edwards

Smith College, Northampton, MA 01063 USA
sedwards@smith.edu

Abstract. An overview of the evidence for an accretion/outflow connection in young stars is presented. Although it is likely that accretion-powered winds play an important role in the angular momentum evolution of forming stars and proto-planetary disks, how this happens is not evident.

Observations of the innermost regions of jets and winds are examined in the light of expectations from current models for MHD wind launching.

The evidence suggests that accretion powered winds are likely launched both radially from the star and via a magneto-rotational disk wind.

1 Introduction

The focus of the JETSET series is on the past, present, and future study of jets from young stars. We are on the threshold of implementing new technologies that will enable significant improvements in angular resolution that promise

S. Edwards: *Stellar Jets*, Lect. Notes Phys. **742**, 3–13 (2008)
DOI 10.1007/978-3-540-68032-1_1

to revolutionize the study of these jets, sharpening our determination of their collimation, kinematics, and excitation and improving our understanding of their origin. It is the purpose of this second JETSET school to attune you to the possibilities that arise from observing stellar jets with high angular resolution. As the introductory speaker, my task is to set the stage for this topic and to remind you of why we study jets.

Collimated outflow in jets and the presence of accretion disks appear to be inseparable phenomena. Although we are not yet certain where or how jets originate in accretion disk systems, we have reason to believe they may play a role in mass and angular momentum transport in the disk [23], in stellar spin-down [24], in heating the disk atmosphere [25], in hastening disk dissipation [31], and in disrupting infalling material from the collapsing molecular cloud core [43]. The study of stellar jets, thus, has ramifications for the entire field of star and planet formation!

In my talk on the beautiful island of Elba, I highlighted some pivotal moments in the study of jets since the initial discovery of their bright bow shocks in the mid-twentieth century by G. Herbig and G. Haro. These included (1) the recognition in 1975 that the emission line ratios of Herbig–Haro nebulae required formation in shocks, postulated by R. Schwartz to derive from strong stellar winds of young T Tauri stars [40]; (2) the discovery in the early 1980s that mass ejection from young stars has a bipolar morphology, manifested by both shock-excited nebulae propagating outward from the star [17, 29] and extended lobes of expanding molecular gas [42]; and [3] the concurrent recognition that HH nebulae are the working surfaces of shocks arising in collimated jets of outflowing material where kinetic energy is thermalized and shock velocities are much smaller than actual space velocities [36]. Since there are numerous reviews of jets in the literature [2, 35, 36], as well as in other contributions to the JETSET school, I will skip the rest of the historical background and the overview of the most recent multiwavelength observations of jets and concentrate here on issues directly related to the accretion/outflow connection and jet launching mechanisms.

2 Probing the Accretion/Outflow Connection

Periodically, it is healthy to revisit the assumptions behind the widespread acceptance of a symbiotic relation between accretion and outflow and to consider ways of improving the precision in determining the efficiency of the ejection process, defined by the ratio of the rate of mass ejection to the rate of disk accretion: $\dot{M}_{\text{eject}}/\dot{M}_{\text{acc}}$. While observations of the full mass range of young stars suggest correlations between outflow and accretion, the underpinnings of this interpretation are provided by the young low mass T Tauri stars (TTS) with small enough extinctions to be optically visible Class II sources. An early suggestion of a connection between outflow and accretion for TTS was a correlation between the luminosity of blueshifted forbidden lines and the excess infrared luminosity [9]. Today the blueshifted forbidden lines are still considered a prime outflow diagnostic, understood to arise in spatially extended bipolar microjets [18, 27, 30]. However, we now know that the infrared excess merely indicates the presence of an optically thick disk but does not directly elucidate its accretion status, since passive reprocessing of stellar radiation by the disk, which may have sizeable flaring or other geometric effects including infalling "envelopes," will also contribute significantly to the infrared excess luminosity [12, 13].

The conclusive demonstration of an accretion–outflow relation for TTS was a correlation between forbidden line luminosities with *accretion* luminosities derived from the optical/ultraviolet emission in excess of photospheric radiation [27]. This emission, known as "veiling," is likely generated in accretion shocks as material is delivered from the disk to the star, presumably via magnetic channelling [5, 6, 26]. Additional evidence that the presence of disk accretion is always accompanied by collimated outflows is furthered by the "weak" TTS, which lack both detectable veiling and forbidden line emission, demonstrating that when there is no accretion of material onto the stellar surface, there is also no outflow.

The evidence for an accretion/outflow connection in the more embedded Class I and Class 0 sources, which cover a wider range of evolutionary states and masses than the TTS, is less firmly grounded. Collimated outflows from these sources are quite dramatic, with spatial extents large enough to resolve considerable structure in high velocity jets in both atomic and molecular gas and to clarify that they are surrounded by slower moving lobes of swept-up molecular gas [7]. Outflow diagnostics such as CO momentum fluxes or H_2 line luminosities provide robust tracers of the outflow kinematics and energetics, which in turn are well correlated with the source bolometric luminosity spanning up to 5 orders of magnitude [14, 38]. While these correlations suggest that the accretion/outflow relation seen in the TTS is also operating here, the bolometric luminosity does not differentiate between photospheric and disk/envelope contributions. Thus, neither can the claim that an accretion disk is present usually be directly verified nor can reliable assessments of disk accretion rates be made. Two directions are being pursued to extend accretion diagnostics beyond the TTS range. One relies on estimating accretion rates from the luminosities of near infrared lines thought to arise in magnetospheric accretion flows, which have been calibrated against accretion rates for TTS [13] and the other is an acquiring sufficiently deep optical observations of embedded sources for veiling excesses to be determined in the traditional way [45]. For the majority of Class I and Class 0 sources, however, we rely on correlations of outflow diagnostics with bolometric luminosity, which only provide indirect evidence that collimated outflows are powered by accretion over the full range of mass and evolutionary status of young stellar objects.

A crucial diagnostic at the heart of the accretion/outflow connection is the ratio $\dot{M}_{\mathrm{eject}}/\dot{M}_{\mathrm{acc}}$ [23]. Even for well-studied TTS in a restricted range of mass and evolutionary state, the ejection/accretion ratio is not well constrained.

When the luminosity from accretion onto the star can be separated from other contributors to the spectral energy distribution, disk accretion rates are probably more reliably known than mass outflow rates. This is inferred from the excellent match between accretion shock models and the observed veiling shortward of 0.5 μ [11], although even then the accretion luminosities are likely underestimated [20, 45]. Uncertainties in mass ejection rates may be as much as an order of magnitude, hampered by difficulties in assessing ionization fractions and filling factors in the shocked gas traced by forbidden lines. New approaches for diagnosing these quantities in jets promise to improve the precision of mass ejection rates, as discussed elsewhere in this volume (articles by Hartigan and Nisini).

Another important consideration in determining the ratio of $\dot{M}_{\mathrm{eject}}/\dot{M}_{\mathrm{acc}}$ is to use *simultaneous* diagnostics of outflow and accretion. The best estimate to date is probably from Hartigan et al. [27], with mass accretion rates revised upward following Gullbring et al. [26], yielding $\dot{M}_{\mathrm{eject}}/\dot{M}_{\mathrm{acc}}$ of 0.1–0.01 (with sizeable scatter) for accreting TTS. Similar techniques have been applied to Taurus Class I sources by White and Hillenbrand who instead found a ratio ~0.9! Since the masses, ages, and mass accretion rates of their sample were similar to those of Class II sources, they suspect a bias in deriving forbidden line luminosities from highly extinct edge-on sources that leads to uncomfortably large estimates of ejection efficiency ratios [1, 45]. This problem requires much more attention in order to improve the precision of $\dot{M}_{\mathrm{eject}}/\dot{M}_{\mathrm{acc}}$, so that we can assess whether it is the same for all systems, or whether its value depends on the evolutionary state of the disk, mass of the star, or other factors.

3 Assessing Wind Launch Mechanisms

Accretion powered outflows are thought to be launched via magnetohydrodynamic processes that depend on a large-scale open magnetic field anchored to a rotating object with collimation provided by the toroidal component focusing the flow toward the rotation axis ("hoop stress") [23]. Three basic steady state MHD ejection scenarios are under consideration, all of which tap energy from a combination of accretion, rotation, and magnetic fields [24]: disk winds emanating over an extended range of disk radii [32], disk winds restricted to a narrow "x" region near the disk truncation radius [41], and stellar winds channeled along field lines emerging radially from the star [37, 39].

3.1 Jet Structure Inside 100 AU

Observational tests that can discriminate among these models require information on the smallest possible spatial scales in order to probe the structure, excitation and kinematics of the jet close to its source. Current capabilities, using HST or AO on 8-m ground-based telescopes provide spatial resolution on the order of 10s of AU for the nearest star formation regions. Combining this capability with techniques such as [18, 44] or reconstructing velocity channel maps from multiple long slit spectra [3] is yielding new information on the collimation and kinematics of jets within 100 AU of the exciting star (see also [28, 33, 34]). Some important new insights include (also see chapter by Dougados) the following.

The onset of collimation: A progressive increase in jet width with distance from the exciting star is inferred in two TTS microjets, where at distances from 15 to 50 AU the FWHM increases from 10 to 30 AU, corresponding to an opening angle of ~20° [18]. At somewhat larger distances (50–200 AU), jet widths in a handful of stars appear to be stable, maintaining FWHM ~30 AU, corresponding to opening angles < 5° [35]. While the data is still sparse, it suggests that the act of collimating a wider flow into a jet occurs around 50 AU.

Structure in the Poloidal Velocity Field: The fastest moving gas in the jet is narrowly confined to the jet axis and is sheathed by slower moving, less extended emission [3]. Although such behavior is compatible with a single extended disk wind where launch velocities decrease with increasing launch radii along the disk, in some objects the two flows appear kinematically distinct when scrutinized with spectroastrometric techniques applied to long slit spectra with high velocity resolution, suggesting there may be two flows with separate origins [38, 39].

Structure in the Toroidal Velocity Field: Velocity shifts perpendicular to the jet axis have been recorded in a few jets within the first few hundred AU of the exciting source [3, 8]. The shifts can be interpreted as rotation in the azimuthal velocity field of the outer, slower moving jet gas [3, 16, 46], which would imply origin in a disk wind carrying at least 2/3 of the disk angular momentum with footpoints in the disk originating from radial extents of 0.5 to 5 AU [46]. However, in at least one case, a resolved TTS disk is rotating in the opposite sense of the inferred jet rotation [9], suggesting that alternate interpretations of the transverse velocity shifts, such as jet precession or axisymmetric instabilities [15], need to be given close consideration.

The above results based on high angular resolution data are certainly tantalizing but must be considered preliminary. Major progress can be made simply by significantly increasing the number of observed microjets with current techniques, thus clarifying the robustness of the phenomena. A complication that needs to be accounted for in interpreting these high angular resolution observations is that jet-forbidden emission is typically confined to shocks in multiple working surfaces that appear as knots with large proper motions. Timescales for the appearance of new knots are only a few years, which can complicate comparison of data taken at different epochs [18, 38]. In the coming decade, the advent of improved angular and velocity resolution with 8-m interferometers and the James Webb Space Telescope (see chapter by Bacciotti, McCaughren and Ray) are likely to resolve the basic questions of where wind collimation into a jet occurs, whether jets include two distinct poloidal flows, and whether their rotation indicates they are a primary source of angular momentum loss in the disk.

3.2 The Inner 0.1 AU

Despite the promise of significant gains in angular resolution from 8-m interferometers and JWST, the crucial region inside 0.1 AU, where the highest velocity winds are launched, will remain out of reach except via deciphering line profiles taken at high spectral resolution. This region, including the very inner disk, the star-disk interface, and the star itself, is heavily influenced by the stellar magnetosphere, which truncates the inner disk and controls the accretion flow onto the star. It is probed via permitted atomic emission lines from the infrared through to the ultraviolet that span a wide range of ionization states. The line profiles, typically with strong and broad emission sometimes accompanied by redshifted absorption below the continuum and/or narrow centered emission components, are usually attributed to formations in magnetospheric funnel flows and accretion shocks on the stellar surface [6, 22], although evidence is growing that there may be a significant contribution from a wind to the broad emission of hydrogen and helium [1, 4, 44]. Definitive evidence for inner winds comes from blueshifted absorption, which until recently was only seen superposed on the broad emission lines of Hα, Na D, Ca (II) H&K, and Mg (II) H&K. While these blueshifted features have been known for decades to signify the presence of a high velocity wind close to the star, they yield little information on the nature or location of the wind-launching region [10].

A surprising new diagnostic of the inner wind region in accreting stars is He I λ10830, which can show remarkable P Cygni profiles, with blueshifted absorption deeply penetrating the continuum, in both accreting TTS and Class I sources [19, 21]. The extraordinary potential for He I λ10830 to appear in absorption derives from the high opacity of its metastable lower level ($2s^3S$), $\sim$21 eV above the singlet ground state, which becomes significantly populated relative to other excited levels owing to its weak de-excitation rate via collisions to singlet states. In a recent survey of classical T Tauri stars, He I λ10830 profiles show subcontinuum blueshifted absorption in $\sim$70% of the stars, in striking contrast to Hα where only $\sim$10% of accreting TTS have blue absorption penetrating the continuum [20]. There is no question that the inner wind illuminated by He I λ10830 derives from accretion, since it is not seen in the weak, non-accreting TTS, and the strength of the combined absorption and emission correlates with the veiling at 1 μ. There is, however, quite a diversity in the morphology of the He I λ10830 blue absorption feature among a sample of accreting TTS that span a wide range of disk accretion rates. Blue absorptions can range from remarkable breadth and depth, where 90% of the 1 μ continuum is absorbed over a velocity interval of 300–400 km/s, to narrow absorption with only modest blueshifts, as illustrated in Fig. 1.

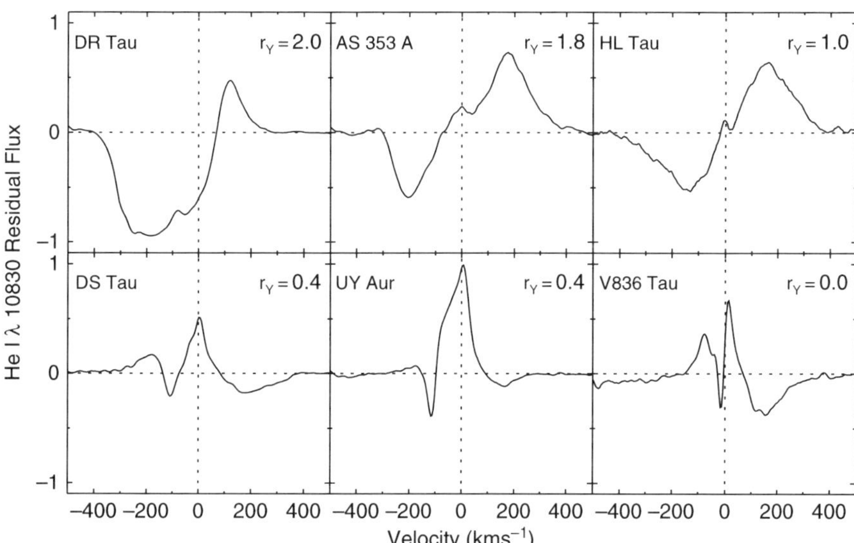

Fig. 1. Illustration of the extremes of blueshifted absorption shown in He I λ10830 profiles of accreting TTS. The upper row shows P Cygni-like profiles with deep and broad blue absorption and the lower row shows examples of narrow blue absorption [20]. The latter profiles also show red absorption from magnetospheric accretion columns. Simultaneous 1 μ veiling, r_Y, is identified for each star

The He I λ10830 profiles offer a unique probe of the geometry of the inner wind. The profile morphology is sensitive to outflow geometry because it is formed under conditions resembling resonance scattering, with only one exit allowed from the upper level and a metastable lower level. Thus it will form an absorption feature via simple scattering of the 1 μ continuum under most conditions. Additionally, helium lines are restricted to form in a region of either high excitation or close proximity to a source of ionizing radiation, which is likely to be within the crucial 0.1 AU of the star where the inner wind is launched.

A recent study takes advantage of these sensitivities to model helium profile formation for two inner wind geometries, a disk wind and a wind emerging radially from the star, using Monte Carlo scattering calculations [21]. Profiles were computed under the assumptions of pure scattering and also with the additional presence of in situ emission. Comparison with the observed He I λ10830 profiles of accreting TTS suggests the subcontinuum blue *absorption* is characteristic of inner disk winds in ∼30% of the stars and of stellar winds in ∼40%. The He I λ10830 *emission* components, however, seem to arise only via in situ emission in stellar winds and/or scattering in stellar winds and magnetospheric accretion funnels, suggesting that the prevalence of stellar winds may be higher than indicated by the morphology of the blue absorption feature. Remarkably, blue absorption suggesting strong stellar winds is found more frequently among stars with the highest disk accretion rates and well-developed microjets, while blue absorption suggesting inner disk winds is more frequently found in stars with lower accretion rates. Furthermore, simultaneous redshifted absorption at He I $\lambda10830$ from magnetospheric accretion flows is rarely seen among the stars with stellar wind profiles but is common among those with lower disk accretion rates and profiles suggesting disk winds (see Fig. 1). A possible explanation for this behavior is a readjustment of the interaction between the stellar magnetosphere and the inner disk depending on the disk accretion rate. Whatever the cause, it appears that conditions for driving significant winds radially outward from the star are enhanced when disk accretion rates are high. What is still unknown is how much this accretion powered stellar wind contributes to the overall mass ejection flux emerging in the jet or to angular momentum loss from the accreting star.

3.3 Now What?

At present, all three of the basic steady-state wind ejection mechanisms (extended disk wind, x-wind, stellar wind) are still viable and evidence is mounting that more than one mass ejection process may contribute to the overall poloidal velocity field in the jet. The actual situation is likely more complicated, as the interaction of a rotating stellar magnetosphere with a disk magnetosphere will lead to non-steady state phenomena, where transient magnetic reconnection events may drive mass loss similar to that found from, for example, coronal mass ejections [24]. Despite the apparent complexity of the problem, it is clear that the basic accretion/ejection process is very robust, as attested to by the plethora of extended collimated jets seen via H_2 in the $4.5\,\mu$ IRAC band with the Spitzer telescope in such very young star formation regions as NGC 1333 currently being investigated by R. Gutermuth and J. Walawender. At this point observational constraints are lagging behind theoretical ideas, but it is likely that well-conceived observational programs carried out with improved high angular resolution cababilities will finally solve the mystery of how jets derive from accretion disk systems.

4 The Future

In the spirit of a talk for young astronomers who will one day assume leadership in the study of jets, let me leave you with the following questions which I hope you will have fun finding answers to!

- How/where are jets launched in accretion disk systems? Is there more than one source of mass outflow? Does $\dot{M}_{\mathrm{eject}}/\dot{M}_{\mathrm{acc}}$ vary with the accretion rate of the disk, with the evolutionary state of the protostar or with the mass of the protostar?

- How much angular momentum do jets carry? Is this angular momentum being removed from the disk, the star, or both?
- Do MHD disk winds enable disk accretion or are they a by-product? What is their effect on the evolution of the accretion disk and the process of planet formation?
- What transpires in an accretion disk system to produce the well-documented semi-periodic discontinuities in mass ejection events on time-scales ranging from years to millenia?
- How do close binaries affect jet formation and the evolution of accretion disks?

References

1. Appenzeller, I., Bertout, C., Stahl, O.: Edge-on T Tauri stars. Astron. Astrophys. **434**, 105 (2005)
2. Bally, J., Reipurth, B., Davis, C.J.: Observations of jets and outlows from young stars. In: Reipurth, B., Jewitt, D., Keil, K. (eds.) PPV. University of Arizona Press, Tucson (2006), p. 215–230
3. Bacciotti, F., Ray, T.P., Mundt, R., Eislöffel, J., Solf, J.: Hubble space telescope/STIS spectroscopy of the optical outflow from DG Tauri: Indications for rotation in the initial jet channel. Astrophys. J. **576**, 222 (2002)
4. Beristain, G., Edwards, S., Kwan, J.: Helium emission from classical T Tauri stars: Dual origin in magnetospheric infall and hot wind. Astrophys. J. **551**, 1037 (2001)
5. Bertout, C.: T Tauri stars – Wild as dust. Astron. Astrophys. **27**, 351 (1989)
6. Bouvier, J., Alencar, S.H.P., Harries, T.J., Johns-Krull, C.M., Romanova, M.M. Magnetospheric Accretion in Classical T Tauri Stars In: Reipurth, B., Jewitt, D., Keil, K. (eds.) Protostars and Planets V, p. 479. University of Arizona Press, Tucson (2006)
7. Cabrit, S.: Constraints on accretion–ejection structures in young stars. In: Bouvier, J., Zahn, J.-P. (eds.) Star Formation and the Physics of Young Stars, p. 147. EDP Sciences, Les Ulis (2002)
8. Cabrit, S., Edwards, S., Strom, S.E., Strom, K.: Forbidden-line emission and infrared excesses in T Tauri stars – Evidence for accretion-driven mass loss? Astrophys. J. **354**, 687 (1990)
9. Cabrit, S., Pety, J., Pesenti, N., Dougados, C.: Tidal stripping and disk kinematics in the RW Aurigae system. Astron. Astrophys. **452**, 897 (2006)
10. Calvet, N.: Properties of the winds of T Tauri stars. In: Reipurth, B., Bertout, C. (eds.) IAU Symposium 182, Herbig–Haro Flows and the Birth of Low Mass Stars, p. 417. Kluwer, Dordrecht (1997)
11. Calvet, N., Gullbring, E.: The structure and emission of the accretion shock in T Tauri stars. Astrophys. J. **509**, 802 (1998)
12. Calvet, N., Hartmann, L., Strom, S.E. Evolution of Disk Accretion In: Mannings, V., Boss, A.P., Russell, S.S. (eds.) Protostars and Planets IV, p. 377. University of Arizona Press, Tucson (2000)
13. Calvet, N., Muzerolle, J., Briceno, C., Hernandez, J., Hartmann, L., Saucedo, J., Gordon, K.: The mass accretion rates of intermediate-mass T Tauri stars. Astrophys. J. **128**, 1294 (2004)

14. Caratti o Garatti, A., Giannini, T., Nisini, B., Lorenzetti, D.: H2 active jets in the near IR as a probe of protostellar evolution. Astron. Astrophys. **449**, 1077 (2006)
15. Cerqueira, A.H., Velazquez, P.F., Raga, A.C., Vasconcelos, M.J., de Colle, F.: Emission lines from rotating proto-stellar jets with variable velocity profiles. I. Three-dimensional numerical simulation of the nonmagnetic case. Astron. Astrophys. **448**, 231 (2006)
16. Coffey, D., Bacciotti, F., Woitas, J., Ray, T.P., Eislöffel, J.: Rotation of jets from young stars: New clues from the hubble space telescope imaging spectrograph. Astrophys. J. **604**, 758 (2004)
17. Cudworth, K., Herbig, G.H.: Two large-proper-motion Herbig–Haro objects. Astrophys. J. **84**, 548 (1979)
18. Dougados, C., Cabrit, S., Lavalley-Fouquet, C., Ménard, F.: T Tauri stars microjets resolved by adaptive optics. Astron. Astrophys. **357**, L61 (2000)
19. Dupree, A.K., Brickhouse, N.S., Smith, G.H., Strader, J.: A hot wind from the classical T Tauri stars: TW Hydrae and T Tauri. Astrophys. J. **625**, L131 (2005)
20. Edwards, S., Fischer, W., Hillenbrand, L., Kwan, J.: Probing T Tauri accretion and outflow with 1 micron spectroscopy. Astrophys. J. **646**, 319 (2006)
21. Edwards, S., Fischer, W., Kwan, J., Hillenbrand, L., Dupree, A.K.: He I lambda10830 as a probe of winds in accreting young stars. Astrophys. J. **599**, L41 (2003)
22. Edwards, S., Hartigan, P., Ghandour, L., Andrulis, C.: Spectroscopic evidence for magnetospheric accretion in classical T Tauri stars. Astrophys. J. **108**, 1056 (1994)
23. Ferreira, J., Casse, F.: Self-collimated stationary disk winds. Astron. Astrophys. **292**, 479 (2004)
24. Ferreira, J., Dougados, C., Cabrit, S.: Which jet launching mechanism(s) in T Tauri stars? Astron. Astrophys. **453**, 785 (2006)
25. Glassgold, A.E., Najita, J., Igea, J.: Heating protoplanetary disk atmospheres. Astrophys. J. **615**, 972 (2004)
26. Gullbring, E., Hartmann, L., Briceño, C., Calvet, N.: Disk accretion rates for T Tauri stars. Astrophys. J. **492**, 323 (1998)
27. Hartigan, P., Edwards, S., Ghandour, L.: Disk accretion and mass loss from young stars. Astrophys. J. **452**, 736 (1995)
28. Hartigan, P., Edwards, S., Pierson, R.: Going slitless: Images of forbidden-line emission regions of classical T Tauri stars observed with the hubble space telescope. Astrophys. J. **609**, 261 (2004)
29. Herbig, G.H., Jones, B.F.: Large proper motions of the Herbig–Haro objects HH 1 and HH 2. Astrophys. J. **86**, 1232 (1981)
30. Hirth, G.A., Mundt, R., Solf, J.: Spatial and kinematic properties of the forbidden emission line region of T Tauri stars. Astron. Astrophys. Suppl. **126**, 437 (1997)
31. Hollenbach, D.J., Yorke, H.W., Johnstone, D. Disk Dispersal around Young Stars In: Mannings, V., Boss, A.P., Russell, S.S. (eds.) Protostars and Planets IV, p. 401. University of Arizona Press, Tucson (2000)
32. Königl, A., Pudritz, R. Disk Winds and the Accretion-Outflow Connection In: Mannings, V., Boss, A.P., Russell, S.S. (eds.) Protostars and Planets IV, p. 759. University of Arizona Press, Tucson (2000)

33. Pyo, T.-S., et al.: Adaptive optics spectroscopy of the [Fe II] outflow from DG Tauri. Astrophys. J. **590**, 340 (2003)
34. Pyo, T.-S. et al.: Adaptive optics spectroscopy of the [Fe II] outflows from HL Tauri and RW Aurigae. Astrophys. J. **649**, 836 (2006)
35. Ray, T., Dougados, R., Bacciotti, F., Eislöffel, J., Chrysostomou, A.: Toward resolving the outflow engine: An observational perspective. In: Reipurth, B., Jewitt, D., Keil, K. (eds.) PPV. University of Arizona Press, Tucson, astro-ph/0605597 (2006), p. 231–244
36. Reipurth, B., Bally, J.: Herbig–Haro flows: Probes of early stellar evolution. Ann. Rev. Astron. Astrophys. **39**, 403 (2001)
37. von Rekowski, B., Brandenburg, A.: Solar coronal heating by magnetic cancellation. Astron. Nachr. **327**, 53 (2005)
38. Richer, J.S., Shepherd, D.S., Cabrit, S., Bachiller, R., Churchwell, E. Molecular Outflows from Young Stellar Objects In: Mannings, V., Boss, A.P., Russell, S.S. (eds.) Protostars and Planets IV, p. 867. University of Arizona Press, Tucson (2000)
39. Romanova, M.M., Ustyuogva, G.V., Koldoba, A.V., Lovelace, R.V.: Propeller-driven outflows and disk oscillations. Astrophys. J. **635**, L165 (2005)
40. Schwartz, R.: T Tauri nebulae and Herbig–Haro nebulae – Evidence for excitation by a strong stellar wind. Astrophys. J. **195**, 631 (1975)
41. Shu, F., Najita, J., Ostriker, E., Wilkin, F., Ruden, S., Lizano, S.: Magneto-centrifugally driven flows from young stars and disks. 1: A generalized model. Astrophys. J. **429**, 781 (1994)
42. Snell, R., Loren, R., Plambeck, R.: Observations of CO in L1551 – Evidence for stellar wind driven shocks. Astrophys. J. 239, L17 (1980)
43. Terebey, S., Shu, F.H., Cassen, P.: The collapse of the cores of slowly rotating isothermal clouds. Astrophys. J. **286**, 529 (1984)
44. Whelan, E.T., Ray, T.P., Davis, C.J.: Paschen beta emission as a tracer of outflow activity from T-Tauri stars, as compared to optical forbidden emission. Astron. Astrophys. **417**, 247 (2004)
45. White, R.J., Hillenbrand, L.A.: On the Evolutionary Status of Class I Stars and Herbig–Haro energy sources in Taurus-Auriga. Astrophys. J. **616**, 998 (2004)
46. Woitas, J., Bacciotti, F., Ray, T.P., Marconi, A., Coffey, D., Eislöffel, J.: Jet rotation: Launching region, angular momentum balance and magnetic properties in the bipolar outflow from RW Aur. Astron. Astrophys. **432**, 149 (2005)

Measurement of Physical Conditions
in Stellar Jets

P. Hartigan

Department of Physics and Astronomy, Rice University, Houston TX 77005-1892
USA
`hartigan@sparky.rice.edu`

Abstract. This article summarizes the physics needed to interpret spectra of stellar
jets, reviews the history of analysis techniques applied to these objects, discusses
some recent results, and considers the prospects for future research in this area.

1 Introduction

Over fifty years have passed, since Herbig [26] first published spectra of two
bright nebulous emission line sources located in the star formation region
NGC 1999. The objects that we now know as the prototype Herbig–Haro
objects HH 1 and HH 2 had also appeared as emission line sources in an
Hα prism survey of the region by Haro [15]. Unlike other emission line nebulae known at the time (such as H II regions), no blue stars existed nearby
that might ionize these nebulae. Moreover, as Herbig noted in his paper, the
emission-line spectra were quite peculiar: forbidden emission lines from low
ionization species such as [S II] and [O I] were unusually strong, and yet high
ionization lines of [O III] and [Ne III] were also present.

In the years that followed Herbig's paper, Herbig–Haro (HH) objects have
played an increasingly pivotal role in clarifying how stars form. We now know
that HH objects represent regions of shocked gas in collimated supersonic
jets which emerge along the rotation axes of accretion disks. The existence
of supersonic outflowing gas from a source where material is supposed to
be accreting onto the central object is at first glance quite surprising, and
the ubiquity of the phenomenon suggests that jets are an essential aspect of
the star-formation process. Because jets are spatially resolved and show clear
proper motions, emission-line images immediately reveal something about the
kinematics of the flows, which in turn helps to constrain models of jets and
accretion disks.

However, if we wish to understand the internal dynamics of jets and obtain a deeper appreciation of how these flows connect with their disks, we
must consider the processes that heat the jet, and we must measure physical

P. Hartigan: *Measurement of Physical Conditions in Stellar Jets*, Lect. Notes Phys. **742**,
15–42 (2008)
DOI 10.1007/978-3-540-68032-1_2

conditions along the outflow, such as the temperature, density, ionization fraction, magnetic fields, and internal velocities. In many ways, HH objects are particularly well suited for such analyses because they emit at least a dozen bright emission lines, all of which are optically thin. Hence, the emitted light escapes freely, and there are no complex radiative transfer issues to consider as there are, for example, in emission line spectra of accretion disks around young stars.

This review consists of three parts. In the next section, I summarize the basic atomic physics that underlies the analysis of emission lines in stellar jets. The techniques are similar to those employed to study other emission nebulae such as H II regions and planetary nebulae, but the relatively low ionization, high densities, and inhomogeneous conditions present behind shock waves in jets pose some unique challenges in the interpretation of the data. In Sect. 3, I present an overview of the history of the analysis of line ratios in HH objects. Extracting physical information from the spectra of HH objects began soon after their discovery and continues to this day, with analysis methods steadily improving. There is a great deal to learn about how science is done by understanding how the incremental advances from different groups have contributed to our present knowledge. In Sect. 4, I review some new results just accepted for publication that deal with magnetic field strengths along jets, and with heating and cooling within 100 AU of the source.

2 Atomic Physics of Stellar Jets

2.1 Observations of Electron Temperature and Electron Density

Connection of Observed Line Fluxes with Level Populations in Atoms

Before we consider the physical processes at work in stellar jets, we should first have some idea as to what we actually observe when we see emission lines from an extended object. Emission lines in stellar jets occur because an atom transitions from an excited bound state to a lower energy bound state. When we speak of an atom in an excited state we really mean that a bound electron in that atom is in a level above the ground state. The atom may come to the excited state in many ways. For example, if the atom is initially in its ground state, a free electron can collide with it and in the process transfer some of the kinetic energy of the collision to the bound electron within the atom, raising the atom to an excited state. Alternatively, if the atom was already in a higher excited state, the collision may de-excite the atom to a lower excited state. Other processes may also leave the atom in an excited state, such as absorbing a photon when the atom is in a lower state, recombining from a higher ionization state, or even as a consequence of charge exchange with another atom.

Regardless of the physical process that leaves the atom in the excited state, if we know the density of atoms in that state we can calculate the amount of light that emerges from a volume of such atoms. In what follows, let us define level 1 as the lower level and level 2 as the upper level. Let n_1 and n_2 refer to the number density of atoms (cm^{-3}) in levels 1 and 2, respectively. Then for a parcel of gas with a number density n_2, the rate of photons emitted from a 1 cm^3 volume is simply $n_2 A_{21}$, where A_{21} is the Einstein A-coefficient for the transition (units s^{-1}). Hence, if we define the volume emission coefficient of the transition j_{21} to be the energy per unit volume, per second, per steradian in the emission line coming from the gas, then

$$j_{21} = \frac{n_2 A_{21} h\nu_{21}}{4\pi},\tag{1}$$

where the units on j_{21} are $\mathrm{erg\,cm^{-3}s^{-1}str^{-1}}$.

The general equation of radiative transfer is

$$\frac{dI}{dl} = -\kappa I + j,\tag{2}$$

where I is the intensity, dl the path length along the line of sight, κ the opacity, and j the volume emission coefficient. Equation 2 is basically a definition of what one means by a volume emission coefficient and opacity. Because the densities are low enough in stellar jets for the gas to be transparent in the considered emission lines (optically thin), we can take κ equal to zero. Hence, to obtain the intensity of the emission line, we simply integrate the volume emission coefficient along the line of sight,

$$I_{21} = \int j_{21} dl.\tag{3}$$

Intensities are independent of distance to the observer. Hence, at the telescope, if a pixel on a CCD subtends a solid angle Ω, then the rate R_{21} of emission line photons stored per second as electrons on the CCD is given by

$$R_{21} = \frac{I_{21}}{h\nu}\Omega A f,\tag{4}$$

where A is the area of the telescope's aperture, and f is an efficiency factor that takes into account light losses in the Earth's atmosphere, telescope optics, and detector. Hence, from an observational point of view, the critical parameter is the number density n_2 of atoms in the upper state. Emission-line ratios are even simpler, depending only on the relative populations of the upper levels responsible for each line and on the ratio of the Einstein A-values. Once we determine n_2 for all levels of interest, the connection to the observations is immediate and the analysis is done.

Statistical Equilibrium

A key realization in the analysis of line emission is that the level populations of atoms in a parcel of gas are in statistical equilibrium, that is, that the rate that these levels are populated equals the rate that they are depopulated. There is a distinct difference between statistical equilibrium and local thermodynamic equilibrium (LTE). In LTE, the levels are populated according to the Boltzmann equation

$$\frac{n_2}{n_1} = \frac{g_2}{g_1} e^{-h\nu_{12}/kT},$$ (5)

whereas in statistical equilibrium, low density gases like those found in stellar jets and in the interstellar medium are mostly in the ground state.

Balancing the rates in and out of the upper state for a 2-level atom where collisional excitation, de-excitation, and radiative decay are the dominant processes, we have

$$n_1 n_e C_{12} = n_2 A_{21} + n_2 n_e C_{21},$$ (6)

where n_e is the electron density, and C_{12} and C_{21} are the rate coefficients for collisional excitation and de-excitation, respectively. Units on rate coefficients are cm^3s^{-1}. Whenever particles of type A interact with those of type B, the rate of the reaction per unit volume is simply $n_A n_B C_{AB}$ $cm^{-3}s^{-1}$. For the simple case where A are point-like particles fixed in space and B are particles that move in one direction with a velocity v and cross section σ, $C_{AB} = \sigma v$. Thus, it is helpful to think of a rate coefficient as simply the product of a cross-sectional area and an impact velocity. If we know how the cross-section varies with velocity, then it is possible to integrate over arbitrary velocity distributions such as occuring in a thermal gas to obtain an average rate coefficient for the reaction.

We can define a critical density to be n_C when the two terms on the right-hand side of (6) are equal

$$n_C = A_{21}/C_{21}.$$ (7)

Notice that the critical density refers to the number density of electrons, not to the number density of the atoms that are the targets for the collisions. In a gas that has the same temperatures of ions, neutrals, and electrons, all the particles have the same av erage kinetic energy, so the electrons will have larger average velocities by a factor of $(m_{atom}/m_e)^{1/2}$. For this reason, collisions between ions and electrons occur at a much higher rate than those between ions and neutrals, the latter being usually negligible unless the gas is almost entirely neutral.

As an aside, it is possible for electrons and ions to have different temperatures. When electrons and ions encounter a shock wave with velocity v_S, some fraction of the incident kinetic energy $(m/2)v_S^2$ for each particle is converted to thermal energy kT. Because ions are heavier, their postshock temperatures are correspondingly higher. Differences between ion and electron temperatures gradually equilibrate behind the shock, a process followed by all modern shock codes. The effect on line emission can be important, for example in supernova shells [28].

Single Ion Analysis: High and Low Density Limits, n_e and T_e

Equation 6 naturally separates densities into two regimes, depending on the electron density. In the low density limit (LDL), $n_e << n_C$ and collisional de-excitation is unimportant. In this limit, every collisional excitation from the ground state to the upper state is followed by radiative decay and the emission of a photon. Alternatively, in the high density limit (HDL), $n_e >> n_C$, so most collisional excitations are followed by collisional de-excitations. In this case, the collisional de-excitations 'quench' excitations that would otherwise produce photons. As far as the dynamics of the gas is concerned, energy is only lost from the system (i.e., the gas cools) when photons escape the gas. Collisional excitations followed by collisional de-excitations do not cool the gas because the net effect is simply to transfer energy from one free electron (the one doing the excitation) to another free electron (the one doing the de-excitation).

We would like to determine which emission line ratios are useful for measuring temperature, electron density, and so on. Consider Fig. 1, which shows two hypothetical three-level atoms A and B. Levels 2 and 3 in atom A have nearly the same excitation relative to ground, so the frequencies of the 2-1 and 3-1 transitions are nearly identical. From (1, 2, 3, and 6), the ratio of the line intensities I_{31}/I_{21} in the LDL simply equals C_{13}/C_{12}, the ratio that collisions from the ground populate these levels. This result makes sense, because in the LDL, all collisional excitations produce a photon when the atom decays back to its ground state. As we shall see below, for closely spaced energy levels, the ratio C_{13}/C_{12} equals the ratio of collision strengths Ω_{13}/Ω_{12} between the respective levels. Collision strengths are independent of the electron density and usually nearly independent of the temperature, so the ratio of the line intensities is a fixed number in the LDL.

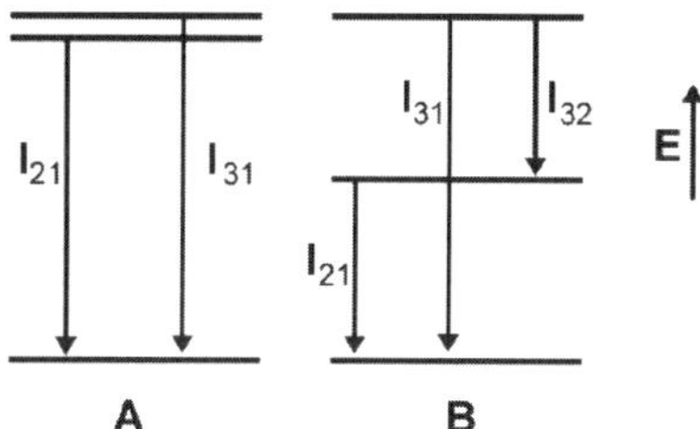

Fig. 1. Energy levels for two hypothetical atoms. The lower, intermediate, and upper levels are defined as levels 1, 2, and 3, respectively for both atoms

In the HDL, the intensity ratio I_{31}/I_{21} approaches a different value, $(C_{13}/C_{12}) \times (A_{21}/A_{31})$. Thus, the line ratio I_{31}/I_{21} changes from one asymptotic value at low densities to a different value at high densities. When the value for the electron density lies between the LDL and HDL, the observed ratio provides an estimate of the electron density. The observed ratio gives an upper limit for the density when the gas is in the LDL and a lower limit when the gas is in the HDL.

In atom B, levels 2 and 3 are at quite different energies, E_2 and E_3 respectively, above ground. In this case, the ratio of the collision rate coefficients C_{13}/C_{12} equals $\Omega_{13}/\Omega_{12} \times \exp(-(E_3 - E_2)/kT)$, so there is now also a temperature dependence to the ratio. This result reflects the fact that more electrons have sufficient energy to excite the atom from the ground level to level 2 than are available to excite from the ground to level 3. In the HDL, the temperature dependence remains, with the values modified by the factor A_{21}/A_{31} as for atom A. Hence, the emission line ratio I_{31}/I_{21} depends on both the electron density and on the temperature. Hence, the ratio of two line fluxes from an ion specifies a fixed curve in (n_e, T_e) space.

Another type of line ratio is one where the transitions share a common upper level, such as I_{32}/I_{31}. From (1 and 3), we see that this ratio simply equals $A_{32}\nu_{32}/A_{31}\nu_{31}$. From a diagnostic standpoint, this ratio would seem to be useless, as it is fixed regardless of the conditions in the plasma. There are, however, some uses for this type of line ratio. First, the ratio is a good check of the instrumental flux calibration when the emission lines are close in wavelength (i.e., the ground state is split; e.g. [N II] $\lambda6583$/[N II] $\lambda6548$). In cases like Atom B in Fig. 1, where $\nu_{31} >> \nu_{32}$, one can use the observed line ratio to estimate the reddening (e.g. [S II] $1.03\,\mu$m/[S II] $\lambda4070$, see Brugel et al. [5]). Alternatively, if one knows the reddening, it is possible to determine the ratio of the Einstein-A values of the transitions experimentally[48], which can then be used in other objects to find the reddening [34, 37].

In rare cases, line ratios from the same upper level may deviate from the ratio of the A-values when one of the transitions is permitted and goes to the ground level [23]. In these so-called resonance lines, the opacity (κ in (2)) becomes large enough for some of the photons to be absorbed by other atoms along the line of sight, which subsequently scatter the photons in a direction away from the observer, lowering their fluxes. This process only occurs with permitted lines, typically at ultraviolet wavelengths.

Multiple Ion Analysis: Ionization Fractions and Abundances

The analysis we have described thus far applies only to emission lines that originate from a single ion. Density and temperature dependencies of line ratios between different ions follow the same general guidelines described above, except that an observed line ratio is no longer just a function of the electron temperature and density but now also depends on the total abundance of each ionic species. If the same element is used (e.g. [N I] $\lambda5200$/[N II] $\lambda6583$), then the ratio depends only on the ionization fraction of that element, while line ratios that mix elements (e.g. [O I] $\lambda6300$/[S II] $\lambda6731$) also depend upon the ratios of the abundances of these elements.

To translate an observed electron density to a total density we need to know the ionization fraction of hydrogen, which is the dominant element. Fortunately, ionization fractions of N and O are tied closely to that of H through charge exchange (see below and Sect. 3), so one can usually obtain the ionization fraction of H in this manner.

Magnetic fields are particularly difficult to measure in stellar jets because the emission lines do not display any Zeeman splitting. It is possible to infer something about field strengths by studying shock waves. We will return to this issue in Sect. 4.

2.2 Physical Processes

Deriving the atomic physics relevant for studies of diffuse nebular gas is beyond the scope of this paper. However, it is useful to summarize the main results that come out of quantum mechanical studies of atoms, because this physics underlies the entire analysis of emission lines. In this section, I provide such an outline, which is not intended to be a step-by-step derivation of all the physics. However, I have tried, wherever feasible, to not simply quote results but instead highlight intermediate steps in the analysis so that the interested student may look for these in physics texts on the subject. There are several good references on atomic physics and line spectroscopy (e.g. [4, 43]).

Compilations of collision strengths, energy levels, A-values, charge exchange cross sections, photoionization cross sections, and photoexcitation cross sections for emission lines of interest exist in the literature [30]. The historical application of these processes to HH research is covered in Sect. 3.

Photoionization

Photoionization processes usually do not dominate the ionization balance within stellar jets. Unlike massive stars, T Tauri stars have relatively cool photospheres and do not emit substantial amounts of ionizing radiation. Some ionizing radiation does occur from hotspots where material falls onto the star, but this radiation is usually absorbed by the surrounding medium before it can propagate significantly into the jet. Ionization events do occur in stellar jets close to the source, but these do not appear to be caused by photoionization (see Sect. 4). However, photoionization is an important process in the cooling zones of shock waves, where, for example, Lyman continuum photons propagate both upstream and downstream of the shock front. These photons may either ionize preshock gas or re-ionize postshock gas.

The physics of photoionization comes down to calculating the photoionization cross section for the atom of interest as a function of frequency. Photoionization cross sections are zero up to an energy threshhold, where they jump to a peak value a_0 (cm^2) and then decline as the frequency increases. Above the threshhold, the photoionization cross section declines as $\nu^{-\alpha}$, where $\alpha \sim 3$ for H. Values for a_0 and α are tabulated in several references [36].

To describe photoionization quantitatively, note that in the presence of a magnetic field the momentum operator $\mathbf{p}$ becomes $\mathbf{p} + e\mathbf{A}/c$, where $\mathbf{A}$ is the magnetic vector potential. We can represent the effect the photon has on the Hamiltonian by using

$$\mathbf{A} = \mathbf{A}_0 \cos(\mathbf{k} \cdot \mathbf{r} - \omega t) = \mathbf{A}_0 \frac{e^{i(\mathbf{k} \cdot \mathbf{r} - \omega t)} + e^{-i(\mathbf{k} \cdot \mathbf{r} - \omega t)}}{2}, \tag{8}$$

where $\mathbf{A}_0$ is a constant. The momentum $(\mathbf{p} + e\mathbf{A}/c)$ enters into the Hamiltonian as a squared term, the photon may also interact with the atom via a $\mu_{\mathbf{S}} \cdot \mathbf{B}$ term, where $\mu_{\mathbf{S}}$ is the magnetic moment of the atom for spin $\mathbf{S}$. Ignoring this spin term for the moment, and discarding the usually negligible $e^2 A^2/c^2$ term that arises from squaring the momentum, we obtain an expression for the interaction terms H_{int} in the Hamiltonian that represent how the radiation affects the state of the atom,

$$H_{\mathrm{int}} = \frac{e}{2mc} \left(\mathbf{p} \cdot \mathbf{A} + \mathbf{A} \cdot \mathbf{p} \right). \tag{9}$$

When, as in this case, the Hamiltonian can be written as the sum of an unperturbed component H_0 and a perturbation H_{int}, one can use the perturbation theory to determine how an atom in an inital state $|\phi_1\rangle$ evolves with time. If the energy eigenstates of the atom are $|\phi_n >$ and the atom is in state 1 at $t = 0$, then the wavefunction at time t is

$$|\psi(t) >= \sum_n |\phi_n >< \phi_n|\psi(t) >. \tag{10}$$

Define $c_n(t) = < \phi_n|\psi(t) >$. Then the probability that the atom will be in state 2 at time t is given by

$$|< \phi_2|\psi(t) >|^2 = |c_2(t)|^2. \tag{11}$$

To determine $c_2(t)$ we must perturb the Hamiltonian for some time interval T with an electromagnetic wave like that of (8). The result, which is not too difficult to show (e.g. [4]), is

$$c_2(t) = \frac{1}{i\hbar} \int_0^t e^{i\omega_{12}t'} H_{21}(t')\mathrm{d}t', \tag{12}$$

where

$$H_{21} = \langle \phi_2|H_{\mathrm{int}}|\phi_1 \rangle \tag{13}$$

is the matrix element between the initial and final states. In this case, the initial state '1' is a bound level, while the final state '2' is unbound, so ϕ_2 is a continuum state (sine wave) that represents the free electron.

The result of this exercise for H-like wavefunctions is that the cross section a_ν (units are cm^2; the ν refers to a cross section at a particular frequency) is

$$a_\nu = \frac{64\pi^4 m e^{10} Z^4}{3^{3/2} c h^6 n^5} \frac{g_\nu}{\nu^3}, \tag{14}$$

where n is the initial level (equals 1 for ionization from the ground state), and g_ν is a Gaunt factor ~ 1 which varies slowly with frequency. Hence, $a_\nu \sim \nu^{-3}$, as noted above. The above expression for a_ν appears complicated, but the cross section at the threshold for ionization from level 1 simplifies to a numerical constant $(64/3\sqrt{3})$ times the fine structure constant $e^2/\hbar c$ times πr_0^2, where r_0 is the Bohr radius.

There is an interesting physical reason why the photoionization cross section drops as ν increases. The dominant term in the interaction Hamiltonian is the electric dipole matrix element $\langle \phi_2 | er | \phi_1 \rangle$, which in position representation is an integral that looks like

$$\int \int \phi_2^*(r, \Omega) er \phi_1(r, \Omega) r^2 \ \mathrm{d}r \ \mathrm{d}\Omega. \tag{15}$$

The integrand for the radial portion is a product of three functions, r^3, $\phi_2^*(r, \Omega)$, which is a sine wave, and $\phi_1(r, \Omega) \sim r e^{-r}$ for the ground state 1s wavefunction. Visualizing the product of these three functions, it is clear that if ϕ_2^* oscillates slowly, then the integrand will retain the same sign over the entire interval of r, where the product $r^4 e^{-r}$ differs significantly from zero. However, when ϕ_2^* oscillates rapidly compared to the rest of the integrand, then the integral tends to cancel and the final result is much lower. Hence, a rapidly oscillating ϕ_2^*, which corresponds to a highly energetic free electron and therefore an energetic photon of large ν, leads to a smaller photoionization cross section. For this reason the photoionization cross section drops with frequency.

Photoexcitation and Pumping

Like photoionization, photoexcitation is a secondary process in stellar jets. The only transitions affected by photoexcitation in a significant way are permitted transitions, where the lower energy level is populated enough to make the transition optically thick. In these so-called resonant lines, photons can be scattered out of the beam and change the line ratios relative to non-resonant lines. This phenomenon has been observed in ultraviolet resonance transitions of [Fe II] in HH 47A with HST spectra [23].

Classically, one can calculate a photoexcitation cross section with a model where the electron is tied to a spring that has some natural resonant frequency ω_0 and damping constant γ, and the electron is subject to a passing electromagnetic wave of frequency ω. The oscillating electron, being accelerated, emits light, and the ratio of the radiated power to the incident flux gives the cross section at that frequency. Integrating over all frequencies, we find

$$\int_0^\infty a_\nu \mathrm{d}\nu = \frac{\pi e^2}{mc} f \ \mathrm{cm^2 s^{-1}}, \tag{16}$$

where the oscillator strength f equals unity in the classical approximation.

The quantum calculation is more involved but follows in an analogous manner the one we outlined for photoionization in the previous section. As in that case, we envision the atom starting out in an intial state $|\phi_1 >$ and seek to determine what the probability is that it will occupy the final state $|\phi_2 >$ after some time t under the influence of the electromagnetic perturbation.

Let us begin by taking the perturbation to be a single plane wave of angular frequency ω represented by an oscillating vector potential as in (8). Taking the electric field and the vector potential to be in the z-direction and the magnetic field in the x-direction, the wave moves in the y-direction so $\mathbf{k} \cdot \mathbf{r} = ky$. Expanding the exponential term in (8),

$$e^{iky} \sim 1 + iky + \cdots \sim 1 \tag{17}$$

we find that the matrix element

$$H_{21} = \langle \phi_2 | H_{\mathrm{int}} | \phi_1 \rangle = \frac{eA(t)}{c} < \phi_2 | \frac{p_z}{m} | \phi_1 > . \tag{18}$$

This is an operator equation, so one uses commutation relations between p_z, z, and the unperturbed Hamiltonian H_0 to obtain

$$p_z = \frac{m}{i\hbar} \left(z H_0 - H_0 z \right). \tag{19}$$

Substituting the expression for p_z into the expression for H_{21} and using $H_0 | \phi_1 > \ = E_1 | \phi_1 >$ and $H_0 | \phi_2 > \ = E_2 | \phi_2 >$ we get

$$H_{21}(t) = \frac{A(t)}{ic\hbar} (E_1 - E_2) \langle \phi_2 | ez | \phi_1 \rangle = \frac{A(t)\omega_{12}}{ic} \langle \phi_2 | ez | \phi_1 \rangle . \tag{20}$$

Using (12) we obtain an expression for $c_2(t)$,

$$c_2(t) = \frac{-A_0 \omega_{12}}{2\hbar c} \langle \phi_2 | ez | \phi_1 \rangle I(t) , \tag{21}$$

where

$$I(t) = \int_0^t e^{i(\omega_{12} + \omega)t'} + e^{i(\omega_{12} - \omega)t'} \, \mathrm{d}t'. \tag{22}$$

The integral is easy to solve as

$$I(t) = i \left[\frac{1 + e^{i(\omega_{12}+\omega)t}}{\omega_{12} + \omega} + \frac{1 - e^{i(\omega_{12}-\omega)t}}{\omega_{12} - \omega} \right] . \tag{23}$$

The first term in (23) is small compared to the second term when $\omega \sim \omega_{12}$. Rewriting the second term as a sine,

$$I(t) = -e^{\frac{-i(\omega_{12}-\omega)t}{2}} \left[\frac{e^{\frac{i(\omega_{12}-\omega)t}{2}} - e^{\frac{-i(\omega_{12}-\omega)t}{2}}}{2i\left(\frac{\omega_{12}-\omega}{2}\right)} \right]. \tag{24}$$

Hence,

$$|c_2(t)|^2 = \frac{A_0^2 \omega_{12}^2}{4\hbar^2 c^2} |\langle \phi_2|ez|\phi_1\rangle|^2 \frac{\sin^2\left(\frac{\omega_{12}-\omega}{2}t\right)}{\left(\frac{\omega_{12}-\omega}{2}\right)^2}. \tag{25}$$

The function $|c_2(t)|^2$ represents the probability that the atom is in state 2 at time t under the influence of a single plane wave with angular frequency ω. Equation 25 has a peak at $\omega = \omega_{12}$, and the peak value increases as t^2, while the width of the peak scales as t^{-1}. Integrating the function over all frequencies yields a result proportional to time t. Hence a uniform distribution of electromagnetic waves across the entire absorption profile results in a constant rate of excitation. For this reason, the rate of excitation must be proportional to the specific mean intensity J_ν (erg cm^{-2}s^{-1}str^{-1}Hz^{-1}). Using contour integration in the complex plane, we find

$$\int_{-\infty}^{\infty} \frac{\sin^2\left(\frac{\omega_{12}-\omega}{2}t\right)}{\left(\frac{\omega_{12}-\omega}{2}\right)^2} d\omega = 2\int_{-\infty}^{\infty} \frac{\sin^2(xt)}{x^2} dx = 2\pi t. \tag{26}$$

In real situations, there is not a single plane wave with a frequency ω but rather an input spectrum with a range of frequencies. To solve this case, consider a perturbation that begins at $t=0$ and extends to time t. With the dipole approximation (17), the perturbation is defined by some electric field $E(t) = E_0 \sin(\omega t)$, or equivalently $A(t) = A_0 \cos(\omega t)$, where

$$E = -\frac{1}{c}\frac{\partial A}{\partial t}, \tag{27}$$

and

$$E_0 = -\frac{\omega A_0}{c}. \tag{28}$$

Equation 12 has a term proportional to $e^{i\omega_{12}t}A(t)$ within the integrand. Taking $A(t)=0$ outside the time interval for the pulse, we can extend the limits to $\pm\infty$. The integral then becomes the Fourier transform $\hat{A}$ of the vector potential,

$$\hat{A}(\omega) = \frac{1}{2\pi}\int_{-\infty}^{\infty} e^{i\omega t}A(t)dt. \tag{29}$$

Combining these results yields

$$|c_2(t)|^2 = \frac{4\pi^2 \omega_{12}^2}{\hbar^2 c^2} |\langle \phi_2|ez|\phi_1\rangle|^2 \left|\hat{A}(\omega)\right|^2. \tag{30}$$

For isotropic radiation,

$$|\langle\phi_2|ez|\phi_1\rangle|^2 = \frac{|\langle\phi_2|er|\phi_1\rangle|^2}{3}, \tag{31}$$

so that

$$|c_2(t)|^2 = \frac{4\pi^2}{3\hbar^2}|\langle\phi_2|er|\phi_1\rangle|^2\left|\hat{E}(\omega)\right|^2. \tag{32}$$

Finally, we must relate $\hat{E}(\omega)$ and the mean intensity J_ν ($\mathrm{erg\,cm^{-2}s^{-1}Hz^{-1}str^{-1}}$). For isotropic radiation, the total energy flux F ($\mathrm{erg\,cm^{-2}s^{-1}}$) in any given direction is

$$\int_0^\infty 4\pi J_\omega \mathrm{d}\omega = F = \frac{c}{4\pi t}\int_0^t E^2(t')\mathrm{d}t' = \int_0^\infty \frac{c\left|\hat{E}(\omega)\right|^2}{t}\mathrm{d}\omega, \tag{33}$$

so that

$$J_\omega = \frac{c\left|\hat{E}(\omega)\right|^2}{4\pi t}, \tag{34}$$

where we have used the fact that $E(t') = 0$ except for $0<t'<t$, $\hat{E}(\omega) = \hat{E}(-\omega)$, and Parseval's theorem

$$\int_{-\infty}^\infty E^2(t)\mathrm{d}t = 2\pi\int_{-\infty}^\infty\left|\hat{E}(\omega)\right|^2\mathrm{d}\omega. \tag{35}$$

Using $J_\nu \mathrm{d}\nu = J_\omega \mathrm{d}\omega$ with $\omega = 2\pi\nu$ we obtain

$$J_\nu = \frac{c\left|\hat{E}(\omega)\right|^2}{2t}, \tag{36}$$

and

$$|c_2(t)|^2 = \frac{8\pi^2 J_\nu t}{3\hbar c}|\langle\phi_2|er|\phi_1\rangle|^2. \tag{37}$$

The Einstein B_{12} value is related to J_ν and $c_2(t)$ through

$$\frac{d|c_2(t)|^2}{\mathrm{d}t} = J_\nu B_{12} \tag{38}$$

Hence,

$$B_{12} = \frac{8\pi^2}{3\hbar^2 c}|\langle\phi_2|er|\phi_1\rangle|^2. \tag{39}$$

The above analysis used the simplest possible expansion of the vector potential in (8). Keeping higher order terms is warranted, if the electric dipole moment matrix element in (8) is zero, in which case one must also include the $\mu_{\mathbf{S}} \cdot \mathbf{B}$ term. Higher order terms in the expansion of (8) lead to electric quadrupole and magnetic dipole transitions (see below).

Radiative Decay

Radiative or spontaneous decay refers to a situation, where an atom in an excited state emits a photon and decays to a lower energy state. We can use the expression for B_{12} in (39) to find the Einstein A-value (units s^{-1}). In LTE, the mean intensity is the Planck function, and the ratio of the level populations n_2/n_1 is given by (5). Statistical equilibrium implies

$$n_2 \left(A_{21} + B_{21} J_\nu \right) = n_1 B_{12} J_\nu. \tag{40}$$

Substituting the Planck function B_ν for the mean intensity J_ν, where

$$B_\nu = \frac{2h\nu^3}{c^2 \left(e^{h\nu/kT} - 1 \right)} \tag{41}$$

we find

$$A_{21} = \frac{2g_1 h\nu^3}{g_2 c^2} B_{12} \tag{42}$$

and

$$B_{21} = \frac{g_1}{g_2} B_{12}. \tag{43}$$

Using (39) for B_{12}, we find

$$A_{21} = \frac{64\pi^4 \nu_{21}^3 g_1}{3 g_2 h c^3} \left| \langle \phi_2 | er | \phi_1 \rangle \right|^2. \tag{44}$$

Although the preceding equations were derived assuming LTE, they consist of relationships between atomic parameters and so are valid outside LTE as well. For example, the A-value simply represents the rate at which an atom in level 2 will decay to level 1. This rate does not depend upon the level populations of other atoms in the gas. Owing to the ν^3 dependence, A-values at infrared wavelengths are much lower than those in the optical and ultraviolet. At ultraviolet wavelengths, $A\sim 10^8$ s^{-1}, while that same line in the far-infrared will have $A\sim 0.1$, even though it is still a permitted line.

The transitions described above have a nonzero electric dipole moment matrix element and are called permitted lines. When this matrix element is zero, there may be nonzero components to the electric quadrupole or magnetic dipole moments that contribute to make the A-value nonzero. These emission lines are called forbidden and are denoted with brackets (e.g. [N II] $\lambda 6583$). At optical wavelengths, the A-values of forbidden lines are typically a factor of 10^5 (magnetic dipole) to 10^8 (electric quadrupole) smaller than those of permitted lines. Nevertheless, forbidden lines are extremely important for nebular gas because they dominate the cooling at typical nebular temperatures of $\sim 10^4$ K (1 eV). At these temperatures, transitions that would produce permitted lines are a factor of ~ 10 higher in energy than kT, making the permitted lines very weak because the upper states are simply not populated.

Emission lines between states with different spins are not allowed in the L–S coupling approximation, but they do occur in nature. Such lines are known as semiforbidden, or intersystem lines, and are denoted with a single bracket (e.g. [C III λ1909). Semiforbidden transitions like [C III λ1909 are useful because they have A-values intermediate between forbidden and permitted lines, and so have high critical densities, but remain optically thin in many cases. Selection rules for the various electric and magnetic multipole radiation components are well documented in the literature [47].

Collisional Excitation

In collisional excitation, a free electron collides with an atom and raises the atom to a higher energy state. The electron emerges from the collision as a free particle with a corresponding lower kinetic energy. Quantum mechanically, this is a scattering process with the incident electron represented as a plane wave, while the exiting electron is a scattered spherical wave with some scattering amplitude and phase shift. It is convenient to define the cross section σ_{12} of such a reaction in terms of a collision strength Ω_{12}, impact velocity v, and statistical weight g_1 of the lower level as

$$\sigma_{12} = \frac{\pi \hbar^2 \Omega_{12}}{m^2 v^2 g_1}. \tag{45}$$

Collision strengths are dimensionless, typically of order unity and have the property that $\Omega_{ij} = \Omega_{ji}$.

The cross section will be zero for incident electron energies that are too low to excite the transition. Often the behavior of the cross section can be quite complex near the threshold excitation of the transition and may exhibit a variety of resonances [6]. These resonances typically disappear when the cross sections are integrated over a Maxwellian distribution of electron velocities, as one encounters in a thermal plasma. When integrated over a Maxwellian, collision strengths are nearly independent of the temperature for most collisions of electrons and ions. Collisional cross sections between electrons and neutral atoms typically vary with temperature roughly as T^α, with $0.5 < \alpha < 1$.

The rate coefficient C_{12} is given by $< \sigma_{12} v >$, where the brackets indicate an average over the distribution of impact velocities. For a Maxwellian, the equation for the rate coefficient becomes

$$C_{12} = \frac{8.63 \times 10^{-6} \Omega_{12}}{g_1 T_{\mathrm{e}}^{1/2}} \mathrm{e}^{-h\nu_{12}/kT_{\mathrm{e}}} \ \mathrm{cm}^3 \mathrm{s}^{-1} \tag{46}$$

Owing to the complexity of resonances, the large number of energy levels, and the need to include relativistic perturbations into the analysis, reliable collision strengths are not always available for all transitions of interest, especially for heavier ions, such as Fe II and Fe III, though availability of these quantities continues to improve gradually with time.

Charge Exchange

An example of a charge exchange reaction is $H^++O\rightarrow H+O^+$. Let the rate coefficient of this reaction be κ_1. The reaction may also proceed in the opposite direction with rate coefficient κ_2. In steady state, where N_X indicates the number density of species X,

$$(N_{H^+}N_O)\,\kappa_1 = (N_H N_{O^+})\,\kappa_2 \tag{47}$$

During the collision, the atoms briefly form an unbound state of the molecule OH^+, which then decays to either H^++O or to $H+O^+$. The ionization state of H is 13.60 eV and for O it is 13.62 eV. Hence, relative to a ground state defined by neutral $H+O$, the state '1' of H^++O has an excitation of $E_1 = 13.60$ eV, while state '2' of $H+O^+$ has an excitation of $E_2 = 13.62$ eV.

In LTE, the ratio of the populations of states 1 and 2 are given by the Boltzmann equation

$$\frac{N_1}{N_2} = \frac{N_{H^+}N_O}{N_H N_{O^+}} = \frac{\kappa_2}{\kappa_1} = \frac{g_1}{g_2}e^{-(E_1-E_2)/(kT)}, \tag{48}$$

where $g_1 = g(H^+)g(O)$ and $g_2 = g(H)g(O^+)$ are the statistical weights of states 1 and 2, respectively. For temperatures of interest ($\sim 10^4$ K), H is all in the ground state, which has $g(H) = 2$, and $g(H^+) = 1$. The ground state configuration of O^+ is $^4S_{3/2}$, which has a statistical weight $g(O^+) = (2*3/2)+1 = 4$. In neutral O, there is a triplet state near the ground, $^3P_{2,1,0}$, with a total statistical weight of $(2*2)+1 + (2*1)+1 + (2*0)+1 = 9$. Hence, $g_1 = 9$ and $g_2 = 8$. Putting all this together we have

$$\frac{N_{O^+}}{N_O} = \frac{8}{9}e^{-0.02\,\mathrm{eV}/(kT)}\frac{N_{H^+}}{N_H} \tag{49}$$

Hence, if the charge exchange cross sections are large enough, this process will tie the ionization fractions of atoms together, weighted by their respective statistical weights and a Boltzmann factor that represents differences in the ionization states of the atoms. Because H atoms are so abundant and He is typically neutral, when charge exchange occurs it almost always involves H. Charge exchange cross sections are particularly large between H and O.

2.3 Isoelectronic Sequences

An isoelectronic sequence refers to a group of atoms and ions that have the same number of electrons. For example, the isoelectronic sequence of 6 electron systems, all of which have ground electronic states of $1s^22s^22p^2$, consists of C I, N II, O III, Fl IV, Ne V, and so on. For the purposes of spectroscopy, atoms and ions within an isoelectronic sequence have identical patterns of energy levels, with the different nuclear charges acting as a scale factor in the binding energies for the levels. The reason for this behavior is that electrons in an unfilled outer shell determine the bound excited states of atoms, and two identical groups of such electrons produce the same sets of energy levels. Electrons become more bound as the number of protons in the nucleus increases, so the transitions shift to shorter wavelengths for higher atomic numbers. For example, wavelengths for the transitions from $^1D_2$ to $^3P_2$ in atoms with $6e^-$ decrease steady from $9849\,\text{Å}$ in C I, to $6583\,\text{Å}$ in N II, to $5007\,\text{Å}$ in O III, to $3426\,\text{Å}$ in Ne V.

Energy levels of light atoms are described well by L-S coupling. Heavier elements may require J-J coupling, but these are rarely of any interest in nebular studies because their abundances are very low. In L-S coupling, energy levels are defined with a notation ^aX_b, where $a = 2S + 1$ is the multiplicity, with S a number representing the total spin, X is the letter S, P, D, F, ... corresponding to L = 0, 1, 2, 3, ... where L is the total orbital angular momentum, and b is the J-value, which ranges from L + S, L + S − 1, ... L − S. For two electrons, one calculates the range of L and S allowed within a particular shell by adding together the individual l_i and s_i ($s_i = 1/2$ for all electrons) in the shell by the rules L = $l_1 + l_2$, $l_1 + l_2 − 1$, ... $|l_1 − l_2|$, and S $= s_1 + s_2$, $s_1 + s_2 − 1$, ... $|s_1 − s_2|$. Rules for more than two active electrons have similar forms.

One complicating factor when the n-values for the electrons are the same, as occurs in the ground configuration, is that electrons are fermions and cannot occupy identical quantum states. Electrons with the same n and l-values are called equivalent electrons, and calculating which levels are excluded on the basis of the Pauli principle can be somewhat involved.

As an example, consider the case of an atom that has all but two of its electrons in filled lower shells ($1s^2$, $2s^2$, etc.), with the two remaining electrons each in a p-shell. Here, $l_1 = l_2 = 1$ (p-states have l = 1), and $s_1 = s_2 = 1/2$ (electrons are spin 1/2 particles). Then L = 2, 1, or 0, and S = 1 or 0. As a result we expect three triplet states, $^3S_1$, $^3P_{2,1,0}$, $^3D_{3,2,1}$, and three singlet states, $^1S_0$, $^1P_1$, $^1D_2$. All these states exist for two non-equivalent p-electrons, for example, in the excited electronic state $1s^22s^22p3p$. However, when the electrons are equivalent, as in the ground electronic state $1s^22s^22p^2$, application of the Pauli exclusion principle shows that the ^3D, ^3S, and ^1P states do not exist (e.g. Appendix B of Eisberg & Resnick [12]).

Two isoelectronic sequences that differ only in the n-value of their outer shells also have the same pattern of energy levels. For example, ions and atoms that are part of an isoelectronic sequence with 14 electrons, including Si I, P II, and S III, all have ground electronic states of $1s^22s^22p^63s^23p^2$ and so have the same pattern of lower levels as do atoms in the 6-electron isoelectronic sequence: a ground triplet state $^3P_{2,1,0}$, a singlet first-excited state $^1D_2$, and a singlet second-excited state $^1S_0$.

Energy levels for outer shells that are more than half-filled are easier to model by considering the equivalent configuration of the same shell with a number of holes equal to the capacity of the shell minus the number of electrons. For example, $1s^2 2s^2 2p^4$, the isoelectronic sequence of 8 electrons (O I, Fl II, Ne III, etc.), is the equivalent of two holes in the p-shell, which holds a maximum of six electrons. The energy levels produced by $2p^4$ are identical to those of $2p^2$, with the exception of the J-values within the ^3P state which are reversed: the ground state in the triplet is now $^3P_2$ instead of $^3P_0$.

With the above considerations in mind, there are five common isoelectronic sequences for abundant light elements (Table 1). For higher-Z elements, once the d-shells become populated the energy level diagrams become very complex, particularly when the shell is about half-filled (e.g. Fe II, which is $3d^5$). Neutral noble gases (Ne, Ar) with a filled outer p-shell exhibit no emission lines of interest in the infrared, optical, or near-UV.

Representative energy level diagrams for each of the five common isoelectronic sequence types are shown in Fig. 2. We now consider how we might use each of the five types to analyze physical conditions in diffuse gas.

s^1: Permitted Resonance Lines

When a single electron lies in an outer s-shell, the atom is hydrogen-like, and the spectrum is dominated by permitted transitions between the various electronic excited states. Permitted transitions from the ground state are resonance lines that may be excited by photons. Because most of the atoms in a diffuse gas are in their ground states, these resonance transitions may become optically thick in HH objects.

Table 1. Ground electronic state configurations for abundant ions

Configuration	Examples	Electrons
s^1	H I, He II,	1
	Li I, C IV, N V, O VI	3
	Na I, Mg II, Al III, Si IV	11
	K I, Ca II, Ti IV	19
s^2	He I	2
	C III, N IV, O V	4
	Mg I, Al II, Si III	12
	Ca I, Ti III	20
p^1/p^5	C II, N III, O IV	5
	Si II, S IV, Fe XIV	13
	Ne II, Na III, Mg IV, Al V	9
	Ar II, K III, Ca IV, Fe X	17
p^2/p^4	C I, N II, O III, Ne V	6
	O I, Ne III, Na IV	8
	Si I, P II, S III, Ar V	14
	S I, Ar III, K IV, Ca V	16
p^3	N I, O II, Ne IV, Na V	7
	P I, S II, Cl III, Ar IV, K V	15

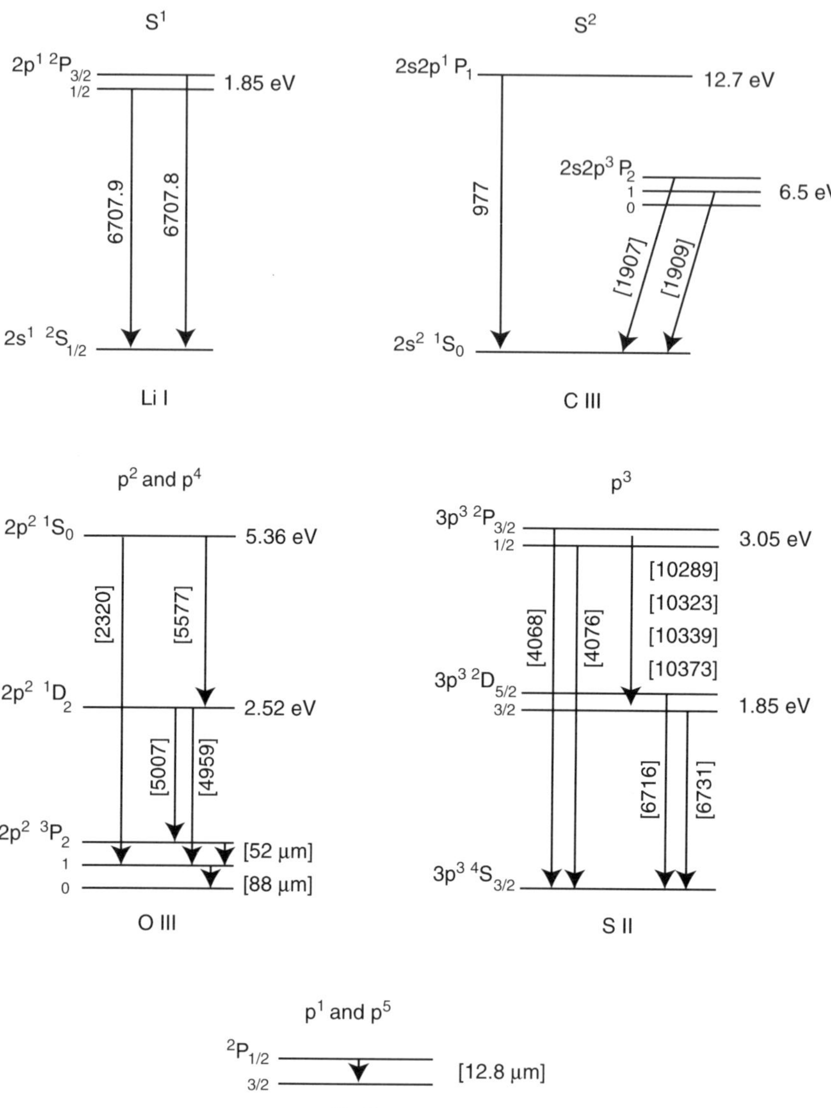

Fig. 2. Representative energy level diagrams for each of the five most common isoelectronic sequence types. Forbidden lines are denoted with two brackets, and semi-forbidden lines with a single bracket. Wavelengths are in Å unless otherwise noted and are applicable only for the specific ion shown

In hydrogen, the first excited state lies 10.2 eV above ground, so the resonant lines are all in the ultraviolet as part of the Lyman series. However, atoms like Na I have a single electron in the 3s shell, which lies only about 2.1 eV below the 3p shell, so the transition falls in the optical as the Na D doublet at 5890 Å and 5896 Å. Likewise, the Li I resonance line, so useful in establishing youth in T Tauri spectra, comes from the 2p–2s transition (Fig. 2) and is a closely spaced red optical doublet at 6707 Å. The resonance lines are all doublets because photons carry one unit of angular momentum, so the upper states of resonance absorption lines that originate from the ground s-shell must all be p-shells. A single electron in a p-shell is a doublet, with a spectroscopic configuration of $^2P_{3/2,1/2}$.

There are many well-known permitted resonance transitions of common s^1 elements, including C IV 1548+1550 Å, N V 1243+1239 Å, O VI 1032+1038 Å, Na I 5890+5896 Å (the "D" lines), Mg II 2796+2804 Å, Al III 1855+1863 Å, Si IV 1394+1403 Å, and Ca II 3933+3968 Å (the "K" and "H" lines, respectively). Many of these lines are seen in emission and in absorption in the spectra of active T Tauri star chromospheres, in classical T Tauri stars caused by accretion activity, and in the interstellar medium as narrow absorption features.

Ca II has an interesting complication in that the 3d energy level falls between the 4s and 4p levels responsible for the H and K lines. Transitions from 4p to 3d are allowed and give rise to the 'infrared triplet' of emission lines at 8498+8542+8662 Å that are bright in chromospherically active stars. The 3d state is metastable and decays back to 4s via a pair of forbidden lines at 7291 Å and 7324 Å. Hence, unlike the majority of forbidden lines, the upper levels of these transitions may be pumped by photons under some circumstances. However, in HH objects the dominant means of excitation for the 3d state appears to be collisions [19]. The main use of the 7291 Å line is to remove its companion line, 7324 Å, from a blend that includes the red quartet of auroral lines from O II, which are important as diagnostics of temperature, density, and ionization fraction as described below.

The 2s state of H, which can be populated, for example, via an excitation from 1s to 3p followed by a decay 3p–2s, is also a metastable state because an electron in the 2s level cannot decay to 1s by emitting a single photon without violating conservation of angular momentum. In a dense gas, collisions couple 2s–2p, which can then decay to 1s, but in a low density gas like that found in HH objects ($n_e \sim 10^4$ cm^{-3}), the only viable way for the 2s state to decay to 1s is by emitting two photons, which is an allowed process if one includes the A^2 term in the perturbation expansion in (8). In 2-photon emission, a virtual energy level appears somewhere between 1s and 2s to facilitate the transitions. The virtual level can occur anywhere, so the resulting spectrum is a continuum that peaks in the ultraviolet around 1400 Å [49].

Two-photon continua have been observed in HH spectra, both in the blue and in the ultraviolet [7, 10]. The continua are strong because HH objects are mostly neutral and are excited by shock waves with enough energy to populate the $n = 2$ level of hydrogen easily. The intrinsic spectral shape of the two-photon continuum is a fixed function that is independent of physical parameters such as density and temperature, so one can easily infer the luminosity of the continuum by observing at a single wavelength and applying a bolometric correction [23].

s^2: Singlets, Triplets, and Semiforbidden Lines

Isoelectronic sequences with two electrons in an outer s-shell have two distinct sets of energy levels: a triplet state where the spins of the two electrons align, and a singlet state where the spins are antiparallel. Such systems are He-like, and have a ground state of $^1S_0$. The $^3S_1$ state does not exist for equivalent s-electrons owing to the exclusion principle, so the lowest energy triplet state is ^3P. Permitted (resonance) lines from higher singlet electronic states (e.g. 3s3p–3s^2) exist in the spectra, and permitted transitions also exist between the various triplet states.

For He I, the ground state 1s^2 $^1S_0$ lies $\sim$20 eV below the first-excited state, so transitions to ground are extreme ultraviolet or even soft X-ray photons. Transitions between the upper states give rise to many optical and infrared lines. For the singlets, these lines include λ2.058 μm (2p–2s), λ5015 (3p–2s), and λ6678 (3d–2p). For the triplets, He I has λ10830 (2p–2s), λ3888 (3p–2s), and λ5876 (3d–2p). He I lines have been used recently by Edwards to probe the inner winds of jets [11].

Transitions also exist between the triplet states and the singlet states that may be either semiforbidden or forbidden. In the example depicted in Fig. 2, the λ1909 line of C III would be an electric dipole line were it not for the change of spin that makes it semiforbidden. The companion line at λ1907 is forbidden (magnetic quadrupole) because the J-value changes by 2. The [C III λ1909 line has been used to probe flows in young stars very close to the sources [14].

p^3: The Density Diagnostic

All p^3 configurations have a ground level and two well-spaced doublets as the upper levels (Fig. 2). The capacity of the p-shell is 6 electrons, so a p^3 configuration is half-filled and there is no clear rule regarding which J-level has a higher energy within the doublets. For example, the $^2P_{3/2}$ state has a higher energy than the $^2P_{1/2}$ state for N I, S II, and Ar IV, but a lower energy state for O II and Ne IV.

As we found in Sect. 2.1, flux ratios of the lines from upper-state doublets are density diagnostics. In the example in Fig. 2, these include the [S II] 6716/6731 ratio, the [S II] 4068/4076 ratio, and various combinations of the four infrared lines around 1 μm. Ratios between transitions that originate from ^2P to those that start from ^2D are sensitive to both density and temperature.

Unfortunately, the higher ionization p^3 configurations in Table 1 are usually not seen in HH jets, which have low ionization, and the [N I] 5198/5200 Å ratio (the analog of [S II] 6716/6731) has a very low critical density and the lines are usually blended because they are so close together in wavelength. Hence, S II is the most important ion for density measurements, though O II is also useful when all the emission lines are bright enough to measure accurately.

p^2/p^4: Temperature and Reddening Diagnostics, Far-IR Lines

The p^2 and p^4 configurations are similar, each with a ^3P ground state, a first excited state of $^1D_2$ a few eV above ground, and a second excited state $^1S_0$ a few eV higher. Transitions between $^1D_2$ and ^3P are referred to as nebular lines, while transitions between $^1S_0$ and $^1D_2$ are auroral lines, and those between $^1S_0$ and ^3P are transauroral lines. Auroral and transauroral lines are strong in the Earth's aurora because these lines have high critical densities; in contrast, nebular lines are bright in astrophysical nebulae where the densities are low enough to mitigate the effects of quenching. Ratios between nebular and auroral, and nebular and transauroral lines are good diagnostics of temperature, while ratios of auroral and transauroral lines give the reddening (Sect. 2.1).

Many of the brightest emission lines in HH objects come from this group of atoms, such as [C I] 9849+9823, [N II] 6548+6583, [O III] 4959+5007, [O I] 6300+6363, [O I] 5577, and [Ne III] 3869+3968. Forbidden transitions between the ^3P levels produce a number of well-known infrared lines, including [C I] 610 μm, [N II] 204 μm, [O I] 63 μm, [O III] 88 μm, and [Ne III] 15.4 μm. In addition, the first excited electronic state (e.g. 2p^{3}3s for the 2p^4 atoms; 2s2p^3 for 2s^{2}2p^2 atoms) gives a doublet level $^5S_{2,1}$ that generates a bright ultraviolet semiforbidden line in the transition to $^3P_1$. Examples include [O I 1302 Å , [C I 2965 Å , [N II 2140 Å , and [O III 1663 Å.

p^1/p^5: An Infrared Forbidden Line, and UV Permitted and Semiforbidden Lines

A single p-electron or hole has a single, closely spaced doublet as a ground state, with $^2P_{1/2}$ level the lowest energy for p^1, and $^2P_{3/2}$ the lowest energy for p^5. Transitions within this doublet include [C II] 156 μm, [N III] 57 μm, [O IV] 26 μm, [Si II] 35 μm, [Ne II] 12.8 μm, and [Ar II] 7.0 μm, all in the infrared. Highly ionized coronal lines of [Fe X] and [Fe XIV] are in the optical at 6375 Å and 5303 Å, respectively.

The first and second excited electronic states for p^1 and p^5 occur when an electron from the s-shell is excited to the p-shell. The resulting configuration has two p-electrons and one s-electron hole. The result gives p^2 ^4P and p^2 ^2D states, which may decay to ground and give rise to both forbidden and semiforbidden transitions. Examples include [C II 2325 Å and C II 1335 Å.

3 Applying Nebular Line Diagnostics to Stellar Jets

Within the first decade after their discovery, Böhm [2] applied standard nebular diagnostic techniques to spectroscopic observations of the brightest HH object, HH 1. The analysis relied upon the ratio of nebular to auroral lines within O I and S II to constrain temperature and density. The results, which still hold today, some 50 years hence, are that a typical electron temperature is $\sim$8000 K and typical electron density $\sim 10^4 \mathrm{cm}^{-3}$. The analysis also included line ratios between different elements, such as O and Ne.

The nature of the heating source of HH nebulae was a significant puzzle until the mid-70s. In his discovery paper, Herbig [26] noted the lack of blue photoionizing stars near HH 1 and HH 2, which seemed to rule out photoionization as a heating source. However, not all HH objects were as isolated from stellar radiation, and in 1974, Strom et al. [50] found an extended emission line source HH 102 in the L 1551 star formation region that was also a reflection nebula. We now know that HH 102 is a rather unusual object and contains a mixture of 'true' HH emission nebulae superposed upon an extended reflection nebula that marks the edge of an outflow cavity. Motivated by HH 102, Strom et al. concluded that all HH objects were reflecting the light of a young star. The reflection nebula hypothesis was able to account for the mostly blueshifted radial velocities seen in HH objects, because circumstellar dust in an infalling envelope will see any emission lines from the star to be blueshifted and will reflect that blueshifted line emission to the observer.

A few years later, Schwartz [45] came up with the modern explanation that shock waves excite HH objects. Schwartz was motivated by the similarity between HH spectra and those of supernova remnants, where a range of ionization states exist. In addition, the emission line analysis showed HH objects to have a low filling factor, as expected for the cooling zones of shocks. A few years later, Schwartz [46] found supersonic velocity dispersions of >100 kms^{-1} in HH objects and noted that large line widths occur when a bow shock forms around a stationary obstacle in the flow. As the 1970s drew to a close, Schmidt and Miller [44] put the reflection nebula hypothesis to rest by noticing that while continuum in HH objects is polarized, emission lines are not, so the emission lines were intrinsic to the object and not scattered light.

The realization that HH objects were shock waves led to a great deal of theoretical work (e.g. by Raymond [41] and Dopita [9] with the goal of explaining the observed emission-line ratios. With the advent of computers, it became possible to model the physical conditions within a cooling zone of a shock and to predict the line emission from such a zone. The computer revolution also greatly facilitated calculations of the collision strengths and A-values needed to address this problem. Shock models of HH objects enjoyed some success, but it was clear that no single shock model could reproduce the observations well.

The first comprehensive work on the emission-line analysis of HH objects is the classic 1981 paper of Brugel, Böhm, and Mannery [5]. This paper stands today as an outstanding analysis of emission lines in HH objects. By combining dozens of emission lines from various elements for several HH objects, the authors made several key discoveries and pointed out some fundamental limitations as well. The new spectra reproduced the earlier measurements of T_e and N_e, but the uncertainties were low enough to show a real intrinsic scatter in the measurements, indicative of a range of temperatures and densities. These observations clearly showed that emission-line analyses are problematic when the cooling zones of the shock waves are not resolved spatially, because in that case a wide range of densities and temperatures contributes to the observed line ratios. Another important result was that the volume filling factors were typically 1% or less, implying a filamentary, clumpy emitting zone. The authors also attempted to determine the role of dust in the flows by measuring refractory abundances, but the uncertainties were high.

When Herbig and Jones [27] presented the first proper motion study of HH objects in 1981, it finally became clear that HH objects represented dense, bullet-like objects in highly collimated outflows. The collimated nature of the flows implied that redshifted flows were rare simply because they propagate into the dark cloud where the extinction is high. The combination of large emission line widths found by Hartmann and Raymond [25] and high-excitation lines of C IV discovered in the first ultraviolet observations of Böhm, Ortolani, and their collaborators [3, 35] gave further support to the notion of HH objects as bow shocks. Very good matches to position-velocity diagrams from long-slit spectra by Raga et al. [39] and to velocity-resolved integrated emission line profiles by Hartigan et al. [24] were possible by incorporating radiative shock models with a bow shock geometry, although the bow shocks were not affixed to stationary cloudlets as Schwartz envisioned but rather resembled those that would accompany a bullet of dense gas moving through the surrounding medium.

The availability of CCDs made deep emission line images possible, and one of the first discoveries from this work (Mundt and Fried [33]) was that HH objects are typically parts of dense jets. Hartigan [16] noted that one of the expectations for shocks in jets is that both a bow shock (in the medium ahead of the shock) and a Mach disk (in the jet) should occur, and that both shocks should be visible, and subsequent observations by Morse, Reipurth, and their collaborators [31, 42] showed this to be the case from both emission line ratios and kinematics. It became clear from these studies that bow shocks in jets typically move into the wakes of previous ejections, a notion published earlier by Dopita [8].

As described in Sect. 2, observations of emission line ratios from p^3 ions like S II give direct measures of the electron density, provided the gas is not in the HDL or LDL. To convert an electron density to a total density, one needs to estimate the ionization fraction in the gas. The correction can be large because HH object jets are mostly neutral.

It is certainly convenient to think of the jet as being filled with some plasma of a fixed density and ionization fraction. However, it is important to realize that the ionization fraction, like the electron density, varies dramatically for shocked flows. As neutral material enters a shock, the temperature rises sharply to the postshock value, and collisions gradually ionize the atoms. While this process occurs, the gas radiates emission lines and cools. Shock models show that the ionization fraction peaks at some value and then gradually declines. Most forbidden emission lines occur in a region where the ionization fraction is declining, so the value that one measures for the ionization fraction depends on where the ions used for the measurement are located in the postshock flow. Using emission lines to constrain the shock models in their 1994 paper, Hartigan et al. [21] found the ionization fractions of the brightest portions of some famous jets to be only a few percent.

Instead of using shock models, one can try to infer physical conditions along the jet by simply using the observed emission line ratios everywhere in the flow, as was done for large apertures in the Brugel et al. 1981 paper. This procedure is the equivalent of modeling the emitting gas as having a constant density, temperature, and ionization fraction over the aperture or pixel of the observed line ratios. As Brugel et al. showed in their paper, analyzing emission lines in this manner is of limited value for systems like HH 1 and HH 2, which are dense, high-shock-velocity objects with unresolved cooling zones whose emission lines span a wide range of ionization states. Similar issues were found by Podio et al. [38] who analyzed both infrared and optical emission lines.

However, there are cases where the cooling zones are resolved, for example, in many HST images. For resolved images, a single density, temperature, and ionization state may characterize the gas well. For these cases, Bacciotti & Eislöffel [1] used an analysis similar to that of Brugel et al. but also used the fact that the charge exchange cross sections were large enough to effectively tie the ionization fractions of N and O to that of H as described in Sect. 2.2. By incorporating charge exchange into the analysis in this manner, one can use the bright red emission lines of [O I], [N II], and [S II] to infer all the physical properties of interest. Results from this analysis show that the ionization fractions range from a few percent in the regions that overlap the Hartigan et al. study (and therefore are in agreement with those results) to tens of percent in other regions. Of particular interest was the result that some jets appear to have ionization fractions that decline as they move away from their sources, suggesting some heating event close to the source.

Even with good measures of ionization fraction and density, estimates of mass loss rates in jets are plagued with uncertainties caused by the inherent clumpy nature of the flows, as I discussed in a recent review [17]. Emission line fluxes increase with the density, so images emphasize dense regions that cool rapidly. There may be cool dense regions within the jet that do not radiate, as well as less dense material that fills a larger fraction of the volume within the jet. The best procedure is probably to measure the mass loss rate as close to the star as possible, before shock waves concentrate the jet into clumps. In any case, because jets are mostly neutral they carry a larger amount of energy and momentum than one would infer by simply using their electron densities, as was done in some early papers [32].

4 Some New Results, Future Prospects

When I presented this lecture in Elba, I included results from two papers that were in the process of being submitted to the Astrophysical Journal. These papers have now been accepted for publication, so I will summarize only the results briefly here.

4.1 Slitless Spectroscopy and a New Analysis Method for Emission-Line Objects

The first paper [18] uses the rather unusual technique of slitless spectroscopy. By opening the spectroscopic slit wide enough to include emission from the entire jet, one obtains an image for each emission line in the jet with a single exposure. The technique is limited to objects that are not resolved spectrally. Using HST over two epochs, we were able to obtain high signal-to-noise images of many emission lines.

Of particular interest to this review is the new analysis technique introduced in the paper, which can be employed for any emission line object. Rather than use a specific emission-line ratio to infer electron densities, another to estimate the temperature, and so on, the paper developes a method where one uses all the emission line ratios, appropriately weighted according to their uncertainties, of all the lines in the analysis. The advantage of this technique is particularly apparent close to the star, where the [S II] 6716/6731 ratio is in the HDL. In the new analysis, the ratio simply provides a lower limit to the density, as it should, and the actual density measurement is constrained more by ratios that involve higher critical density lines such as [O I] 6300. The optimal solution for the electron density, temperature, and ionization fraction are those that minimize a quadratic form that resembles a chi-square, although the statistical distribution is more complex because the line ratios are not necessarily independent of one another.

The analysis yields several intriguing maps of the physical parameters in the HH 30 jet. Since the data were taken in two epochs separated by two years, it is possible to follow proper motions in the jet and observe how the physical parameters vary with time. Results include a drop of ionization with distance, an increase of density along the axis of the flow, and what appear to be heating events that ionize the flow within ~100 AU of the source and move outward with the flow. By observing such phenomena, we can begin to connect the dynamics of jets with the accretion disks that drive them.

4.2 Magnetic Fields in Jets

Magnetic fields are particularly difficult to constrain from emission line ratios because the lines do not show Zeeman splitting. However, one can estimate field strengths by knowing enough about the shock. For example, given the preshock density (constrained by the observed line fluxes) and shock velocity (determined by the line ratios and observed shape of the bow shock) a shock model predicts what the electron density should be in the cooling zone. Observed electron densities are at least an order of magnitude lower than the predicted value, indicative of some other source of pressure. Even a small magnetic field in the preshock gas translates into a significant pressure in the postshock region and will lower the observed electron density there.

Results from this analysis by Morse and collaborators [31] are that magnetic fields in front of bow shocks in jets are low, $\sim 15\,\mu$G. Although one might expect magnetic fields to be strong in jets if they are driven by magnetized accretion disks, these observations indicate that fields play only a minor role in the dynamics at large distances. The only field measurement close to the source seems to be that of Ray et al. [40], who found strong fields of the order of a few Gauss at distances of tens of AU from the source.

Can we reconcile a jet being strongly magnetized at the source but only weakly magnetized at large distances? For a magnetized disk wind the field will be mostly toroidal at large distances and will scale as r^{-1}, while the density drops as r^{-2} for a collimated jet. Hence B$\sim n^{0.5}$, so the magnetic signal speed should remain constant with distance and a flow that is strongly magnetized (e.g. $V_{\mathrm{Alfven}} \sim 100$ km s^{-1}) should remain so at large distances. In fact, such a strong field would prohibit any weak shocks from forming in the jet, but we know shock velocities in jets are ~ 30 kms^{-1} from proper-motion measurements as well as emission-line studies [21, 22].

In order to study this issue, we performed the first pulsed magnetized simulations aimed at connecting physical conditions at large distances with those present near the source. The main results, to appear in print this year [20], show that B$\sim$n in shocks and rarefactions, so the global structure of a pulsed jet follows some intermediate power $\alpha \sim 0.85$, where B$\sim n^{\alpha}$. Because $\alpha > 0.5$, the magnetic signal speed drops with distance, on average, and the jet gradually becomes more hydrodynamic instead of MHD at larger distances from the source. Essentially the velocity perturbations sweep the field into a few dense magnetized blobs that are separated by low field regions.

4.3 Future Research

Future research of emission lines in HH objects and jets will probably divide scientifically into two broad categories. On one hand, there is a great deal of interest in connecting the jets to their launching points in the accretion disks that surround young stars. The process of accretion and collimated outflow occurs elsewhere in the universe (interacting binaries, black holes, AGN's, and perhaps even planetary nebulae), but the phenomenon is easiest to study in young stars because HH jets radiate emission lines and are spatially resolved. Many unanswered questions remain in this area, including the role that magnetic fields play in driving the jet, the importance of time variability and instabilities, how mass is loaded from the disk into the jet, and the transfer of angular momentum from the disk to the outflow.

A second broad area is the study of the fluid dynamics within HH flows. Many questions remain here as well, especially with regard to the role magnetic fields play in the dynamics of colliding clumps, the efficiency of entrainment along the edges of jets, and the physics of the shock waves that couple the atomic outflow with that of the molecular gas. It may even be possible to create laboratory analogs of astrophysical jets [13, 29]. Being able to create jets in a laboratory setting is an attractive prospect, especially if the setup allows one to control the strength and geometry of magnetic fields in a collimated supersonic flow.

I close with a plea to everyone working in this field to try to obtain time-resolved spectra and images of HH jets over many epochs. Only with such data sets will we be able to really compare motions and shock waves in real jets with their numerical and laboratory analogs. Time-resolved data are the only way to determine the degree to which accretion is tied to outflow. The connection between accretion and outflow is perhaps the most fundamental aspect of the jet phenomenon, but our understanding of it remains superficial.

References

1. Bacciotti, F., Eislöffel, J.: Astron. Astrophys. **342**, 717 (1999)
2. Böhm, K.-H.: Astrophys. J. **123**, 379 (1956)
3. Böhm, K.-H., Böhm-Vitense, E., Brugel, E.: Astrophys. J. **245**, L113 (1981)
4. Bransden, B., Joachain, C.: Introduction to Quantum Mechanics. Wiley, New York (1989)
5. Brugel, E., Böhm, K.-H., Mannery, E.: Astrophys. J. Suppl. **47**, 117 (1981)
6. Chen, G., et al.: Phys. Rev. A **74**, 2709 (2006)
7. Curiel, S., Raymond, J., Wolfire, M., Hartigan, P., Morse, J., Schwartz, R., Nisenson, P.: Astrophys. J. **453**, 322 (1995)
8. Dopita, A.: Astron. Astrophys. **63**, 237 (1978)
9. Dopita, M.: Astrophys. J. Suppl. **37**, 111 (1978)
10. Dopita, M., Binette, L., Schwartz, R.: Astrophys. J. **261**, 183 (1982)
11. Edwards, S., In: Jets from Young Stars II, F. Bacciotti, E.Whelan & L. Testi (eds.), Lect. Notes Phys, **742**, 3–13 (2008)
12. Eisberg, R., Resnick, R.: Quantum Physics of Atoms, Molecules, Solids, Nuclei, and Particles. Wiley: New York (1974)
13. Foster, J., et al.: Astrophys. J. **634**, L77 (2005)
14. Gomez de Castro, A., Verdugo, E.: Astrophys. J. **597**, 443 (2003)
15. Haro, G.: Astrophys. J. **55**, 72 (1950)
16. Hartigan, P.: Astrophys. J. **339**, 987 (1989)
17. Hartigan, P.: Astrophys. Space Sci. **287**, 111 (2003)

18. Hartigan, P., Morse, J.: Astrophys. J., **660**, 426 (2007)
19. Hartigan, P., Edwards, S., Pierson, R.: Astrophys. J. **609**, 261 (2004)
20. Hartigan, P., Frank, A., Varniere, P., Blackman, E.: Astrophys. J. **661**, 910 (2007)
21. Hartigan, P., Morse, J., Raymond, J.: Astrophys. J. **436**, 125 (1994)
22. Hartigan, P., Morse, J., Reipurth, B., Heathcote, S., Bally, J.: Astrophys. J. **559**, L157 (2001)
23. Hartigan, P., Morse, J., Tumlinson, J., Raymond, J., Heathcote, S.: Astrophys. J. **512**, 901 (1999)
24. Hartigan, P., Raymond, J., Hartmann, L.: Astrophys. J. **316**, 323 (1987)
25. Hartmann, L., Raymond, J.: Astrophys. J. **276**, 560 (1984)
26. Herbig, G.H.: Astrophys. J. **113**, 697 (1951)
27. Herbig, G.H., Jones, B.: Astrophys. J. **86**, 1232 (1981)
28. Laming, J., Raymond, J., McLaughlin, B., Blair, W.: Astrophys. J. **472**, 267 (1996)
29. Lebedev, S., et al.: Astrophys. J. **616**, 988 (2004)
30. Mendoza, C.: IAU Symposium 103, Planetary Nebula, p. 143, Reidel, Dordrecht, 1983
31. Morse, J., Hartigan, P., Cecil, G., Raymond, J., Heathcote, S.: Astrophys. J. **399**, 231 (1992)
32. Mundt, R., Brugel, E., Bührke, T.: Astrophys. J. **319**, 275 (1987)
33. Mundt, R., Fried, J.: Astrophys. J. **274**, L83 (1983)
34. Nisini, B., Caratti o Garatti, A., Giannini, T., Lorenzetti, D.: Astron. Astrophys. **393**, 1035 (2002)
35. Ortolani, S., D'Ordorico, S.: Astron. Astrophys. **83**, L8 (1980)
36. Osterbrock, D.: Astrophysics of gaseous nebulae and active galactic nuclei, p. 37. University Science Books, Mill Valley (1989)
37. Pesenti, N., Dougados, C., Cabrit S., O'Brien, D., Garcia, P., Ferriera, J.: Astron. Astrophys. **410**, 155 (2003)
38. Podio, L., Bacciotti, F., Nisini, B., Eislöffel, J., Massi, F., Giannini, T., Ray, T.: Astron. Astrophys. **456**, 189 (2006)
39. Raga, A., Böhm, K.-H.: Astrophys. J. **308**, 829 (1986)
40. Ray, T., Muxlow, T., Axon, D., Brown, A., Corcoran, D., Dyson, J., Mundt, R.: Nature **385**, 415 (1997)
41. Raymond, J.: Astrophys. J. Suppl. **39**, 1 (1979)
42. Reipurth, B., Heathcote, S.: Astron. Astrophys. **257**, 693 (1992)
43. Rybicki, G., Lightmann, A.: Radiative Processes in Astrophysics. Wiley: New York (1979)
44. Schmidt, G., Miller, J.: Astrophys. J. **234**, L191 (1979)
45. Schwartz, R.: Astrophys. J. **195**, 631 (1975)
46. Schwartz, R.: Astrophys. J. **223**, 884 (1975)
47. Shevelko, V.: Atoms and Their Spectroscopic Properties, p. 41. Springer-Verlag, Heidelberg (1997)
48. Smith, N., Hartigan, P.: Astrophys. J. **638**, 1045 (2006)
49. Spitzer, L., Greenstein, J.: Astrophys. J. **114**, 407 (1951)
50. Strom, S., Grasdalen, G., Strom, K.: Astrophys. J. **191**, 111 (1974)

Part II

Adaptive Optics, and Space Observations
Spectro-astrometry

Basic Concepts and Parameters
of Astronomical AO Systems

S. Esposito and E. Pinna

INAF - Osservatorio Astrofisico di Arcetri, L.go E. Fermi n.5 - 50125 Firenze,
ITALY
esposito@arcetri.astro.it

Abstract. The paper describes the basic concepts and parameters of astronomical adaptive optic (AO) systems. In particular, the paper introduces and discusses the main parameters and error sources that determine the performances of an AO system. From this discussion, the current limitations of AO systems are derived. Following this, the laser-generated reference star method is described, being the best technique to solve the main limitation of Astronomical AO system, namely the limited sky coverage. Then advantages and disadvantages of such technique are given. Using the considered matter, the case of LBT, the first 8 m class telescope using a deformable secondary mirror, is described. The LBT AO system performances in the single telescope case are briefly outlined. Then some concepts of optical interferometry are reported. They are used to analyze the definition of homothetic interferometer, the LBT case, and explain its differences from the Michelson stellar interferometer configuration. Finally a short description of the LBT interferometric instruments is given.

S. Esposito and E. Pinna: *Basic Concepts and Parameters of Astronomical AO Systems*, Lect. Notes Phys. **742**, 45–78 (2008)
DOI 10.1007/978-3-540-68032-1_3

Key words: Atmospheric wavefront aberration, Adaptive Optics, Laser Guide Star, Homotetic interferometer, Large Binocular Telescope

1 Introduction

High angular resolution observations became more and more important in modern Astrophysics in all the classical research fields from Solar System studies to Cosmology. In this framework, the development of techniques that allow us to reach diffraction limited performance using the present telescopes are of high importance. Among all the techniques applied to ground-based telescopes to reach a diffraction limited performance, the most powerful one is the so-called *Adaptive Optics* or AO. This technique aims to measure the wavefront perturbation due to refraction index fluctuations of the atmosphere and then to correct it quickly to follow the turbulence evolution. In this kind of system, there are three fundamental elements, as shown in Fig. 1; the wavefront sensor that measures the instantaneous wavefront aberration, the wavefront corrector that corrects the phase fluctuation introducing different optical paths for different rays, and the control system, or the so-called wavefront computer that computes from the wavefront sensor data the commands to be applied to the wavefront corrector. The wavefront corrector is usually

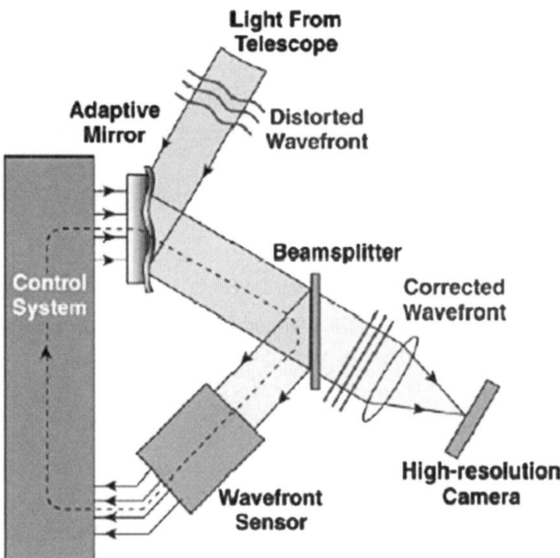

Fig. 1. A block diagram of an AO system. The figure reports the deformable mirror, the wavefront sensor, and the wavefront computer and highlights the two different optical paths of the light from the reference source and the scientific target, respectively

a deformable mirror[1]. Finally the system needs a reference star to measure the wavefront aberration. This object has to be quite bright as we will see in the next sections. Hence, in some cases it cannot be the scientific target itself because it is too dim. In this situation, the reference star has to be angularly close enough to the scientific target so that the achieved wavefront measurement can be successfully used to correct the image of the scientific object. The adaptive optics system having the architecture presented in Fig. 1 was introduced for the first time by H. Babcock in 1953 [1]. Following this, authors like Tatarsky [27] and Fried [10] developed the basis of the theory of image formation in random media, which has subsequently been directly applied to compute the performance and limitation of AO systems. The practical realization of an AO system in Astronomy began later, at the end of the 1980s, with the construction of the Come-On system [25] for the ESO 3.6 m telescope on La Silla. From then on, all the 8 m class telescopes like VLT, Keck, Gemini, Subaru, and LBT have built and planned many different types of AO systems. Today, the majority of professional telescopes of the 4 m and 8 m class have an adaptive optics system and related diffraction limited instruments.

This lecture can be divided in to two parts. The first part aims to describe the basic concepts and limitations of contemporary AO systems (Sects. 2,3,4,5, and 6). The second part (Sects. 7 and 8) briefly introduces the AO systems of LBT telescope [17] for the single telescope first and then in interferometric mode. Here, the aim is to describe what should be available in terms of angular resolution at LBT in the next few years. Also as optical interferometry is becoming increasingly important with time, we outline some basic features of interferometric beam combinations, in particular in relation to the concept of homothetic interferometers. Homothetic interferometers allow one to divide interferometers into two classes, the Michelson stellar interferometer b[22] (like the VLTI) and the Fizeau interferometers (like the LBT). It is clear that an AO system is now a fundamental sub-system of any instrument of an 8–10 m class telescope and indeed of any extremely large telescope (ELT) presently under design. Hence, we believe that it is crucial for the next generation of young astronomers to have a thorough knowledge of the basic principles of AO.

2 Opening Notes on Image Formation in Astronomical Telescopes

We will summarize in the following some basic results on image formation theory, as achieved following the linear theory of optics and Fourier Optics in particular. For the rest of the paper, we will consider all the astronomical objects as incoherent sources so that the superposition principle applies to

---

[1] By deformable mirror (DM), we mean an electro-optical device having a reflecting surface with an adjustable shape.

the object intensities and not to the object electric field. Having said that, we start our work from the basic equation reported below

$$I(\boldsymbol{\alpha}) = \int O(\boldsymbol{\beta})\mathrm{PSF}(\boldsymbol{\beta} - \boldsymbol{\alpha})\mathrm{d}\boldsymbol{\beta}, \tag{1}$$

where $\boldsymbol{d} = \lambda\boldsymbol{f}$ while $\boldsymbol{\alpha}$ and $\boldsymbol{\beta}$ are two bi-dimensional vectors giving the 2D coordinates of a point in the telescope focal plane expressed as angle, $O(\boldsymbol{\alpha})$ and $I(\boldsymbol{\alpha})$ are the object and image intensity distributions, respectively, $\mathrm{PSF}(\boldsymbol{\alpha})$ is the telescope's so-called Point Spread Function or the image of a point like source. This last quantity is easily determined using Fourier optics and can be expressed as:

$$\mathrm{PSF}(\boldsymbol{\alpha}) = \mid \mathrm{FT}\,[W(\boldsymbol{\alpha})]\mid^2 \tag{2}$$

In writing this equation, we assumed that no aberration is present and so the function W, the so-called telescope pupil function, is simply one inside the circle of radius R and zero outside[2]. For this circular aperture telescope, the PSF is usually called the Airy disk that has an FWHM equal to

$$\theta = \lambda/D,$$

where λ and D are the operating wavelength and the telescope diameter. A cut passing through the center of the PSF is represented in Fig. 2. The meaning of (1) can be clarified by looking at the same formula in the Fourier space, as is usually done in Fourier Optics. Taking the fourier transform of both members of (1), we obtain the following relationship

$$\widetilde{I}(\boldsymbol{f}) = \widetilde{O}(\boldsymbol{f}) \cdot \widetilde{P}(\boldsymbol{f}) \tag{3}$$

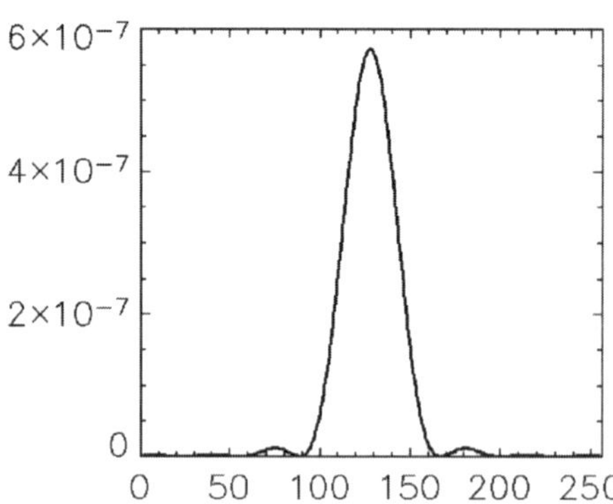

Fig. 2. The central cut of the Point Spread Function of a diffraction limited circular lens

---

[2] We will assume for all the following a circular telescope with no obstruction.

where $\boldsymbol{f}$ is the bidimensional coordinate in the Fourier plane, and P indicates now for simplicity the PSF function. Taking the modulus of left and right members of above equation gives:

$$\mid \widetilde{I}(\boldsymbol{f}) \mid = \mid \widetilde{O}(\boldsymbol{f}) \mid \cdot \mid \widetilde{P}(\boldsymbol{f}) \mid \tag{4}$$

The modulus of $\widetilde{P}$ is defined in optics as the Modulation Transfer Function (MTF) and can be expressed considering the equation for the PSF and the Fourier transform auto-correlation theorem in the following way:

$$\mathrm{MTF}(\boldsymbol{f}) \propto \int W(\boldsymbol{r})W(\boldsymbol{r} + \lambda\boldsymbol{f})\mathrm{d}\boldsymbol{r} \tag{5}$$

It is easy at this point to see that the MTF acts like a filter that attenuates (and de-phases) the Fourier components of the intensity distribution of the considered object. If the MTF is close to zero or zero at a given spatial frequency, that frequency will not be present in the object image. Moreover, it is clear to see that when $\lambda\boldsymbol{f} > D$ the MTF is zero so that the maximum spatial frequency transmitted by the telescope is D/λ. A graphical representation of the MTF can be found considering that the integral of (5) is given by the over-position area of the two circles of diameter D displaced by a quantity $f\lambda$, as reported in Fig. 3. In the case where no aberrations are present, the MTF function can be analytically derived. [The mathematical expression is quite complicated and can be found in several Optics textbooks [15].] However the diffraction-limited MTF behavior is very simple and almost linear, as reported in Fig. 4. In this case, it easy to see that the MTF retains a value different from zero for all the x axis values or all the spatial frequencies in the range $0, D/\lambda$. When some aberrations are present, the system MTF is degraded, and higher spatial frequencies are attenuated or no longer even transmitted.

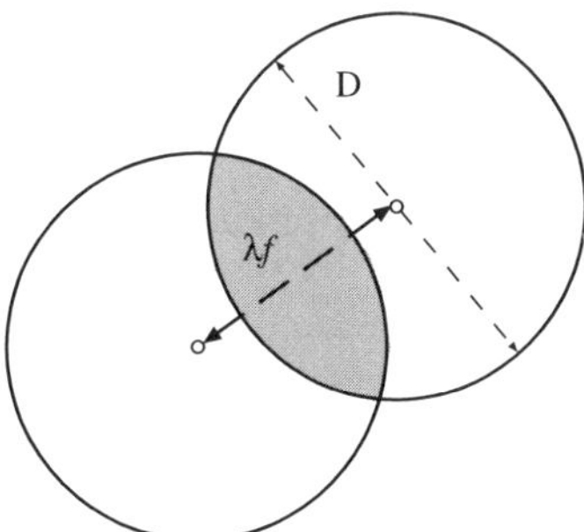

Fig. 3. The geometrical arrangement of the autocorrelation integral of (5). The frequency f here is an angular frequency so that $f\lambda$ is a length. The maximum f where the MTF is different from zero is clearly $f = D/\lambda$

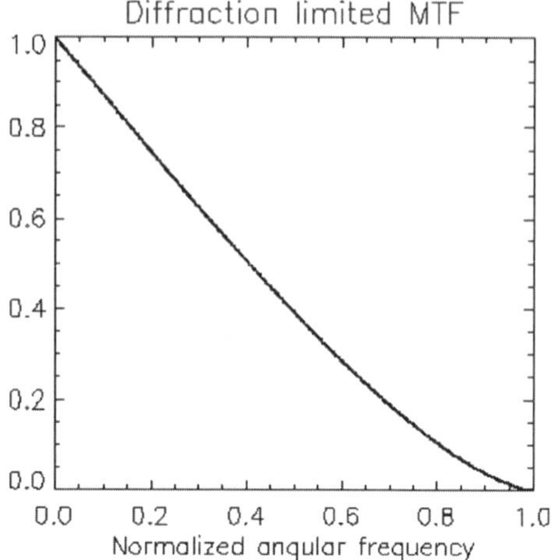

Fig. 4. The analytical representation of the MTF as function of angular frequency normalized at D/λ

3 Atmospheric Turbulence and Wavefront Aberrations

We mention in the introduction that the performance of ground-based telescopes is limited by the atmospheric refraction index fluctuations. Refraction index fluctuations originate when a turbulent flow of air is present. This turbulence, coupled with the fact that the atmospheric temperature is changing with altitude, creates a mix of air particles of different temperatures, which in turns generate the refraction index fluctuations. As we see, the wavefront passing through the atmosphere is disturbed or more precisely aberrated by these fluctuations. In this section, we will quantify the wavefront aberration due to atmospheric turbulence introducing the phase structure function and the Fried parameter r_0 starting from the velocity fluctuation in a turbulent flow.

3.1 Turbulent Flow and Atmospheric Turbulence

The velocity fluctuation distribution in a turbulent flow was studied by Kolmogorov in 1941 [19] and later applied in the branch of optical waves propagation through turbulence by Tatarsky in 1967 [27]. Briefly, we will consider a laminar flow of a fluid having speed $\boldsymbol{v}$, typical scale L, and kinematic viscosity ν. The flow geometry is very simple and is reported in Fig. 5. In this situation, the fluid dynamics states that when the Reynolds number of the flow is larger then 10^3 the flow becomes turbulent. The fluid Reynolds number is defined as

$$R_{\mathrm{e}} = \boldsymbol{v}L/\nu$$

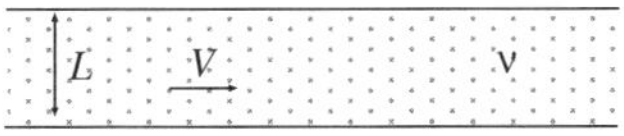

Fig. 5. A sketch of the laminar flow introducing the symbols needed in the following analysis

To understand better the meanings of the R_e number we can write it in the following way:

$$R_e = (v^3/L)/(v^2/L^2\nu) \tag{6}$$

Now we note that the numerator and denominator of the above fraction represent the kinetic energy per unit time and mass and the energy dissipated by viscous friction (per unit time and mass). So it is understandable then when the dissipated energy is much less than the energy received by the fluid (per unit of time and mass) the laminar regime breaks down and we enter the turbulent regime. This is always the case for the free atmosphere were the kinematic viscosity value is $1.5 \times 10^{-7} [\mathrm{m^2 s^{-1}}]$. Substituting this value in the Reynolds number expression, we find that even a wind speed of $1\,\mathrm{ms^{-1}}$ will generate a Reynolds number of the order of 10^6 giving rise to a fully developed turbulent flow. As already mentioned, it was Kolmogorov who devised a simple but effective approach to the description of the velocity fluctuation in a turbulent flow. The Kolmogorov theory is usually called the vortexes cascade from the simple concept that is the theory basis. This concept is explained in Fig. 6, where the vortexes cascade is graphically illustrated together with the relevant quantities of the process. Briefly, we start with the turbulent flow generating vortexes on a scale equal to the laminar flow initial scale L. This scale is usually called the outer scale of turbulence.

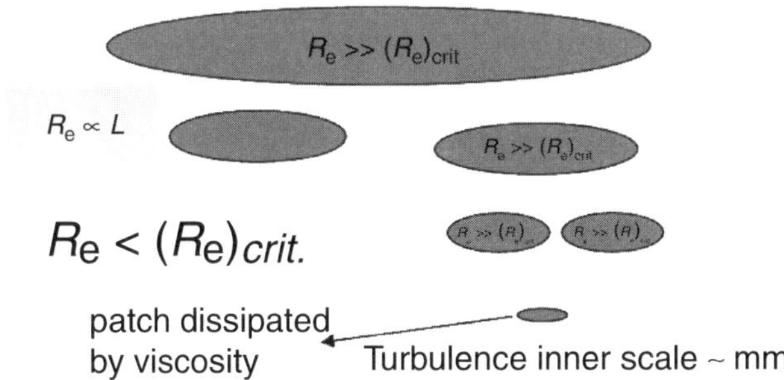

Fig. 6. Graphical representation of the vortex cascades. The turbulent flow generates a vortex of linear dimension L (*outer scale*).The Reynolds number R_e of this vortex is greater than the critical one $(R_e)_{\mathrm{crit.}}$ and the vortex is split into another vortex of a smaller linear dimension. If this vortex still has $R_e \gg (R_e)_{\mathrm{crit.}}$, the process is iterated, generating the vortex cascade. The cascade stops when the the vortex energy is dissipated by viscosity. The linear dimension of the smallest vortex is called *inner scale*

These vortexes will have some velocity fluctuation and will have a Reynolds number greater than the critical one. As a result of this condition, they vortexes will split in to other vortexes having a smaller scale and other velocity fluctuations. This process will stop when the small vortexes have a Reynolds number lower then the critical one. At this point, they will be dissipated by viscous friction in the fluid and will not split anymore. The spatial scale where this happens is called the turbulence inner scale and is of the order of a few mm. Using dimensional consideration and the above given description of the turbulence regime, Kolmogorov was able to express the so-called structure function of the velocity fluctuations that we report below

$$D_v(r) = C_v^2 r^{2/3},\tag{7}$$

where r is the distance between the two considered points, and C_v^2 is a constant that accounts for the strength of the velocity fluctuations.[3] Tatarsky applied this expression in the computation of the phase perturbation experienced by an electromagnetic wave propagating in an inhomogeneous medium (a medium where the refraction index is not constant in space), introducing the concept of a physical quantity that is a conserved passive additive [27]. The relevant example is the atmospheric temperature. This quantity does not change in any way the fluid dynamics, and its values are not modified by atmospheric turbulence. For example, a colored dye added to a turbulent flux would be a passive conserved additive and would be useful for tracing the velocity fluctuations. Tatarsky demonstrates that the fluctuation distributions of any such quantity in a turbulent flow are described by the same structure function of the velocity fluctuations. Using this result, he described the temperature fluctuation in the turbulent atmosphere by the following structure function

$$D_T(r) = C_T^2 r^{2/3},\tag{8}$$

where r is the distance of the two considered points, and C_T^2 is a constant accounting for the strength of the temperature fluctuations. The next step in Tatarsky's work is to consider the expression of the atmospheric refraction index as a function of pressure P and temperature T. This common expression is reported below for easy reading.

$$n = 1 + 10^{-6} \times 79 \left(1 + \frac{7.5 \times 10^{-3}}{\lambda^2}\right) \frac{P}{T}\tag{9}$$

From this equation, it is easy to find out the expression of refraction index fluctuation as a function of temperature. In the following, we will neglect the fluctuations due to pressure changes because pressure fluctuations are usually

---

[3] We report here the mathematical definition of the structure function $D_x(\rho)$ of a random function $x D_x(\rho) = \langle (x(r + \rho) - x(r))^2 \rangle$. The quantities r and ρ can be one, two, or three dimensional quantities. The last case is the case of the velocity fluctuation structure function.

much less in percentage than the temperature fluctuations. With this in mind we find that

$$dn = -79\frac{P}{T^2}10^{-6}dt \tag{10}$$

Combining (8) and (10) we find the expression for the refraction index fluctuation structure function reported below

$$D_n(r) = C_n^2 r^{2/3} \tag{11}$$

In this formula, the quantity C_n^2 is called the constant of refraction index fluctuations structure function and is a measurement of the strength of the atmospheric turbulence in the considered conditions.

3.2 The Phase Structure Function

To compute the phase perturbation experienced by a plane wave propagating through a medium with refraction index fluctuations expressed by (11), we will considered the geometry given in Fig. 7.

Even in this case, the phase fluctuations will be described using a structure function but the phase structure function is given as a two-dimensional function instead of a three-dimensional function (used for the refraction index structure function). In particular, the phase structure function measures the phase difference between two points at distance ρ when both points are located in a single plane perpendicular to the propagation direction of the considered plane wave. This plane is usually taken as containing the telescope

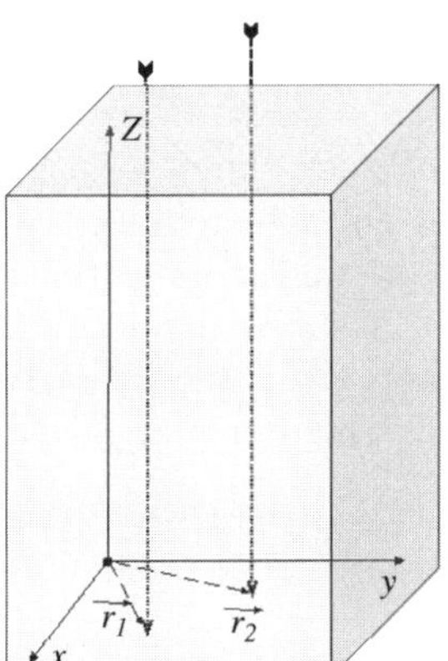

Fig. 7. The geometry used to compute the phase structure function. The grey volume represents the turbulent atmosphere. The integral in (12) is done along the vertical lines showed in the picture. The bottom plane represents the telescope entrance pupil plane

entrance pupil. Considering that the expression of the phase $\phi(r)$ in the plane x,y of the figure is given by

$$\phi(\boldsymbol{r}) = \frac{2\pi}{\lambda} \int_0^{+\infty} n(\boldsymbol{r}, z)\mathrm{d}z \tag{12}$$

then the phase structure function is defined as

$$D_\phi(\rho) = \left\langle [\phi(\boldsymbol{\rho}) - \phi(\boldsymbol{\rho} + \boldsymbol{r})]^2 \right\rangle \tag{13}$$

The analytical expression for the phase structure function was found by Fried in 1965 [10] and is given by

$$D_\phi(r) = 6.88(r/r_0)^{5/3} \tag{14}$$

In this equation, the quantity r_0 is the so-called Fried parameter that is defined as follows, where λ and γ are the considered wavelength and the line of sight zenith angle.

$$r_0 = 0.566\lambda^{6/5} \cos(\gamma) \left[\int_0^h C_n^2(h)\mathrm{d}h \right]^{-3/5} \tag{15}$$

An interesting thing can be inferred from looking at (14). It is easy to see from here and from the r_0 definition that the phase fluctuation is inversely proportional to the considered wavelength so that the optical path defined as the phase fluctuation times the wavelength is achromatic. This result is quite important in the study of the AO systems.[4]

3.3 The Taylor Hypothesis of Frozen Turbulence

The theory considered so far gives the description of the spatial distribution of the phase fluctuations at the telescope entrance pupil in a turbulent atmosphere. However, until now we have not taken into account the dynamic evolution of such a pattern. A simple model describing the dynamics of the turbulence that is usually applied in the case of AO system computations is called the Taylor model [28]. This model makes two assumptions:

- the refraction index fluctuation distribution is constant with time
- the fluctuation distribution is translated on the telescope aperture by the wind

From this model, the turbulence evolution can be derived by simply translating the phase perturbation over the telescope aperture, at the speed of

---

[4] This result is due to the fact that we neglect the wavelength dependance in (9). and is usually allowed because at optical wavelength the neglected term accounts for some percentage of the optical path difference fluctuation.

wind. The basis of the Taylor model can be identified considering the previous discussion on the Kolmogorov theory of vortex cascade. Let us consider again the case of fully developed turbulence. In this case, the energy per unit of time and volume of a given velocity fluctuation is given by $E_l = (v_l)^2$. Then, as seen already, the energy dissipated by viscous friction at the scale l per unit of time and volume is $\left[(v_l)^2/l^2\right]\nu$. Taking the ratio of these two quantities gives the lifetime of a velocity fluctuation of scale l. Numerically we find

$$\tau_l = l^2/\nu \tag{16}$$

This expression shows that the lifetime is smaller for smaller inhomogeneity. In particular, a quadratic law relates the size of the inhomogeneities and their lifetime. So, if we compute the lifetime for inhomogeneities having linear dimension r_0 we find $\tau_{\min} = 25[\mathrm{s}]$. This time is actually much larger than the time taken for a given phase perturbation pattern to pass over a telescope having a diameter of $10\,\mathrm{m}$, even with a very moderate wind of some meters per second. Hence, the pattern can be considered stable and the Taylor hypothesis holds. At this point, it is easy to estimate the characteristic time of the turbulence evolution that results

$$\tau_{\mathrm{atm}} \simeq r_0/v_{\mathrm{wind}} \tag{17}$$

More precise computations of this quantities gives a value of

$$\tau_{\mathrm{atm}} = 0.32 r_0/v_{\mathrm{wind}}$$

We refer the reader to the Beckers [2] paper for a detailed description of this parameter.

4 Effect of Atmospheric Turbulence on the Long Exposure PSF

It is important to be able to determine the telescope MTF when atmospheric perturbations are present. In other words, we want to compute the MTF of a telescope looking through the turbulent atmosphere. In the astronomical case, the exposure time is usually much longer than the atmospheric correlation time; thus, the relevant MTF is the time-averaged MTF. Assuming ergodicity conditions, we will substitute the time average with a statistic average. For this computation, we will have to use the phase structure function defined already. The results reported in this section have been obtained by Fried in 1966 [11]. The expression for the MTF in the case of an integration time much longer then the correlation time of refraction index fluctuations is given by

$$\langle \mathrm{MTF}(f) \rangle \propto \left\langle \int W(\boldsymbol{r}+\boldsymbol{d})W(\boldsymbol{r})\mathrm{d}\boldsymbol{r} \right\rangle = \int P(\boldsymbol{r}+\boldsymbol{d})P(\boldsymbol{r})\mathrm{d}\boldsymbol{r} \left\langle \mathrm{e}^{(\phi(\boldsymbol{r}+\boldsymbol{d})-\phi(\boldsymbol{r}))} \right\rangle,$$

$$\tag{18}$$

where $\langle and \rangle$ represent the expectation value of the given quantity a. Now, considering the central limit theorem the above expression can be rewritten as

$$\langle \mathrm{MTF}(f) \rangle \propto T_0(d/\lambda) \cdot e^{-0.5 D_\phi(d)} \tag{19}$$

In this way, the system MTF has been written as the product of the unperturbed telescope MTF T_0, already reported in Fig. 4 and a damping factor that depends exponentially on the phase structure function which in turn is a function of the Fried parameter or of the turbulence strength. In Fig. 8, we report some cases of MTF for a D/r_0 value of 30, 20, and 10. These are typical cases for an 8 m telescope at visible and NIR wavelengths, respectively. In the picture, we assume that the atmosphere is corrected using an AO system with 30 actuators on the telescope diameter. The assumption is made here that all the wavefront perturbations of scale larger then the actuator pitch of $d_{\mathrm{act}} = D/30$ are perfectly corrected. This is reflected in the behavior of the structure functions that saturate for distances larger then d_{act}. The left side picture is reporting the structure function values for the uncorrected and corrected cases as a function of the different values of D/r_0. On the right side picture we report the corresponding MTF computed using (19).

5 AO System Fundamental Parameters

In this section, we will describe the main parameters of an AO system, in order to give an estimate of the AO systems' current performances and limitations.

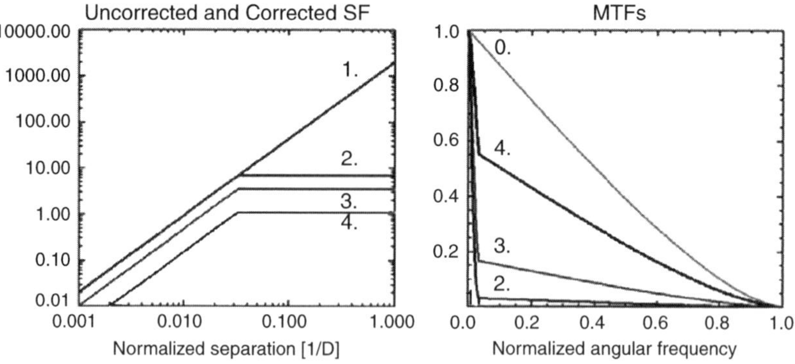

Fig. 8. The left part of the figure reports the corrected phase structure functions for the cases $D/r_0 = 30, 20, 10$ (2., 3., and 4. respectively), together with the uncorrected one (1.). The effect of the wavefront correction is clearly visible in the saturation of the structure functions. The right side reports the corresponding MTFs computed from (19), with the diffraction limited MTF as reference (0.)

In Fig. 1, the basic layout of an AO system is reported to clarify where the relevant parameters come into play. We start by looking at the picture from the deformable mirror, going in a counterclockwise direction.

5.1 Number of Actuators and System Cycle Time

The relevant parameter for the deformable mirror is the number of degrees of freedom, usually given by the number of actuators. The aim of this device is to introduce in the incoming wavefront an optical path difference equal and opposite to the perturbation induced by the atmosphere at a given time. The sum of the two perturbations will give as a result a plane wave. As mentioned, the spatial scale of the atmospheric perturbation is given by the r_0 value. Hence, in order to have the proper resolution on the corrective device, we need to have an actuator grid with pitch smaller then r_0. In this case, the total number of actuators can be estimated to be of the order of $(D/r_0)^2$. The second parameter considered is the time required to compute and apply the correction. As we find in Sect. 3.3, the correlation time of the atmosphere is given in the Taylor hypothesis as $\tau \simeq r_0/v$, where v is the considered wind velocity. To be able to apply an efficient correction, the cycle time of the system has to be smaller than τ so that we can write $t_{\mathrm{cicle}} < r_0/v$.

5.2 Wavefront Perturbation Spatial Sampling

Let us consider now the spatial sampling required to properly measure the wavefront aberration. As stated previously, the characteristic length of the atmospheric perturbation is given by r_0. Assuming this, the wavefront spatial sampling has to be of the order of r_0. A commonly used wavefront sensor in AO system is the so-called Shack-Hartmann sensor [20] which uses a lenslet array to obtain the wavefront slopes in several patches of the telescope entrance pupil. A sketch of the sensor is reported in Fig. 9. This will help the reader to have a better understanding of the basic parameter involved in the following calculations. As shown in the above picture, the lenslet array is optically conjugated to the telescope exit pupil. The array has a number of lenslets of the order of $(D/r_0)^2$. Each of the spots produced by the lenslet array is focused on a CCD detector. The intensity pattern of each spot is recorded for any given wavefront measurement. The spot displacement with respect to the zero or to the unperturbed position is measured by computing the intensity of center of gravity

$$x_{\mathrm{c}} = \sum_{i=1}^{N} I_i x_i \Big/ \sum_{i=1}^{N} I_i \tag{20}$$

where I_i and x_i are the intensity and the position of the i-th pixel considered. Moreover it is easy to show in geometrical optics approximation that

$$x_c = f_{\mathrm{sh}}\overline{\partial w(x,y)/\partial x}, \tag{21}$$

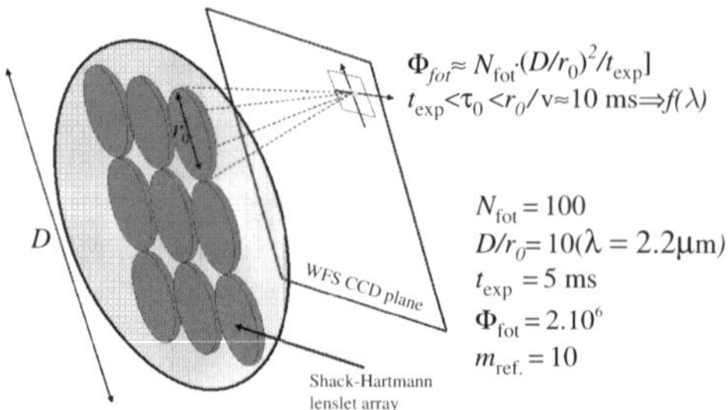

Fig. 9. A simple design of a Shack-Hartmann sensor reporting the basic optical elements of the device. The grey circle is the image of the entrance pupil of the telescope. The single lenslet dimension is chosen to be equal to the scaled length of r_0 on the pupil image

where $w(x, y)$ is the wavefront perturbation and, $f_{\rm sh}$ is the Shack-Hartmann lenslet focal length. The overbar denotes an average of the first derivative over the considered sub-aperture. Thus, the center of gravity is proportional to the wavefront slope in the considered sub-aperture defined by a particular lenslet. The measured data are used to write a linear system of finite difference equations relating the wavefront slopes to the phase difference on the various sub-apertures [12]. This system is usually solved by least square methods. The error in the reconstructed wavefront depends on the total number of photons received by each sub-aperture. This photon number is given by

$$\Phi = N_{\rm fot} \left(D/r_0\right)^2 /t_{\rm exp} \propto r_0^{-3} \tag{22}$$

in order to have 100 photons, per sub-aperture, per integration time[5]. Assuming $D/r_0 = 10$, $t_{\rm exp} = 0.005$, we need an overall flux $\phi = 2e6$ or a star of visual magnitude 10. This number shows that the reference star to be used for wavefront sensing is relatively bright and is not always available in the surroundings of the faint astronomical object we want to observe. Finally, we note that the flux dependance on the third power of r_0 tells that the flux requirements became very strong when the correcting wavelengths got shorter and shorter. A great advantage is usually obtained by performing the wavefront sensing at optical wavelengths and the adaptive correction in the NIR (usually 1–5μm). Doing so, the wavefront sensor can use a visible CCD camera (much faster and less critical than the IR detector). The adaptive correction is done in the J, H, and K bands, where r_0 is larger and the requirements for all the parameters like number of actuators, sub-apertures, and time response, are all relaxed with respect to the visible band. This strategy is allowed

---

[5] We assume here that 100 photons per sub-aperture per integration time are enough to have a good SNR in the center of gravity measurement. It can be shown that the accuracy in this measurement is given by $\delta x_{\rm c} = \theta/\sqrt{N}$, where θ and N are the lenslet spot full width half maximum and the received number of photons.

by the achromaticity of the optical path, as demonstrated in Sect. 3.2. This means that the correction done by the adaptive mirror, driven by an optical wavefront sensor, is valid for all the wavelengths. An adaptive Optics system working in this configuration is called polychromatic AO system. Finally, it is important to stress that for a polychromatic system the r_0 we referred to above estimating the number of actuators and the number of sub-apertures is computed at the correcting wavelength (in the NIR).

5.3 Reference Source and its Angular Distance from the Scientific Object

The main effect that limits the usefulness of an AO system as described up to now is the so-called angular anisoplanatism. The angular anisoplanatism issue was analyzed by Fried [13] and later by other authors. We have already mentioned that because the reference star has to be considerably bright, the scientific target is rarely used as a reference star. Instead, a closely positioned star of high enough magnitude is used. To see how close the reference star has to be, we consider Fig. 10. In the figure, the telescope entrance pupil and a single turbulent layer are represented. As clearly shown, the two rays crossing the same point of the entrance pupil and coming from the scientific and reference objects, respectively, do cross the turbulent layer in two different points. In the sketch (Fig. 10), the reference source is located at an angular distance $\theta = r_0/h$ from the on-axis scientific target. The phase perturbations relevant to our discussion are also shown in Fig. 10 . In particular, Φ_1 is

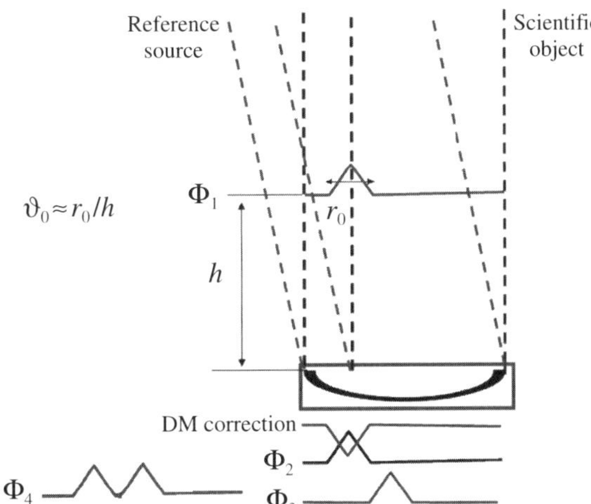

Fig. 10. The figure reports the geometrical arrangement to quantify the wavefront sensing error due to the angular distance between the reference source and the scientific target. The Φ_i are the different phase perturbations involved in the discussion in the text

the phase perturbation on the considered layer at high h; Φ_2 is the phase perturbation experienced by the scientific object placed on-axis, while Φ_3 is the one experienced by the off-axis reference star. The DM is driven to introduce the correction following the phase peak of Φ_3. The result of this correction is represented as Φ_4. This error is called named it such angular anisoplanatism, after Fried [13]: Now we have found as a coherence distance of the atmospheric perturbation the Fried parameter r_0. With this in mind, it is easy to realize that the maximum angle θ allowed between the reference and the scientific object is given by

$$\theta \simeq r_0/h \tag{23}$$

this particular value of θ called isoplanatic angle and is usually indicated as θ_0. The effect of angular separation between reference and scientific target is illustrated in Fig. 23. Substituting real values for r_0 and h, like $0.2\,\text{m}$ and $10\,\text{km}$, we find $\theta = 4\,\text{arcsec}$. This angle is quite small and usually does not allow us to find the 10 mag star we need. It is exactly this problem that limits the sky coverage of an adaptive optics system or the fraction of sky that can be observed efficiently.

5.4 An Estimate of the AO Systems Sky Coverage

The sky coverage (SC) is defined as the percentage of the sky that can be observed using the AO system. This sky fraction is given by: $SC = \pi\theta_0^2 \cdot N^*_{m_0}/4\pi$, where θ_0 is the isoplanatic angle, $N^*_{m_0}$ is the number of stars having magnitude $m < m_0$. Finally m_0 is the AO system limiting magnitude identified using (27). Computing the SC requires a model for the star density as a function of the celestial coordinates. We summarize here all the equations obtained in previous sections and used to quantify the actual sky coverage of an AO system. The formulae we identified are:

$$N_{\text{act}} = (D/r_0)^2 \tag{24}$$

$$\tau_0 \simeq r_0/v \tag{25}$$

$$N_{\text{sub}} = (D/r_0)^2 \tag{26}$$

$$\Phi_{\text{ref}} = vD^2/r_0^3 \tag{27}$$

$$\theta_0 \simeq r_0/h \tag{28}$$

Combining these basic equations J. Beckers [2] compiled a table that we present in Fig. 11. In this table, the various AO system parameters are identified as a function of the values of r_0, τ_0, and θ_0. Figure lists the sky coverage of the AO system as a function of the correcting wavelength. The sky coverage in the K band is 14% but it goes down to 5% in J band. In the visible wavelength regime the situation gets dramatically worst. This is mainly due to the third power dependance of the needed reference flux from r_0 which in

Limiting V magnitude for polychromatic wavefront sensing and sky coverage at average Galactic latitude for different spectral bands[a]

Spectral band	λ (μm)	r_o (cm)	τ_o (sec)	τ_{det} (sec)	V_{lim}	θ_o (arcsec)	Sky coverage (%)
U	0.365	9.0	.009	.0027	7.4	1.2	1.8 E-5
B	0.44	11.4	.011	.0034	8.2	1.5	6.1 E-5
V	0.55	14.9	.015	.0045	9.0	1.9	2.6 E-4
R	0.70	20.0	.020	.0060	10.0	2.6	0.0013
I	0.90	27.0	.027	.0081	11.0	3.5	0.006
J	1.25	40	.040	.0120	12.2	5.1	0.046
H	1.62	55	.055	.0164	13.3	7.0	0.22
K	2.2	79	.079	.024	14.4	10.1	1.32
L	3.4	133	.133	.040	16.2	17.0	14.5
M	5.0	210	.21	.063	17.7	27.0	71
N	10	500	.50	.150	20.4	64	100

[a] Conditions are: 0.75 arcsec seeing at 0.5 μm; $\tau_{det} = 0.3\ \tau_o = 0.3\ r/V_{wind}$; $V_{wind} = 10$ m/sec; $H = 5000$ meters; photon detection efficiency (includes transmission and QE) = 20%; spectral bandwidth = 300 nm; SNR = 100 per Hartmann-Shack image; detector noise = 5 e$^-$.

Fig. 11. A table taken from the J. Beckers's review published in 1993 [2]. The situation is now changed due to the use of laser-generated reference stars

turn is almost linearly proportional to $\lambda^{6/5}$. In the V band the sky coverage is of the order of some parts per million. As already mentioned this was the big limitation of the astronomical AO system at the time Beckers was writing. The solution to this problem was proposed in 1985 and is described in the next section.

6 Solving the Sky Coverage Problem: The Artificial Reference Sources

As we have seen, the main limitation of an astronomical AO system, as described till now, is the limited sky coverage. This limit is due to the relatively small number of stars bright enough to be used as a reference source. To solve this problem, in 1985, R. Foy and A. Labeyrie [8] proposed the use of an artificially generated reference star. To generate this kind of source, the authors proposed the use of the resonant backscattering from atoms and molecules in the free atmosphere. A favorable one was the sodium atom. Sodium atoms accumulate at a height of $\sim$90 km.

The resonant backscattering of such atoms can be excited using laser light tuned at the wavelength of 589 nm. This produces photons (which return to the telescope) from a given location in the sky. In addition, the reference source location is determined by the laser pointing direction and can be moved at will, and placing the reference source close to the scientific object solves the sky coverage problem. This type of laser system is called a laser guide star (LGS) and is now a reality for most 8–10 m class telescopes. It will also be a key issue for the ELTs of the future. However, even if the use of LGSs increase substantially, the resultant SC will still not be the 100% desired by the astronomers. This is due to two effects briefly explained below.

6.1 The Tilt Indetermination Problem

To explain the problem of tilt indetermination in the signal obtained by an artificially generated reference star, let us consider the geometry of the laser launch and measurement. There are several possible geometries for the laser launch. However, the tilt indetermination problem holds, even if in slightly different terms, for all possible launching and measurement geometries. We will consider in our discussion one of the most commonly used geometries, where the laser beam is launched from behind the telescope secondary mirror. This situation is represented in Fig. 12.

The laser beam is focused on the sodium layer at an altitude of 90 kms. The backscattered light is received by the full aperture of the telescope. The

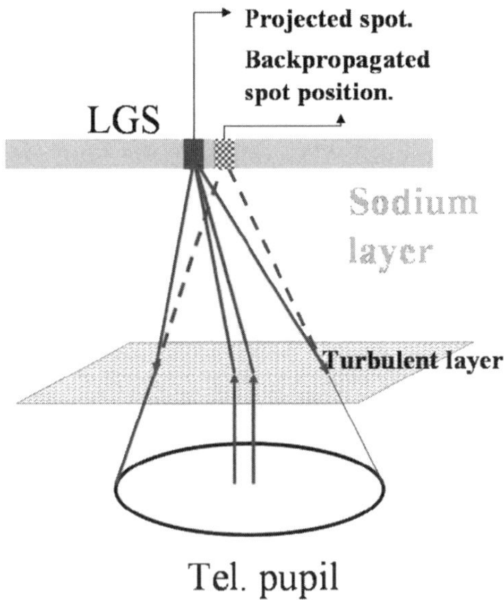

Fig. 12. The above sketch depicts the laser beam launched from the telescope to the Na layer to produce the artificial reference source. During the upcoming path, this beam crosses the turbulent layer, introducing a tilt, which in turn generates a displacement of the artificial star in the sodium layer. In this way, the tilt detected on the LGS is the sum of the tilt experienced by the laser before the LGS generation and the one experienced when the LGS beam is received by the telescope aperture crossing again the turbulent layer

global tilt of the wavefront should be detected as the laser spot displacement in the telescope focal plane. However, the spot displacement measured is due to two different terms. The first one is the tilt that the beam experiences in going up to the 90 km altitude layer. This term is the tilt term due to the tilt perturbation on the scale of the launching telescope, usually of the order of 30–40 cm. The second one is the tilt experienced by the wavefront that propagates from the sodium layer down to the entrance pupil of the telescope. So, the spot displacement is the sum of two tilts and is not a correct measurement of the tilt of a natural guide star that crosses the atmosphere only once. Using the LGS tilt to correct the tilt of the scientific object seriously degrades the telescope angular resolution. To solve this problem several solutions have been proposed [6, 9, 23], but the core solution is to measure the tilt term of the wavefront perturbation using a natural guide star. In this case, we are back to the initial problem of finding a reference star. However, there are two favorable circumstances that mitigate this problem. First, the isoplanatic angle for the tip tilt measurement is larger than the usual isoplanatic angle, allowing a larger sky region to be used for reference star searching. Second, the tilt measurement is made with the full telescope aperture so that the the star magnitude can be fainter than the standard magnitude needed for wavefront sensing described in Sect. 5.2. Overall, the sky coverage for AO systems with a laser guide star is usually improved and ranges between 30% and 90% [26]. These numbers have to be compared with sky coverage values of some percent reported in Fig. 11.

6.2 The Focus Anisoplanatism Problem

There is a second problem that has to be taken into account when a laser guide star is used as a reference star in an AO system. Because the reference source is located at a finite altitude over the telescope, the waves reaching the telescope are spherical. These spherical waves do not sample the same atmospheric volume as the plane wave coming from the star, thus implying that the aberration measured using the laser light is not exactly the same as that experienced by the light of the scientific object. This effect was first studied by Fried [14], who named it focus anisoplanatism, as it arises from a difference in focus of the reference and the scientific object. The geometrical situation is reported in Fig. 13, where a section of two sampled volumes is reported for clarity. As shown in Fig. 13, the perturbations experienced by a ray hitting a given point of the primary mirror are different depending on the ray starting from a source at finite altitude or from a celestial object assumed to be placed at infinity. [It is clear from what we stated in the previous sections that when the distance d, shown in Fig. 13, is larger than r_0, the focus anisoplanatism error becomes significant].

Finally we note that due to the finite altitude some of the turbulence is not sampled at all by the laser light. Following the analysis of this error given by D. Fried, we find that the wavefront error due to this effect can be expressed

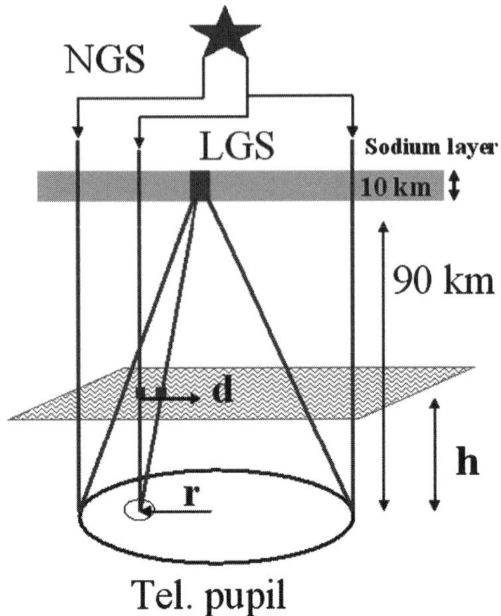

Fig. 13. The above sketch illustrates how the finite altitude of the Na layer affects the wavefront detection. The LGS is seen by the telescope as an object at a finite distance. Due to this, two rays hitting the same telescope pupil point **r**, but coming from the LGS and natural guide star (NGS) respectively, cross the turbulent layer at different points at distance d

by

$$\sigma^2 = (D/d_0)^{5/3}, \tag{29}$$

where d_0 is a parameter that scales with the wavelength as r_0 having quite a complicate expression [14]. The value of this parameter for a standard astronomical site is about 4 m in the infrared. Hence, a single laser guide star can be used to perform AO observations with a focus anisoplanatism error of some radians squared in the near infrared.

7 LBT: An Adaptive Telescope

To show an application of adaptive optics to an 8 m class telescope, we will briefly describe the AO system of the Large Binocular Telescope (LBT). The LBT with AO can be used in single dish or interferometric mode. The LBT is a binocular telescope made up of two 8.4 m primary mirrors. The mirrors are placed on the same mount and can work as two separate units or as an interferometer (the center to center distance of the two primary mirrors is 14 m). A drawing of the telescope is presented in Fig. 14. It is also interesting to

Fig. 14. The Large Binocular Telescope mechanical drawing, showing the two primary mirrors supported by a single mount. Cylinders between the two primary mirrors represents the adaptive intruments of the LBT. From bottom to top they are LUCIFER, LBTI, and NIRVANA

note that LBT is possibly the first 8 m telescope to have an adaptive secondary mirror [24]. In other words, both the secondary mirrors of the LBT will have 672 actuators and can provide adaptive correction for all the telescope focii. We will considered in the following all the instruments that are located at the Bent Gregorian focal stations. Those stations and instruments are represented by the cylinders in Fig. 14.

7.1 LUCIFER: The Single Dish Imager and Spectrograph

We begin our AO system description with LUCIFER[6] [21], the NIR imager and spectrograph placed at the front Bent Gregorian focal station of LBT. This instrument can work in seeing limited or diffraction-limited conditions with a Field of View of 4×4 or 0.5×0.5 arcminutes, respectively (some instrument characteristics, particulary for these modes, are reported in Table 1).

---

[6] **LBT NIR** spectroscopic **Utility** with **Camera** and **Integral-Field Unit** for **Extragalactic Research**.

Table 1. Some basic characteristics of LUCIFER's imaging and spectroscopic modes

LUCIFER	
Detector	HAWAII-2 (2048 × 2048)
Bands	Z J H K
MOS mode	30 position cold mask
Diffr. limited mode	
FoV	0.5 × 0.5 arcmin
Scale	0.015 arcsec/pixel
Slit length	≤ 0.5 arcmin
R (K)	20600 (0.137 arcsec slit)
R (H)	28200 (0.100 arcsec slit)
R (J)	37100 (0.076 arcsec slit)

In the diffraction limited case, the atmospheric turbulence correction will be provided by the First Light AO system (FLAO). This system is mainly composed of the adaptive secondary mirror [24], a pyramid wavefront sensor [7], and a custom electronic system for the real-time computations. Various numerical simulations [5] were done to assess the performance of LUCIFER when the AO system is in operation.

In Fig. 15, we report some simulation results that quantify the limiting magnitude in the imaging mode. In particular, the plot quantifies the point source magnitude needed in the H band, to achieve a SNR = 3 in 30 min of integration, assuming as main noise the sky background brightness. The H band sky background is assumed to be 13.5 mag per arcsec square. Note that the table contains different curves showing the performance as a function of the angular separation between the scientific object and the reference star magnitude. The diffraction- and seeing-limited cases are reported as straight lines for reference. Finally, the number of pixels used to detect the source is optimized, and the top small table shows that the optimal size of the detection region is of 75 mas. This is about 2.4 times the FWHM of the LBT telescope in the H band or the diameter of the first zero of the diffraction-limited PSF (which contains about 80% of the incoming energy).

It is interesting to note that for very bright reference stars the performances are worst when the reference star is located very close to the scientific object (supposed on axis). This is due to the noise created by the photons from the reference star in the scientific target image. The curves for the on-axis reference star assume that the reference and the scientific object are the same, meaning that there is no noise contribution from the reference source. The number of the considered table can be scaled to give the detection limit in terms of surface brightness assuming a spatial resolution of 75 mas. To scale the achieved results, the magnitudes of the y axis in Fig. 15 have to be subtracted by 5.6 mag. A similar kind of plot has been calculated in the case of the diffraction-limited spectroscopic observation with LUCIFER and are

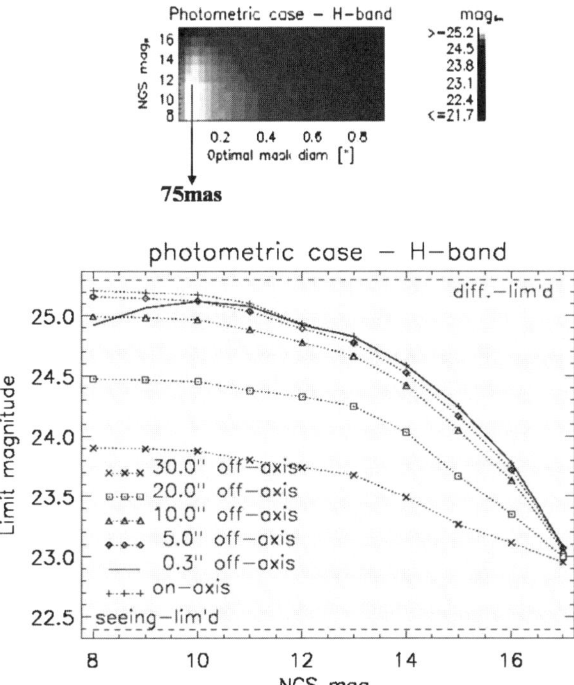

Fig. 15. Bottom: a plot of the limiting magnitude for detection of a point source against the sky background as a function of the reference star magnitude. **Top**: a 2D representation of the optimal window width (x axis) for source detection as a function of reference star magnitude (y axis)

reported in Fig. 16. In this case, the plot quantifies the point source magnitude needed to achieve an SNR $= 3$ in the H band, in the spectroscopic bin of LUCIFER. The spectral resolution element is 0.06 nm set by the dimension of the slit $\simeq 0.1$arcsec. Again, the main noise contribution is the sky background brightness assumed to be 13.5 mag. Even in this case, different curves quantify the performance as a function of the relative distance between the scientific object and the reference star. The curves for a diffraction-limited telescope and for a seeing-limited telescope are reported for comparison. The y axis magnitudes can be scaled to evaluate the detection limit for surface brightness [mag/arcsec^2] by subtracting 5.9 magnitudes. Doing so, the plot represents the surface brightness, required to achieve a SNR $= 3$ in a spectral bin as a function of the NGS magnitude.

7.2 The Adaptive Homothetic Interferometer

At this point, we enter the description of the most challenging part of the LBT operation, the interferometric operations. Before going further, we go

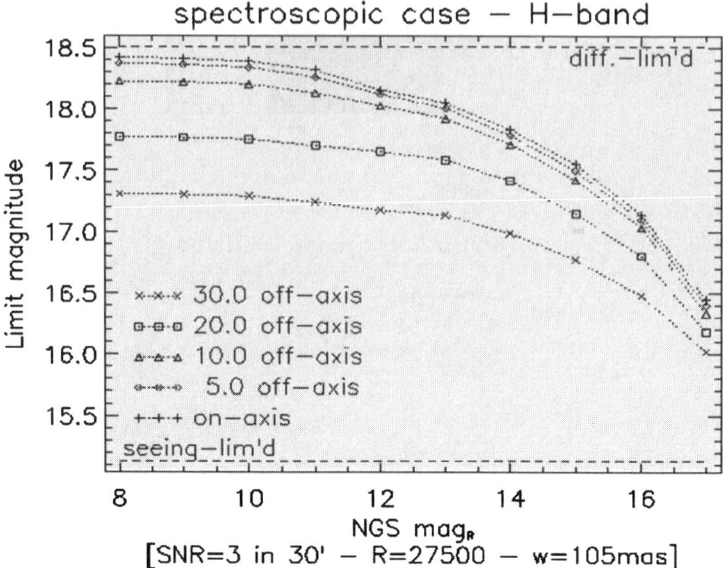

Fig. 16. A plot of the limiting magnitude to reach an SNR of three in the spectroscopic bin (0.06 nm) against the sky background as a function of the reference star magnitude

through some basic concepts of interferometry to explain the definition of a homothetic interferometer. Let us consider a single lens as shown in Fig. 17. The main characteristic of the ideal lens is that the optical path length from surface S to focal point P along each ray of the given wavefront is the same. This condition holds for every direction in the FoV. For this reason, the rays from any object in the FoV arrive at the corresponding focal point in phase

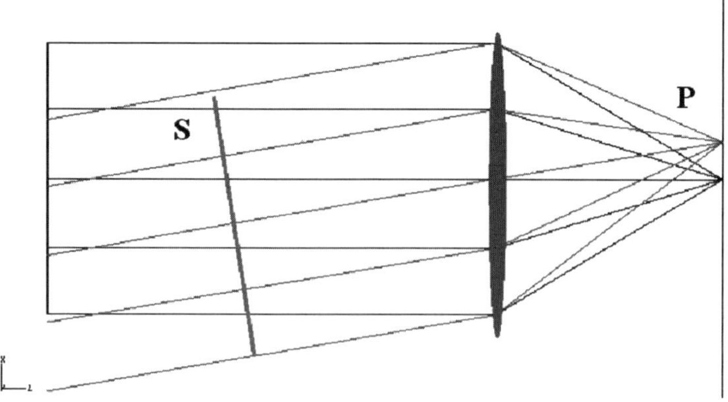

Fig. 17. A figure showing the basic behavior of a lens accordingly to paraxial optics

and positively interfere. The lens glass thickness is calculated to compensate the natural optical path difference of the incoming rays for all the FoV. In this way, all the points located at different positions in the FoV have the same image or, in other words, have the same Point Spread Function, as shown in Fig. 2. Let us consider now the same lens but with a diaphragm covering it. The diaphragm covers the entire lens, except for two small holes of diameter D, placed at a separation b. This separation b is almost equal to the lens diameter (Fig. 18). In this situation, the PSF width is slightly smaller then the initial one (Fig. 19). This is because some portion of the lens having a smaller baseline with respect to the two-holes distance has been rejected. The image we get in the focal plane when $D \ll b$ is usually referred to as Young fringes pattern. In the left picture of Fig. 19, the PSF of a single hole is reported together with the sinusoidal behavior of the interference of the two chief rays from the given sub-apertures. The period of this last curve is λ/b. The final PSF of the system is obtained by considering the two curves and is represented in the right side of the picture. Going off-axis, the Young fringes have a maximum in the location of the image position given by geometrical optics. From a qualitative point of view, this happens because the phase delay between the chief rays of the two sub-apertures is zero. This is due again to the optical path compensation performed by the lens. Finally let us consider now the case of the Michelson stellar interferometer [22] that is represented schematically in Fig. 20. In this case, the light coming from a scientific object is received using two mirrors of diameter D separated by a distance b, with no optical power (M1 and M2 in the sketch). Then, the two beams are reflected and redirected to a mirror with optical power, a lens (L1) in our design, to be focused in the interferometer focal plane.[7] The baseline of this setup is the

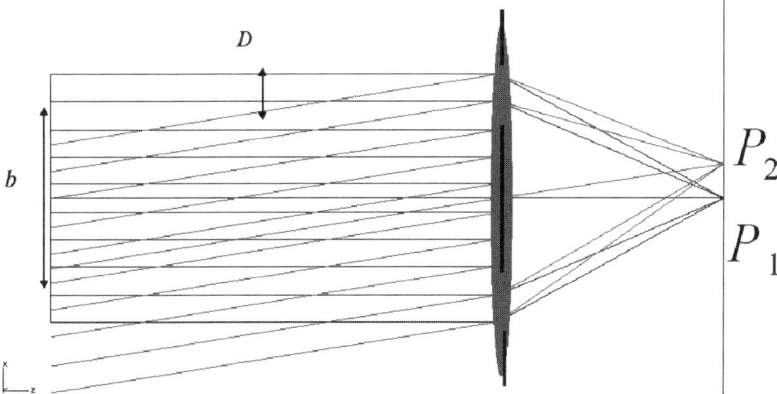

Fig. 18. The optical system made up by a lens and a two-holes diaphragm

---

[7] Note that the big lens L2 in our representation is shown only to explain the system behavior and is not part of the setup.

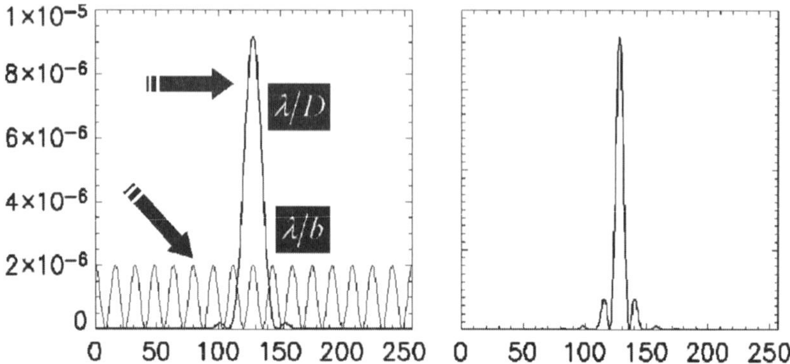

Fig. 19. The elements involved in determining the PSF of the two sub-aperture system. **Left**: The PSF of the single aperture of diameter D superimposed to the sinusoidal behavior of chief ray interference for a couple of apertures spaced by a distance b. **Right**: the two sub-aperture system PSF

distance between the two flat mirrors and can be much larger than the mirrors diameter D, as is presented in the figure. Following the reasoning we did for the two aperture case, we could say that the system PSF should have a width that is much smaller then the PSF of the used mirrors. In particular, the ratio

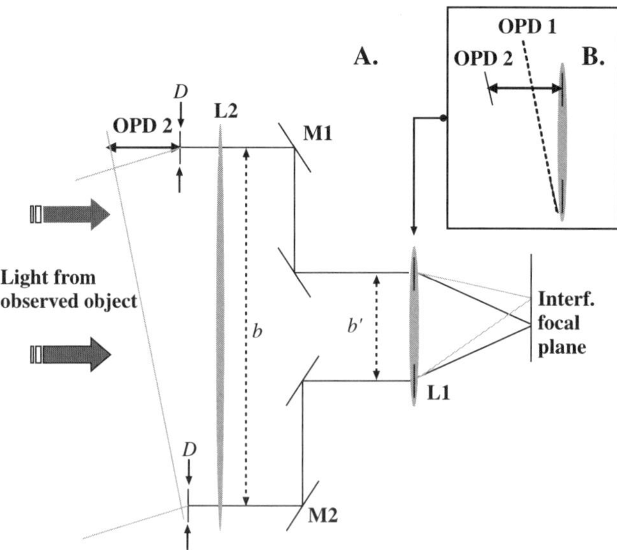

Fig. 20. A schematic representation of the Michelson stellar interferometer. Two sources are represented in the figure to compare the chief rays' behavior for an on-axis object and an off-axis object

between the two PSFs should be D/b. The PSF of this interferometric setup is reported in Fig. 21. The left PSF is achieved for an on-axis object, and the center PSF is achieved for an off-axis object. In both cases, a noticeable gain is achieved in angular resolution because of the small fringe spacings in the PSF. As anticipated, the fringe spacing is a factor D/b smaller then the PSF of the single lens of diameter D (left plot of Fig. 21). It is usual, however, for the PSF shape to change with field of view positions. This means that an image composed of several points will be reimaged through the mentioned system using a different PSF per each point of the FoV. In this case, the image we obtain is no longer a good reproduction of the observed object intensity pattern. Why did this problem not occur with the lens or with the lens and two sub-aperture we described before? As already mentioned, in the lens case the lens glass compensates for the geometrical difference in optical path so that all the rays from a given source arrive in phase at the focal point. In this case, something different happens. A simple idea of the system behavior is found by considering the small inset B in Fig. 20. The two wavefront portions, coming from an off-axis object and sampled by the two mirrors, arrive on the lens with an optical path difference highlighted in the inset by the solid arrow. This is the optical path difference OPD 2, which the big lens L2 would have compensated for, to keep the PSF stable in the field of view. However, the optical path that the lens L1 compensates (at that particular position in the field of view) is different from the one considered above and corresponds to the optical path difference OPD 1 identified by the dashed line.[8] For this reason, the two chief rays from the considered mirrors do not interfere positively in the case of an off-axis source. In other words, the sinusoidal pattern we introduced in Fig. 19 is shifted, with respect to the mirrors PSFs, because of the uncompensated optical path difference. This effect changes the PSF shape with respect to the on-axis case. It is easy to see that this effect is

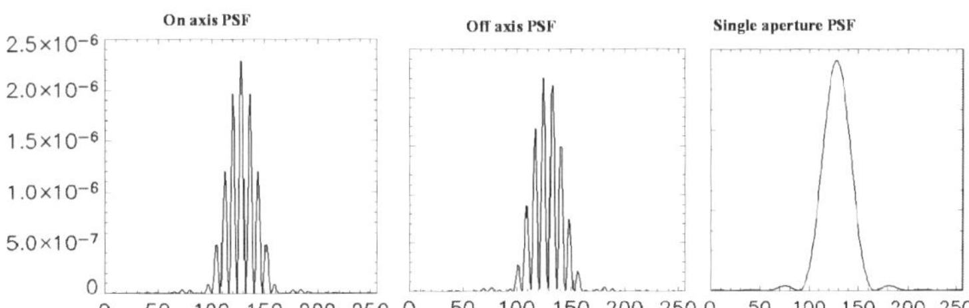

Fig. 21. Left: the on-axis interferometric PSF. The PSF maximum is in the center of the profile. Center: the off-axis PSF. The maximum of the PSF is not in the center of the profile. **Right**: the PSF of the single aperture of diameter D

---

[8] The dashed line is obtained by considering the wavefront surface perpendicular to the initial propagation direction, identified by the off-axis object position in the sky.

not present on the on-axis object, where for symmetry reasons the sinusoidal pattern remains centered on the mirrors PSF.

Now, what is the difference of the present setup from the lens and the two sub-apertures case we already considered? In this interferometric case, the entrance pupil (the two mirrors of diameter D placed at distance b) is not purely scaled on the re-imaging lens L1. In fact, the mirror diameter remains the same while the mirror separation is scaled from b to b'. This violation to a pure scaling or re-imaging of the entrance pupil of the interferometer generates the optical behavior, where the PSF is not constant with the field of view. An interferometer where the pupil is purely scaled in all the optical train is called homothetic and creates real images of the source. However, no scaling is perfect, and so it is important to understand what is the limit that can be tolerated, when an interferometer is used to produce real images. Let us consider the left picture of Fig. 22, where a homotethic configuration is sketched. Note that the two lenses have been added before the folding mirrors to collimate and compress the beams. In formulae requiring the homotheticity means using the symbols defined in figure:

$$\frac{b}{b'} = \frac{D}{D'} \tag{30}$$

In this case, the chief rays are in phase for all the points of the field of view. Now, suppose that the interferometer aperture is correctly scaled in the plane of the refocusing lens L1 (as shown in the right picture of Fig. 22). However, one of the two apertures is linearly displaced in this plane by an amount ε. Then let us consider the case of a wavefront having a tilt with an OPD amplitude of $N\lambda$ over the interferometric pupil. This is the case of an off-axis object placed at N times the angle λ/b. At the re-imaging lens the wavefront has a slope of $N\lambda/b'$. The displacement of one side of the aperture, the one represented as displaced on the right side in the picture, introduces a phase delay with respect to the perfect case equal to

$$N\lambda/b'\varepsilon$$

as graphically shown in the picture. If we require that the displacement of the sub-aperture does not perturb the image we have to require

$$N\lambda/b'\varepsilon << \lambda \tag{31}$$

or

$$\varepsilon << b'/N \tag{32}$$

This last equation states the precision required in the positioning of the two apertures when the homothetic condition is required on a field of view of $N\lambda b$. In other words, the equation states that the precision in the re-imaging process has to increase as the required field of view. In the LBT case, for example,

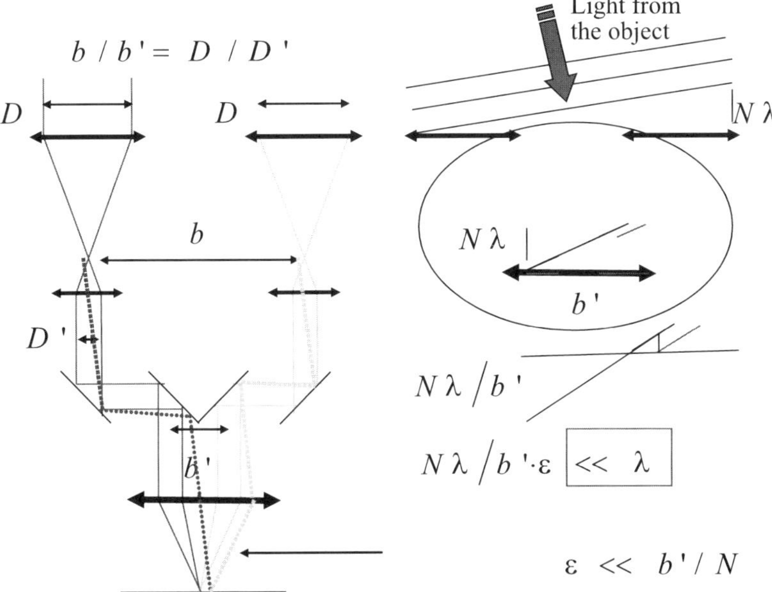

Fig. 22. A figure showing a homothetic interferometer. The right side of the picture is the basic homothetic arrangement while the left picture outlines the main elements used in the homothetic condition computations

the infrared instrument with interferometric capability called NIRVANA has an infrared camera mounted in the interferometric focal plane. Because the field of view of the camera is 15×15s arcsec, and the interferometer resolution λ/b is about 18 milliarcsec, the quantity N is equal to about 750, and b/N is found to be, for a 100 mm pupil spacing on the reimaging optics, equal to 0.13 mm. Requiring that the aperture displacement be much smaller leads to displacements of the order of 0.01 mm or 10 μm. Thus in the pupil re-imaging process the image centering has to have an accuracy of 10 μm. Many other effects can produce the same loss of homotheticity such as, pupil rotations, the pupil magnification, and so on. For all these effects, a similar computation can be done to show what is the limit for a rotation or magnification error when a certain homotheticity is required.

8 Interferometric Instrument at LBT

To complete our description of the adaptive instruments of LBT, we report in this section about the two interferometric instruments of the telescope. Remember that the LBT has two primary mirrors of 8.4 m with a center to center distance of 14 m, achieving the maximum equivalent resolution of a 23 m aperture. The opto-mechanical design of LBT allows one to obtain a homothetic interferometer in both the interferometric focal stations (top and central cylinders in Fig. 14). All interferometric instruments use adaptive optics to improve performance.

We will start with the European interferometric instrument, briefly mentioned above, called NIRVANA [16]. The instrument objective is to realize interferometric images adaptively corrected on a large field of view (15 × 15 arcsec). Images are taken in the NIR 1–2.2 µm. The instrument is built in collaboration with the Istituto Nazionale di Astrofisica in Italy, the Max Planck Institute fur Astronomie in Heidelberg, The Max Planck Institute for Radioastronomy in Bohn, and the University of Cologne. The main sub-systems of NIRVANA are the beam combining optics, the Multi Conjugate Adaptive Optics system (MCAO) having four deformable mirrors and six wavefront sensors, the fringe tracker camera to sense the differential piston of the two subapertures, and finally an NIR camera having a field view of 15 × 15 arcseconds. A drawing of NIRVANA is reported in Fig. 23, where some details of the MCAO sub-systems are visible. The instrument assembly is shown, in the bottom left inset, mounted in the structure of LBT as a large box between the primary mirrors. A simulated image obtained with NIRVANA is presented in Fig. 24. Here a stellar field of 15 × 15 arcseconds is shown. The particular shape of the PSF with the three central fringes, similar to the PSF we already discussed in the case of the simple lens, is evident. Even in this case, the number of side lobes is due to the ratio between the aperture diameter, 8.25 m, and their separation, 14 m. The interferometric images of LBT taken with the telescope in a given orientation with respect to the celestial coordinates will contain this feature so that the full resolution of the interferometer

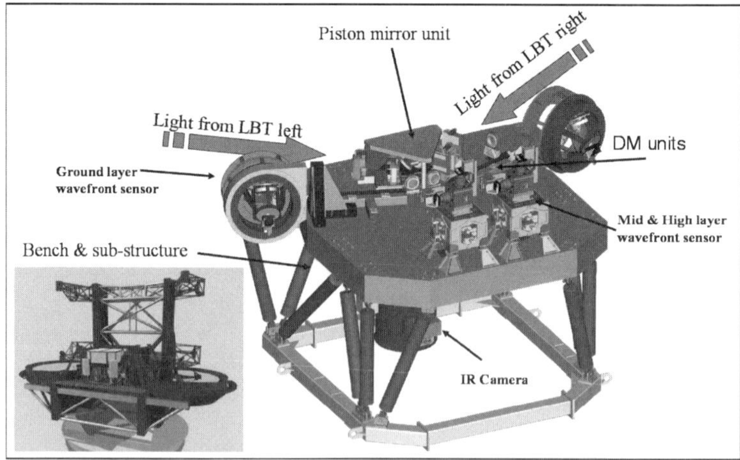

Fig. 23. A drawing of the NIRVANA instrument as mounted in the azimuth platform. The small box in the bottom left corner shows the instrument in the telescope structure, while the large figure shows the main components of the AO system of NIRVANA

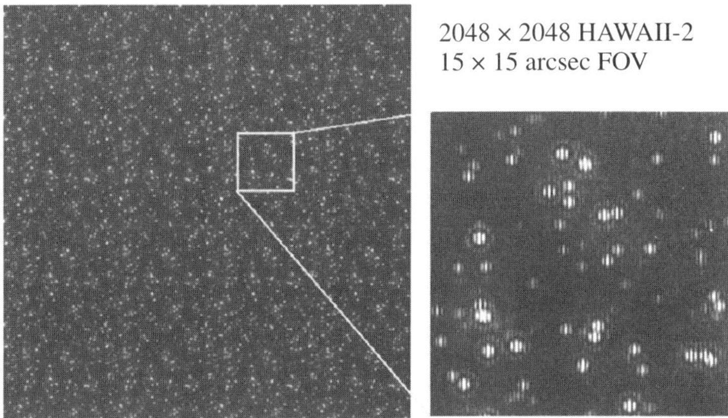

Fig. 24. A simulated image of a 15 × 15 arcsecond stellar field as seen from the LBT interferometric instrument NIRVANA. The three fringes of the interferometric PSF of LBT are clearly visible

is achieved only in the fringe direction. However, taking multiple images with the telescope having the baseline projection in a different orientation with respect to the celestial coordinates will allow the user to have the maximum resolution in all directions. For more details on the image processing due in the NIRVANA case the reader is referred to the [4].

The second interferometric instrument of LBT is the American interferometer called LBTI (*Large Binocular Telescope Interferometer*) [18]. This instrument is realized by the Steward Observatory. In Fig. 25, a mechanical sketch of the instrument is shown. The two input focal planes of LBT telescopes (left and right units) are located close to the top boxes labeled UBC (Universal Beam Combiner). The principal aim of this instrument is to provide sensitive nulling interferometric observations of nearby solar-like stars at $2 - 20\,\mu$m. In particular, LBTI will be used in collaboration with NASA to perform initial observations for target selection of the U.S. spatial planet finder mission.

The nulling mode uses LBT as a Bracewelll interferometer [3] to produce a dark fringe in the center of the field of view (Fig. 26). This acts as a coronograph allowing masses very close to the central star to be detected. However, to produce a high contrast nulling interferometer, an adaptive optics correction is required. Two AO units will be used, placed at the two interferometric focal stations. The deformable mirrors used will be the adaptive secondary mirrors of the two telescopes. The nulling optics will be placed in the dewar labeled NIL, while the nulling camera called NOMIC will be mounted in the corresponding dewar. The arrangement of such a dewar is reported in Fig. 25. Finally, LBTI is able to provide a beam combination compatible with wide-field homothetic (imaging) interferometry in the thermal infrared 1–$30\,\mu$m. The imaging cameras for that are indicated in Fig. 25 as Fizeau imagers.

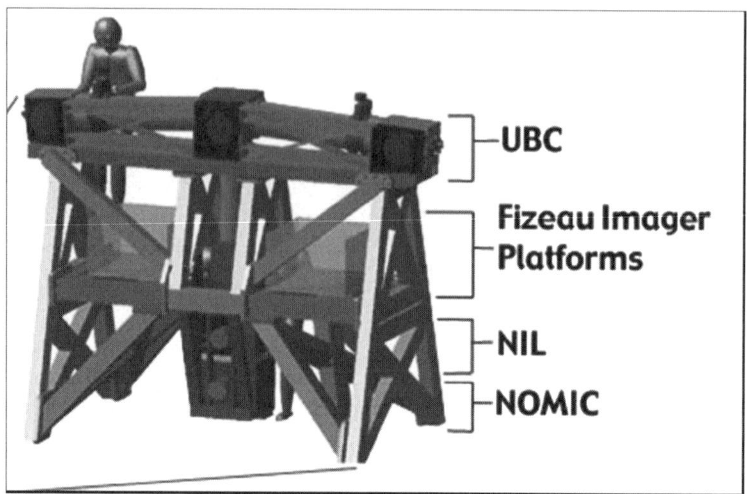

Fig. 25. A sketch of the opto-mechanical arrangement of the LBTI instrument. The left and right side of the structure are close to the two LBT focal planes of the interferometric station

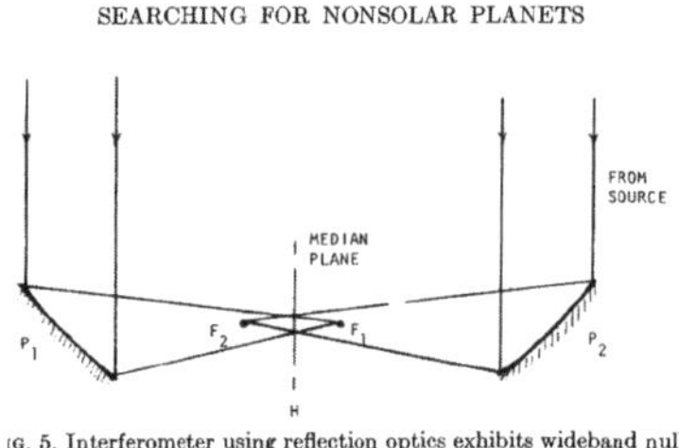

Fig. 26. A sketch of the nulling interferometer principle, where two beams interfere having a phase delay of π. This delay is introduced by using a dedicated optical system

9 Conclusion

The paper has introduced the basic concepts and limitations of modern Astronomical AO systems. The most relevant parameters of such AO systems have been outlined. Using the expression identified for such quantities, the main limitation of Adaptive Optics, namely, the small sky coverage has been analyzed. The present day solution to such a problem, the use of laser guide star, has been described. Then, the diffraction limited instruments that will be operational on the LBT telescope have been introduced. This should provide the reader with an idea of what will be available in the next few years at the LBT telescope in terms of high angular resolution capabilities. Finally, we consider it very important for the astronomers of the next future to understand performance and limitations of AO systems that today appear to be fundamental to the sub-systems of any 8–10 m class telescopes and the key subsystem of any Extremely Large Telescope presently under design.

Acknowledgments

Some of the figures presented in this paper are taken from available papers and presentations made by the groups developing the instruments LUCIFER, NIRVANA and LBTI. The authors thank all these groups for then contributions.

References

1. Babcock, H.W.: The possibility of compensating astronomical seeing. PASP **65**, pp. 229–236 (1953)
2. Beckers, J.M.: Adaptive optics for astronomy – Principles, performance, and applications. Annu. Rev. Astron. Astrophys. **31**, 13–62 (1993). DOI 10.1146/annurev.aa.31.090193.000305
3. Bracewell, R.N.: Detecting nonsolar planets by spinning infrared interferometer. Nature **274**, pp. 780–781 (1978)
4. Carbillet, M., Correia, S., Boccacci, P., Bertero, M.: Restoration of interferometric images. II. The case-study of the Large Binocular Telescope. Astron. Astrophys. **387**, 744–757 (2002). DOI 10.1051/0004-6361:20020389
5. Carbillet, M., Riccardi, A., Femenía, B., Fini, L., Esposito, S.: Preliminary results of simulations for the adaptive optics system of the large binocular telescope. AstroTech Journal, Mem. Soc. Astr. It. **2** (1999)
6. Esposito, and, S.: Techniques to solve the tilt indetermination problem: methods, limitations, and errors. In: Bonaccini, D., Tyson, R.K. (eds.) Adaptive Optical System Technologies. Proceedings of the SPIE, vol. 3353, pp. 468–476 (1998)
7. Esposito, S., Tozzi, A., Puglisi, A., Fini, L., Stefanini, P., Salinari, P., Gallieni, D., Storm, J.: Development of the first-light AO system for the large binocular telescope. In: Tyson, R.K., Lloyd-Hart, M. (eds.) Astronomical Adaptive Optics Systems and Applications. Proceedings of the SPIE, vol. 5169, pp. 149–158 (2003). DOI 10.1117/12.511503
8. Foy, R., Labeyrie, A.: Feasibility of adaptive telescope with laser probe. Astron. Astrophys. **152**, L29–L31 (1985)
9. Foy, R., Migus, A., Biraben, F., Grynberg, G., McCullough, P.R., Tallon, M.: The polychromatic artificial sodium star: A new concept for correcting the atmospheric tilt. Astron. Astrophys. Suppl. **111**, 569–578 (1995)
10. Fried, D.L.: Statistics of a geometric representation of wavefront distortion. J. Opt. Soc. Am. (1917–1983) **55**, 1427–1435 (1965)
11. Fried, D.L.: Optical resolution through a randomly inhomogeneous medium for very long and very short exposures. J. Opt. Soc. Am. (1917–1983) **56**, 1372–1379 (1966)
12. Fried, D.L.: Least-square fitting a wave-front distortion estimate to an array of phase-difference measurements. J. Opt. Soc. Am. (1917–1983) **67**, 370–375 (1977)

13. Fried, D.L.: Anisoplanatism in adaptive optics. J. Opt. Soc. Am. (1917–1983) **72**, 52–61 (1982)
14. Fried, D.L., Belsher, J.F.: Analysis of fundamental limits to artificialguide-star adaptive-optics-system performance for astronomical imaging. J. Opt. Soc. Am. **11**, 277–287 (1994)
15. Goodman, J.W.: Introduction to Fourier optics. Introduction to Fourier optics. McGraw-Hill series in electrical and computer engineering; electromagnetics, 2nd ed., McGraw-Hill, New York (1996). ISBN: 0070242542 (1995)
16. Herbst, T., Ragazzoni, R., Andersen, D., Boehnhardt, H., Bizenberger, P., Eckart, A., Gaessler, W., Rix, H.W., Rohloff, R.R., Salinari, P., Soci, R., Straubmeier, C., Xu, W.: LINC-NIRVANA: A Fizeau beam combiner for the large binocular telescope. In: Traub, W.A. (ed.) Interferometry for Optical Astronomy II. Proceedings of the SPIE, vol. 4838, pp. 456–465 (2003)
17. Hill, J.M., Green, R.F., Slagle, J.H.: The large binocular telescope. In: Stepp, L.M. (ed.) Ground-Based and Airborne Telescopes. Proceedings of the SPIE, vol. 6267, pp. 62670Y (2006). DOI 10.1117/12.669832
18. Hinz, P.M., Angel, J.R.P., McCarthy Jr., D.W., Hoffman, W.F., Peng, C.Y.: The large binocular telescope interferometer. In: Traub, W.A. (ed.) Interferometry for Optical Astronomy II. Proceedings of the SPIE, vol. 4838, pp. 108–112 (2003)
19. Kolmogorov, A.N.: The local structure of turbulence in incomressible viscous fluids for very large Reynolds numbers. Compt. Rend. Acad. Sci. (SSSR) **30**, 301 (1941)
20. Malacara, D.: Optical shop testing. In Malacara, D. Wiley Series in Pure and Applied Optics. Wiley, New York (1992)
21. Mandel, H., Appenzeller, I., Bomans, D., Eisenhauer, F., Grimm, B., Herbst, T.M., Hofmann, R., Lehmitz, M., Lemke, R., Lehnert, M., Lenzen, R., Luks, T., Mohr, R., Seifert, W., Thatte, N.A., Weiser, P., Xu, W.: LUCIFER: A NIR spectrograph and imager for the LBT. In: Iye, M., Moorwood, A.F. (eds.) Optical and IR Telescope Instrumentation and Detectors. Proceedings of the SPIE, vol. 4008, pp. 767–777 (2000)
22. Michelson, A.A., Pease, F.G.: Measurement of the diameter of alpha Orionis with the interferometer. Astrophys. J. **53**, 249–259 (1921). DOI 10.1086/142603
23. Ragazzoni, R.: Propagation delay of a laser beacon as a tool to retrieve absolute tilt measurements. Astrophys. J. **465**, L73–L75 (1996). DOI 10.1086/310139
24. Riccardi, A., Brusa, G., Salinari, P., Busoni, S., Lardiere, O., Ranfagni, P., Gallieni, D., Biasi, R., Andrighettoni, M., Miller, S., Mantegazza, P.: Adaptive secondary mirrors for the large binocular telescope. In: Tyson, R.K., Lloyd-Hart, M. (eds.) Astronomical Adaptive Optics Systems and Applications. Proceedings of the SPIE, vol. 5169, pp. 159–168 (2003). DOI 10.1117/12.511374
25. Rousset, G., Fontanella, J.C., Kern, P., Gigan, P., Rigaut, F.: First diffraction-limited astronomical images with adaptive optics. Astron. Astrophys. **230**, L29–L32 (1990)
26. Sandler, D.G., Stahl, S., Angel, J.R.P., Lloyd-Hart, M., McCarthy, D.: Adaptive optics for diffraction-limited infrared imaging with 8-m telescopes. J. Opt. Soc. Am. **11**, 925–945 (1994)
27. Tatarsky, V.I.: Wave Propagation in a Turbulent Atmosphere, p. 548. Nauka, Moskow (1967)
28. Taylor, G.I.: The spectrum of turbulence. Proc. R. Soc. Lond. A **164**, 476 (1938)

IR Spectroscopy of Jets: Diagnostics and HAR Observations

B. Nisini

INAF-Osservatorio Astronomico di Roma
nisini@oa-roma.inaf.it

Abstract. In this contribution, I will review the diagnostic capabilities of near- and mid-IR emission lines in the study of jets from Young Stellar Objects (YSOs). I will then present recent applications applied to high angular resolution (HAR) observations of jets in young embedded sources.

1 Introduction

The first observations of outflows in the infrared date back to the 1980s, with the pioneering H_2 2.12 μm works (e.g. [2, 54]), where observations were done with single channel detectors coupled with CVF filters. Since then, the progress in IR observations has been enormous, thanks to the development of sensitive IR arrays with progressively larger dimensions and to the higher

B. Nisini: *IR Spectroscopy of Jets*, Lect. Notes Phys. **742**, 79–104 (2008)
DOI 10.1007/978-3-540-68032-1_4   © Springer-Verlag Berlin Heidelberg 2008

spatial resolution now provided by the 8-m class telescopes with Adaptive Optics (AO) system. Not only have ground-based NIR observations seen a boost, thanks to technology development, but in the last decade, the ISO and Spitzer missions have also shown the potential of mid-and far-IR spectroscopic observations, opening a new window for the study of warm gas at low excitation. The new frontier in high angular resolution IR observations is now represented by the IR interferometry, and projects like the VLTI and CHARA are expected to make a new revolution in our understanding of jet physics.

The fast improved quality of the data and the access to a large wavelength coverage have also fostered the development of new and more sophisticated tools for data interpretation. As a consequence, during the last years a large variety of models for the interpretations of IR observations of outflows have been developed, which run from line predictions in shocks to synthetic images from disk-wind models and numerical simulations.

In this contribution, after an introduction on the advantages and complementarity given by observations of jets in the IR with respect to the optical and on the limitations imposed by the atmosphere and technology, I will review the properties of the main emission lines at near- and mid-IR wavelengths and how they can be used to infer the main jet physical parameters. Finally, I will present some recent results obtained by HAR observations in jets from embedded young sources. NIR AO-assisted observations in T Tauri jets will be reviewed in the Dougados contribution to this book. Other useful reviews on IR observations of jets and outflows can be found in [3] and [45].

2 Advantages and Limitations of IR observations

Observing in the infrared presents a number of advantages, in addition to being the only way to probe species and physical domains that are not observed at other wavelengths.

The most immediate advantage is the limited interstellar extinction, which allow the observer to diagnose physical conditions of jets embedded in dense environments, such as those of young protostars. Given $A_K \sim 0.1 A_V$, in the infrared one can ideally observe regions ten times more embedded than at visual wavelengths. In practice, this advantage is limited by technical problems, which make IR observations less sensitive than in the optical (see below). IR observations, in addition, make it possibile to diagnose gas at low excitation, not emitted at optical wavelength, and thus to reveal physical conditions that would be otherwise unproven. Indeed, gas excited at temperatures < 5000 K and densities $>10^4$ cm^{-3} cools down only through atomic and molecular IR lines.

Important lines from abundant atoms and ions such as [C I] , [S I] , [S II] $\lambda6731$, [N I] , [O I] $\lambda6300$, [Fe II]   are found in a wide range of wavelength spanning near- to far-IR (see Sect 3.2 and Table 1), while H_2 has a near-IR

Table 1. Bright forbidden lines in the near- and mid-IR spectral range

Species	Transition	λ^a	E_{up}	n_{cr}^b
		μm	K	cm^{-3}

Near-IR lines

Species	Transition	λ^a	E_{up}	n_{cr}^b
[Fe II]	$^4D_{7/2} - ^4F_{9/2}$	1.644	$1.14\,10^4$	$7.2\,10^4$
[Fe II]	$^4D_{7/2} - ^6D_{9/2}$	1.257	$1.14\,10^4$	$7.2\,10^4$
[Fe II]	$^4D_{7/2} - ^6D_{7/2}$	1.321	$1.14\,10^4$	$7.2\,10^4$
[Fe II]	$^4D_{5/2} - ^4F_{9/2}$	1.533	$1.21\,10^4$	$5.5\,10^4$
[Fe II]	$^4D_{3/2} - ^4F_{7/2}$	1.599	$1.25\,10^4$	$4.3\,10^4$
[N I]	$^2P_{3/2} - ^2D_{5/2}$ $^2P_{1/2} - ^2D_{5/2}$	1.040	$4.12\,10^4$	$3.0\,10^6$
[N I]	$^2P_{3/2} - ^2D_{3/2}$ $^2P_{1/2} - ^2D_{3/2}$	1.041	$4.12\,10^4$	$3.0\,10^6$
[S I]	$^1D_2 - ^3P_2$	1.08	$1.32\,10^4$	$7.0\,10^5$
[S I]	$^1D_2 - ^3P_1$	1.13	$''$	$''$
[S II]	$^2P_{3/2} - ^2D_{3/2}$	1.029	$3.52\,10^4$	$1.7\,10^6$
[S II]	$^2P_{3/2} - ^2D_{5/2}$	1.032	$''$	$''$
[S II]	$^2P_{1/2} - ^2D_{3/2}$	1.034	$3.51\,10^4$	$2.3\,10^6$
[S II]	$^2P_{1/2} - ^2D_{5/2}$	1.037	$''$	$''$
[C I]	$^1D_2 - ^3P_1$	0.983	$1.45\,10^4$	$1.1\,10^4$
[C I]	$^1D_2 - ^3P_2$	0.985	$''$	$''$

Mid-IR lines

Species	Transition	λ^a	E_{up}	n_{cr}^b
[Fe II]	$^4F_{9/2} - ^6D_{9/2}$	5.34	2692	840
[Fe II]	$^4F_{7/2} - ^4F_{9/2}$	17.9	3494	$4.1\,10^4$
[Fe II]	$^6D_{5/2} - ^6D_{7/2}$	26.0	553	$1.6\,10^4$
[S I]	$^3P_2 - ^3P_1$	25.2	570	$4.2\,10^4$
[Si II]	$^2P_{3/2} - ^2P_{1/2}$	34.8	411	$1.2\,10^2$
[Ne II]	$^2P_{3/2} - ^2P_{1/2}$	12.8	1115	$5.4\,10^4$

^{a}vacuum wavelength.
^{b} critical density for collisions with electrons at $T_e = 10^4$ K.

spectrum widely used to study the molecular jets usually associated with young Class 0 and I protostars. The far-IR, on the other hand, contains the rotational transitions of the most abundant molecules in non-dissociative shocked gas, such as CO and H_2O.

Observations in the IR, however, suffer several limitations. At ground, atmospheric transmission is above 50% only in few windows below $20\,\mu m$. Longer wavelengths can be observed only with telescopes mounted on aircraft or satellite missions. In addition to the absorption, the atmosphere emits in the infrared, introducing a strong background that needs to be removed from the scientific observations. Below 2.2 μm, the main contribution to the sky emission comes from the forest of strong OH roto-vibrational lines. Above $2.5\,\mu m$ there is thermal emission both by the atmosphere and by the telescope with its warm optics that dominates the background. Chopping and beam-switching techniques are required to limit the small scale spatial and temporal fluctuations of this background. Because of that, observations above 2.5 μm from ground remain limited in sensitivity and are much more efficiently done from space.

The main limitation on the spatial resolution that can be reached in the IR comes from diffraction. For a 8 m telescope at a wavelength of $2\,\mu m$, the diffraction disk corresponds to an angular diameter of $\sim \theta'' = 0.5\lambda_{\mu m}/D_m = 0.12''$, while at $10\,\mu m$ the angular diameter is $0.9''$. The diffraction limit at a given wavelength can be compared to the seeing disk, whose size is described as λ/r_0, where r_0, the Fried parameter, is the length over which the incoming wave-front is not significantly disturbed by motions in the atmosphere. r_0 depends on the wavelength as $r_0 \propto \lambda^{6/5}$, with the seeing disk varying as $\theta \propto \lambda^{-0.2}$. At 2 μm, in good atmospheric conditions, the seeing is of the order of 0.4–$0.5''$, which means that the resolution of an 8-m diffraction-limited IR instrument is about a factor of 4 better than a seeing-limited instrument. Adaptive Optics (AO) devices can, therefore, improve the spatial resolution of a near-IR device on a large telescope, but this improvement becomes less and less important at longer wavelengths or with telescopes 4 m or less. For example, the diffraction-limited NICS camera on HST (2.4 m) has a resolution similar to an 8-m seeing-limited telescope on ground (Fig. 1).

Furthermore, the big limitation of the so-far operational AO systems is the need of optical guiding stars in a relatively small field of view of a dozen arcsec. In practice, this implies that the target source should be relatively bright in the R band, making it impossible to use these devices to observe embedded YSOs. In the study of jets, this is a significant limitation, since bright IR lines are usually observed more frequently in jets from very young and optically invisible protostars. At present, the only instrument making use of a guiding star IR sensor is NACO working at VLT; it, however, requires stars with K magnitude greater than 8, making limiting its use to only few targets. In the near future, the use of artificial laser guide stars will alleviate this problem, although long exposures will still need a natural star for tip-tilt motions not corrected with laser (see the lecture by Esposito in this volume).

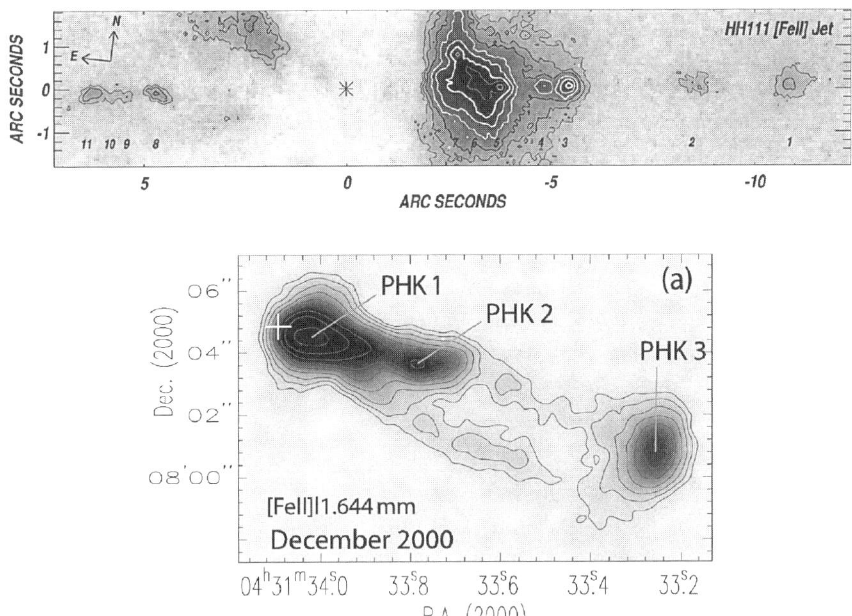

Fig. 1. Top: Diffraction-limited HST-NICMOS image of the HH 111 jet central region obtained through a filter sensitive to [Fe II] emission at 1.64 μm [44]. **Bottom:** Seeing-limited continuum subtracted Subaru image of the L1551 jet in a narrow band filter centered on the [Fe II] 1.64 μm line [42]. The spatial resolution attainable by the two instruments is comparable owning to the difference in their sizes

3 Diagnostics with IR Lines

3.1 Optically Thin and Collisionally Excited Lines

Before describing the main diagnostic IR lines, it is instructive to give a brief account of how the main physical parameters are derived from the analysis of forbidden lines excited in jets. A more detailed and rigorous account is found in the chapter of this book by Hartigan. The interpretation of forbidden lines is made simple by the fact that in the jet physical conditions the lines are usually optically thin and excited by collisions. In ionized jets, atoms and ions are excited by electron collisions, while in neutral and molecular regions atomic and molecular hydrogen are the more important collision partners.

The intensity per unit area and solid angle of a line connecting two levels j, i can be written as

$$I_\nu = \frac{h\nu}{4\pi} A_{j,i} \int f_j n \, \mathrm{d}s \,, \tag{1}$$

where ν is the frequency of the line, $A_{j,i}$ is the Einstein coefficient for spontaneous radiative rate, $f_j = n_j/n$ is the fractional population of the level j,

n is the particle density of the emitting species, and the integration is taken along the line of sight. The partition of ions between the different levels can be determined by assuming that the population have reached steady-state under collisional excitation:

$$n_i \left(\sum_j n_0 \gamma_{i,j} + \sum_{j<i} A_{i,j} \right) = \sum_j n_j n_0 \gamma_{j,i} + \sum_{j>i} n_j A_{j,i}, \qquad (2)$$

where $\gamma_{i,j}$ is the collisional rate coefficient for the transition between i and j and n_0 are the collision partners. The $\gamma_{i,j}$ coefficients (that have units of $au^3\ s^{-1}$) represent the velocity-averaged collisional cross sections of the colliders and depend on the temperature:

$$\gamma_{i,j} =< \sigma_{i,j} v >= \frac{4}{\pi} \left(\frac{\mu}{2kT} \right)^{3/2} dv(\sigma_{i,j}(v)v)v^2 \exp(-\mu v^2/2kT) \qquad (3)$$

Therefore, the dependence on the temperature in the (2) is implicit in the $\gamma_{i,j}$ coefficients. The sum is taken over all the levels that are populated: In nebular conditions, such as those found in jets, only the levels of the lowest multiplets are excited, and (2) in most of the cases reduces to a system of three or five equations. A significant exception, important in the IR, is Fe that has several multiplets at energies less than $30,000$ K and thus many lines excited in nebular conditions (see Sect. 3.3). One important parameter that influences the diagnostic capability of a single line is the *critical density*, which is given by

$$n_{cr} = \sum_{j<i} A_{i,j} / \sum_{j \neq i} \gamma_{i,j}. \qquad (4)$$

For densities much larger than the critical density of a given level, collisions dominate over radiative rates, and the level population is given by its thermal equilibrium value,

$$f_i = \frac{n_i}{n} = \frac{g_i \exp(-E_i/kT)}{\sum_j g_j \exp(-E_j/kT)}, \qquad (5)$$

where g_j is the degeneracy of the state j and E_j its excitation energy. In this limit, the level population does not depend on the total density anymore, and the line intensity is proportional only to the column density $N(cm^{-2}) = \int n\, ds$. On the other hand, if the density is lower than the critical density, the collisional de-excitation from level i is negligible (i.e. sub-thermal excitation) and f_i can be approximated as

$$f_i = \frac{n_0}{A_{i,j}} \sum_{j \geq i} \gamma_{0,j}, \qquad (6)$$

where $\gamma_{0,j}$ is the collisional rate coefficient to state j from the ground state, and the summation is taken over all states j whose decay leads eventually to population of state i. In this situation, the line intensity is proportional to $\int n_0 n \, ds$.

One important consequence of the above discussion is that the the ratio of lines from the same species, but with different critical densities is sensitive only to the density of the collider, provided the lines have a similar excitation energy. This is usually the case if one considers pairs of lines coming from the same multiplet, as the [S II] $\lambda\lambda$ 6716,6731 that are normally adopted to derive the electron density in atomic jets. We will see in the following that similar pairs of lines can be found in the IR that probes different density environments. On the other hand, the ratio of two lines probes the gas temperature if the lines originate from levels with different excitation energy; this happens if one considers transitions from different multiplets of the same species. It derives that if a single spectrum contains two transitions coming from the same multiplet, and one transition comes from a different multiplet, one can use the ratios among these transitions to infer both temperature and density from a single observed species. This happens, for example, with the iron spectrum, as is shown in Fig. 2 and discussed in details in Sect. 3.3.

3.2 IR Lines Excited in Jets

Figure 3, showing the near- and mid-IR spectra of the Herbig–Haro object HH 54, illustrates the main lines that are excited in YSOs jets. In Table 1, the brightest atomic and ionic lines falling in the IR are also listed, together with their main parameters. Atomic lines are usually observed in neutral or single

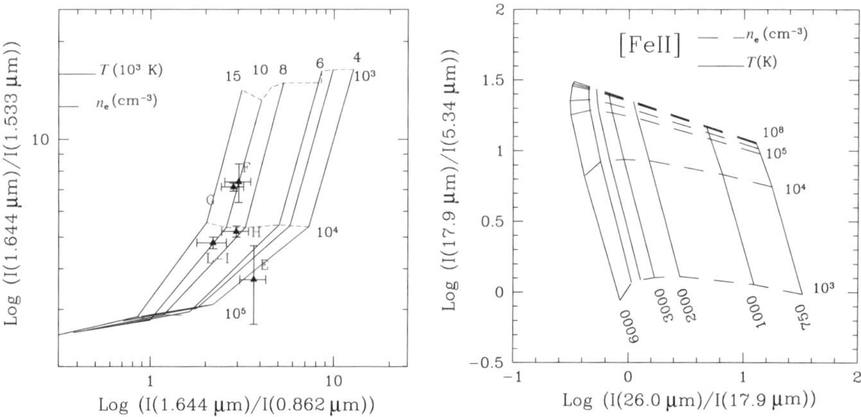

Fig. 2. Diagnostic diagrams for n_e and T_e determination based on different [Fe II] lines. On the left is a diagram constructed combining the near-IR density sensitive ratio 1.64 μm /1.53 μm and the temperature sensitive ratio between the 1.64 μm and the optical 8620 Å line. The symbols indicate the line ratios observed in the HH 1 jet knots (from [37]). On the right, a diagram based on bright mid-IR [Fe II] lines is shown

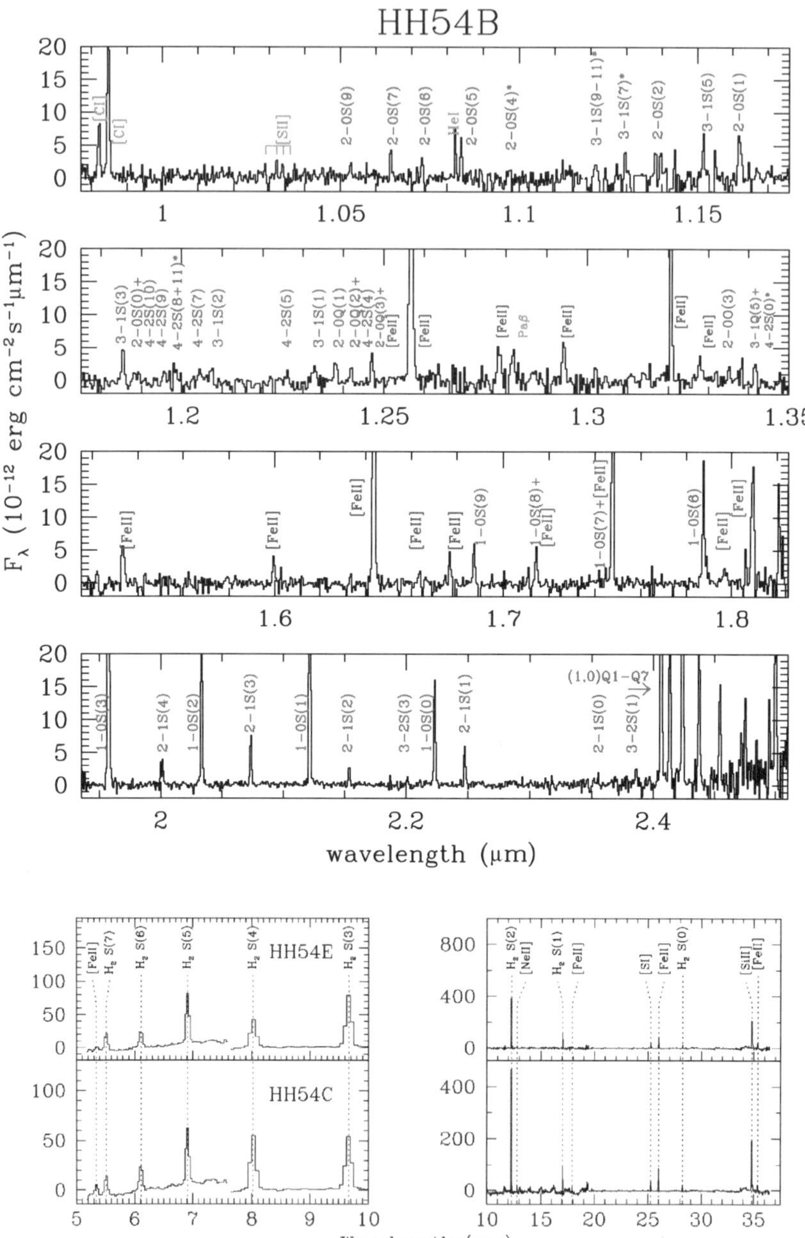

Fig. 3. Spectra of different knots in the Herbig-Haro object HH 54 obtained in the near-IR with ISAAC on VLT (**top**, [22]) and in the mid-IR with Spitzer (**bottom**, [34])

ionized form, typical of a low excitation nebular spectrum. Most of these lines, both in the near- and in the mid-IR, are from [Fe II] fine-structure transitions. The iron lines originate in the same gas component which gives rise to the optical spectrum, tracing gas at low ionization with temperatures in the range 8000–20 000 K and densities 10^3–10^5cm^{-2}. Until very recently, the only [Fe II] lines detected in jet spectra were the bright 1.25 and 1.64µ transitions (i.e. [23]). With the use of sensitive instruments, the complete Fe$^+$ IR spectrum has been observed, and these lines are now widely used to infer physical parameters in ionized flows. A full account of the diagnostic capability of [Fe II] lines is given in Sect. 3.3.

In the Z and J band, there are additional ionic lines that can be used to constrain the gas excitation conditions. These include [S II] and [N I] transitions consisting of four closely spaced lines for each species ([S II] at 1.029, 1.032, 1.034, and 1.037µm, and [N I] at 1.0400, 1.0401, 1.04100, and 1.04104 µm) having excitation energies of $\sim$35 0000 and 40 0000 K, respectively, and thus observed only in the most excited jet knots. The [S II] 1.03 µm lines can be in principle used in combination with the [S II] $\lambda\lambda$6716,6731 ($T_{\rm ex}$ $\sim$15 000 K) to estimate the electron temperature, following what is described in Sect. 9. For this purpose, however, one needs to be sure that the IR and optical observations cover the same jet area and have been inter-calibrated to give a meaningful line ratio (see e.g. [37]). Similarly, the [N I] lines at 1.04 µm could be used with the [N II] $\lambda\lambda$6583,6548 optical lines to determine the ionization fraction in the jet beam. Additional bright lines are those of [C I] at 0.9824 and 0.9850 µm. These transitions are observed in low excitation conditions, and the [C I]/[S II] ratio is a very good indicator of excitation in HH objects. In the mid-IR, other bright atomic transitions are the fundamental lines of [Si II] and [S I] at 34.8 and 25.2 µm, respectively, and the [Ne II] at 12.8 µm. While the

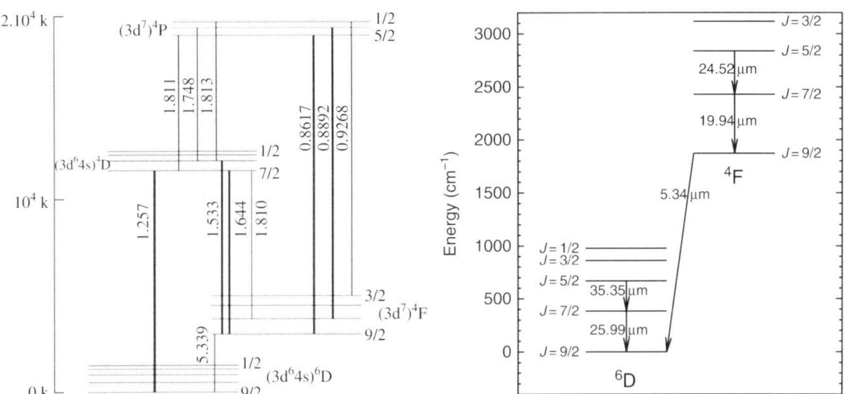

Fig. 4. Fe$^+$ energy level diagrams. **Left:** The 16 first levels are shown with the prominent near-IR and optical lines indicated (from [40]). **Right:** Zoom on the first low-lying 8 levels with the bright mid-IR lines evidenced (from [34])

former probe poorly excited neutral gas (having excitation temperatures in the 100–500 K range), the [Ne II], having neutral neon and ionization potential of 21.6 eV, is an indicator of high excitation conditions, like those rising in high velocity and dense shocks (i.e. [26]). In highly excited and dense gas, several H I recombination lines can also be detected in the near-IR, i.e. the Paβ at 1.28 µm, the Brγ at 2.16 µm and the complete Brackett series in the H band. Such lines that have critical densities in excess of 10^{10} cm^{-3} require density conditions higher than the Hα line to be sufficiently populated and thus are more often detected only in a spatially not resolved region around the driving source.

As can be seen from Fig. 3, plenty of H_2 molecular lines are detected in the same region giving rise to the ionic transitions. The atomic and molecular components come, however, from distinct regions in the jet, which are often not spatially resolved. H_2 transitions are excited in a gas component characterized by molecular gas at temperatures not greater than 3000–4000 K. Such a component can be present either in the external layers of an atomic jet beam, at the region of the interface with the ambient medium, or in the wings of bow shocks, where the velocity of the shock is not high enough to ionize the gas. The diagnostic capabilities of molecular hydrogen will be discussed in details in Sect. 3.4.

Important coolants in dense outflows can also be found in the far-IR, especially the [O I] 63, 145 µm lines and pure-rotational lines of abundant molecules such as CO, H_2O and OH. These species have been so far observed by the Infrared Space Observatory instrumentation at a very poor spatial resolution limited by the diffraction of the 60 cm telescope (i.e. $\sim$80"). In the near future, the Herschel satellite, with its 2.5 m telescope, will be able to improve in resolution which will however be still very limited by diffraction. An account of the diagnostic capabilities of the FIR lines and of the results obtained on jets by the ISO satellite can be found in [36].

3.3 Diagnostic with [Fe II] lines

In the cold molecular clouds, iron, like other refractory species such as C and Si, is locked into dust grains and thus its gas-phase abundance is extremely low. Along jet beams and in shocked regions, sputtering and photo-evaporation processes can destroy all or part of the dust grains, releasing Fe in gas-phase [14, 29]. Given its rather low ionization potential (7.9 eV), iron is almost all single ionized in the physical conditions typical of jets environments, and, due to its particular electronic configuration, gives rise to an extremely rich optical and IR line spectrum, mainly responsible for the gas cooling in dense atomic jets. The near-and mid-IR [Fe II] lines originate from transitions involving the first 16 fine structure levels (see Fig. 4). In the near-IR, the brightest lines are those connecting the fine structure levels of the ^4D term with the levels of the ^4F and ^6D terms. These levels have very similar excitation temperatures, ranging from $\sim$11 000 to $\sim$12 000 K, and thus their ratio is poorly sensitive to the gas temperature. On the other hand, having different critical densities (between 10^4 and $10^5 \mathrm{cm}^{-3}$), line ratios can be effectively used to diagnose the electron density, as explained in Sect. 9. Several ratios can be used for this purpose, the most useful being the $1.533/1.644\,\mu\mathrm{m}$, $1.60\,\mu\mathrm{m}/1.644\,\mu\mathrm{m}$, and $1.67\,\mu\mathrm{m}/1.64\,\mu\mathrm{m}$. It is important to note that owing to their large critical density, such lines need an electron density higher than $10^3 \mathrm{cm}^{-3}$ to be efficiently excited, thus they cannot be used as an alternative for diagnostics in low density jets, such as those often found in T Tauri stars.

To measure the gas temperature from [Fe II] line ratios, one has to consider also transitions originating from the upper ^4P term that have excitation energies of $\sim$19 000 K. Such transitions, however, either fall in regions of poor atmospheric transmission, such as the 1.811 and 1.813 μm lines or fall in the optical range, such as the bright 0.8617 μm line. A combined optical/IR analysis is, therefore, required to simultaneously derive n_e and T_e from the [Fe II] lines (see [37, 41]), which are a powerful diagnostic since they rely on ratios from lines of a single species, thus unaffected by any assumption on the abundance (Fig. 5, left panel).

It is, however, important to keep in mind that the excitation conditions traced by IR [Fe II] lines are different from those traced by bright optical lines such as [S II] $\lambda\lambda$ 6716,6731 and [N II]$\lambda\lambda$6583,6548 (see Hartigan in this volume). This has been demonstrated by recent studies combining optical and IR diagnostics of the same jets: [Fe II] lines probe gas denser and colder than optical transitions and are likely originated in a cooling layer located at a distance from the shock front larger than that generating the optical lines, where the compression is higher and the temperature is declining (see Fig. 5 right panel).

Another very important information that can be retrieved from the NIR [Fe II] lines is the amount of visual extinction toward the emitting gas component. This is possible by using pairs of optically thin lines originating from the same upper level but separated enough in wavelength. In this case, the theoretical intensity ratio depends only on frequency and transition probabilities (cfr 1) and not on physical conditions in their emission region. Their observed ratio can be therefore written as

$$I_{\lambda_1}/I_{\lambda_2} = A_{\lambda_1}\lambda_2/A_{\lambda_2}\lambda_1 10^{-E(\lambda_1-\lambda_2)/2.5}; \qquad (7)$$

hence, the color excess at the wavelength of the considered lines ($E(\lambda_1 - \lambda_2)$) can be easily measured from the observed ratio. The visual extinction A_V can be then retrieved from the color excess adopting an extinction law. Suitable ratios are the $1.64\,\mu\mathrm{m}/1.25\,\mu\mathrm{m}$ and $1.64/1.32\,\mu\mathrm{m}$, which involve the brightest [Fe II] transitions in the near-IR range. One needs, however, to be aware that the goodness of this A_V determination depends on how precise by the transition probabilities are known. In fact, given the complex atomic structure of Fe^+, the computed radiative transitions are still subject to a certain degree of uncertainty, and different references report values that can differ up to

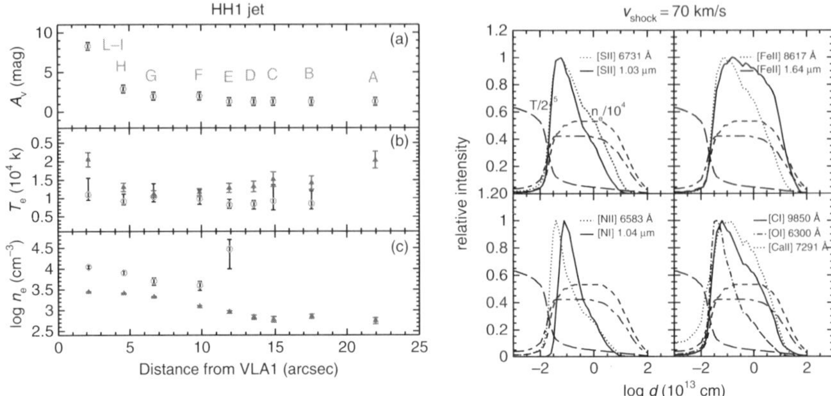

Fig. 5. (Left) Physical parameters (A_v, T_e, and n_e) derived along the HH 1 jet employing [S II] $\lambda6731$,[O I] $\lambda6300$ and [N I] optical lines *(filled triangles)* and near-IR $\lambda6300$ [Fe II] lines *(open circles)*; **(Right)** Relative intensity profiles of several optical and IR lines as a function of the distance from the shock-front for the 70 kms^{-1} shock model by [24]. From [37]

20%. The two more updated calculations of [Fe II] fine structure radiative rates are from [38] and [43]. The intrinsic ratios of 1.64 µm/1.32 µm and 1.64 µm/1.25 µm lines differ in adopting one reference or the other. In general, adopting radiative rates from [38] higher extinction values, up to a factor of two, with respect to the value estimated from the [43] rates are derived. In addition, even taking the same set of theoretical rates, the A_V that one gets from the 1.64 µm/1.32 µm ratio is usually lower than the value measured from the 1.64/1.25 observed ratio (see Appendix B in [37]). Recently, [49] have empirically estimated the radiative rates of most of the NIR transitions from a high S/N spectrum of P Cygni: their derived values differ from both the [43] and the [38] theoretical coefficients but are closer to the latter. Comparing the A_V values derived adopting different [Fe II] lines and radiative rates with the A_V estimated by other observational means in the same region (e.g. H_2 ratios or Balmer decrement), it can be empirically shown that the more reliable values of A_V are obtained by adopting the 1.64 µm/1.25 µm theoretical ratio from [43] or the 1.64 µm/1.32 µm ratio from [38].

A different diagnostic application of the IR [Fe II] lines examines the possibility of measuring the amount of gas depletion in jets, thus setting constraints on the degree of grain destruction by shocks. For this purpose, one has to measure the absolute gas-phase Fe abundance in the investigated region and compare it with the solar elemental abundance. Any difference in these two values indicates that part of the iron is still locked on grains and that dust is still present in the jet beam. To quantitatively derive the gas-phase elemental abundance, however, one has to compare the [Fe II] emission with the emission of a non-refractory species of known abundance that originate in the same region. A very direct method, suggested by [39], is to consider the ratio between [Fe II] 1.25 μm and [P II] at 1.18 μm. Phosphorus is a non-refractory species and its 1.18 μm line has about the same excitation conditions as the [Fe II] 1.25 μm line. Under these conditions, the expected ratio should be about [Fe/P]/2~56 assuming solar abundance for the two species. Measured values indicate Fe gas-phase abundances lower than solar in a number of cases (HH 1, HH 34, and HH 111, [37, 41]). In HH 1, the Fe depletion is higher in the inner and denser regions of the jet, where the high density gas has been less reprocessed by shocks. The fact that Fe can still be partially depleted in jet beams makes the derivation of the mass flux using this tracer and assuming solar abundance only a lower limit of the real value. In spite of that, mass flux values estimated in this way in a few Class I objects result in equal or even larger values than the mass flux derived from the luminosity of optical tracers, [S I] in particular [37, 41].

In the mid-IR, important [Fe II] lines involve transitions between the ^4F and ^6D multiplets and include the fundamental line at 25.98 μm as well as the 17.94, 35.35, and 5.339 μm among the brightests. These transitions probe gas at lower electron density (10^3–10^4cm^{-3}) and temperatures (i.e. 500–5000 K) with respect to the gas observed in NIR lines (see Fig. 4). Therefore, they are potentially a very important diagnostic tool to investigate embedded atomic jets at low excitation, such as those expected from young Class 0 protostars. The 5.339 μm line has been predicted to be one of the most important shock coolants under a wide range of shock conditions [24]. These lines are inaccessible from ground-based telescopes and have been so far detected only by the ISO and Spitzer instruments, with a very limited spatial resolution ranging from 4″ (Spitzer at 5 μm) to 20″ (ISO-SWS at 20 μm). Examples of observations of mid-IR lines by these instruments can be found in [13, 34, 52]. The JWST mission, with its MIRI instrument, will provide access to these important lines with a larger spatial resolution of $\sim 1''$.

3.4 Diagnostic with Molecular Hydrogen

Due to its homopolar nature, molecular hydrogen has no dipole moment and thus all ro-vibrational transitions within the electronic ground state are quadrupolar with low A_{ij} rates. The molecule exists in two, almost independent states, namely, ortho-H_2 (H nuclei with parallel spins) and para-H_2 (H nuclei spins antiparallel). There are no radiative transitions between ortho- and para-H_2 but ortho–para conversion may occur through proton exchange reactions between H_2 and H^0 and H^+. The selection rules for quadrupole transitions are such that $\Delta v = 0, \pm n$ and $\Delta J = 0, \pm 2$, with v and J being the vibrational and rotational quantum numbers, respectively. Thus the ground state line at 28 μm corresponds to the transition $J = 2 \rightarrow 0$ and has quite a high excitation temperature of 500 K, meaning that H_2 can be excited and observed only in regions where the normally cold molecular cloud gas has been sufficiently heated. In star forming regions, H_2 is mainly excited either by radiative pumping into its electronic excited states, followed by a decay which populates by cascade all the vibrational levels of the electronic ground state, or by collisions between H_2 molecules or with neutral H. The first process occurs in the presence of far-UV radiation, such as in PDR regions and gives rise to strong optical and IR fluorescent emission [4], while the second is most typically found in shocks, and it is the main excitation mechanism observed in jets and HH objects. The diagnostic of H_2 is indeed extremely important in the study of jets from young protostars, where the dense jet material is mainly molecular and at low excitation and thus can be probed only with abundant molecular tracers. In fact, H_2 is the only abundant shock molecular coolant that can be observed from ground, therefore its study is important also for the understanding of the energetics involved in transferring energy and momentum from the jet to the ambient medium.

The H_2 pure rotational lines, which probe temperatures in the range 300–1000 K, fall in the mid-IR spectral band and can be observed from space. The near-IR spectral range is instead very reach of rotational transitions originating from the v=1–4 vibrational levels. Lines coming from the v=1 level probe temperatures of $\sim$2000 K, but temperatures up to $\sim$4000 K can be evidenced with the higher vibrational levels lines which fall in the J band (e.g. [21]). One of the most studied lines, though not necessarily the brightest one, is the 1–0 S(1) line at 2.12 μm[1] which is often mapped with IR-arrays equipped with narrow-band filters centered at its central wavelength. Such maps have been widely used to search for molecular jets in large areas of sky [16, 50].

The H_2 transitions being quadrupolar, their radiative rates are two or three orders of magnitudes weaker than the typical rates for molecular transitions: Thus their critical densities are low, typically 10^2–$10^3 \mathrm{cm}^{-3}$ at $\sim$2000 K, and they are quickly thermalized. As we see in Sect. 9, the emission from optically thin lines in LTE depends only on the gas temperature; hence the ratio of different H_2 transitions can be used as a thermometer for the gas in the outflow. A simple way to derive the temperature from a set of observed H_2 lines is through the construction of excitation, or Boltzmann, diagrams. In a thermal distribution, the column density $N(v, J)$ of a given level (v, J), with respect to the total H_2 density, is given by the Boltzmann distribution:

$$\frac{N(v, J)}{N(H_2)} = \frac{g_{v,J}}{Q}\, \mathrm{e}^{-E(v,J)/kT_{\mathrm{ex}}} . \tag{8}$$

Here Q is the partition function, and the statistical weight $g_{v,J}$ is the product of the nuclear spin statistical weight, which, in equilibrium, has values of 1 and 3 for para- and ortho-H_2, respectively, and the rotational statistical weight, which is $(2J + 1)$.

---

[1] The notation of H_2 transitions is such that the change in vibrational state is written first (e.g. 1–0) and the rotational change second with the following code: S denotes a change of ΔJ=-2, Q corresponds to ΔJ=0, and O to ΔJ=+2, while the number in parenthesis indicates the final rotational level.

Column densities in the different levels can be derived from the measured intensities $I_{v,J}$ adopting (see. (1))

$$I_{v,J} = \frac{h\nu}{4\pi}\ A_{v,J}\ N(v,J)\,.\tag{9}$$

Combining these expressions, one derives that

$$ln\frac{N(v,J)}{g_{v,J}} = ln\left(\frac{I_{v,J}\ 4\pi}{g_{v,J}h\nu A_{v,J}}\right) = -\frac{E(v,J)}{kT_{ex}} + ln\frac{N(H_2)}{Q}\,,\tag{10}$$

i.e. for a uniform temperature the natural logarithm of the $N(v,J)/g_{v,J}$ measured ratios should fall on a straight line when plotted versus the excitation energy $E(v,J)$, with a slope proportional to T_{ex}^{-1}. In addition, the intercept to zero of this straight line gives the total column density of the H_2 molecules involved in the emission (see Fig. 6). Being confident of the approximation of single temperature thermalized gas, one can use the Boltzmann diagram also to get a measure of the visual extinction. For this purpose, several transitions coming from the same upper level, and well separated in wavelength, can be used, as we saw in the previous section. Very often such a procedure is adopted using the ratio between pairs of $v{=}1$ transitions, such as the 1–0 $S(i)/1$–0 $Q(i{+}2)$, which all lie in the K band. Such extinction estimates are, however, not very accurate since (i) these lines are not sufficiently separated in wavelength, their ratio is not sufficiently sensitive to extinction variations; (ii) the 1–0 Q(i) lines fall in the 2.3–2.5 µm spectral range, where the atmospheric transmission is very poor, so their fluxes are usually affected by large errors. If the extinction is not too high, several bright pairs of lines lying between 1 and 1.35 µm can be more efficiently used for the extinction determination (see

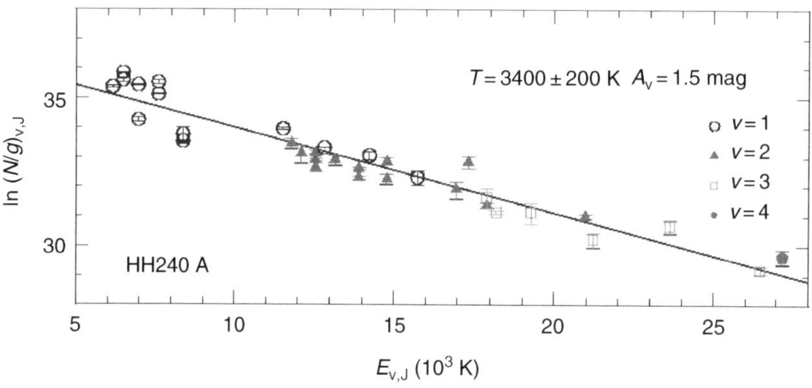

Fig. 6. Boltzmann excitation diagram of the HH 240A object in L1641. Symbols refer to lines from different vibrational levels observed with SofI at NTT, while the straight line is the LTE fit through the data at a single temperature of 3400 K (from [35])

e.g. [5, 21]). Once temperature, extinction, and column density are known, one can use these parameters to measure the total mass flux in the jet: This can indeed be directly inferred in spatially resolved molecular jets from the relationship

$$\dot{M}_{\text{flux}} = 2\mu m_{\text{H}} N(H_2) A \, dv_t/dl_t \,, \tag{11}$$

where μ is the mean atomic weight, m_{H} the proton mass, A the area of the emitting region sampled by the slit, dl_t the projected length, and dv_t the tangential velocity that has to be measured from proper motion studies.

However, the assumption of a single temperature emission is an approximation that can be usually applied only if a limited set of lines with similar excitation conditions are considered, e.g. the v=1–2 lines in the K band. When transitions from several vibrational levels are involved, one may start seeing a temperature stratification, depending on the temperature profile of the post-shocked regions traced by the involved lines or owing to the presence of shocks with different strengths in the beam. Temperature stratifications are evidenced as a curvature in the Boltzmann diagram: Such a curvature is usually very marked when pure-rotational lines are included, as shown in Fig. 7, where the near-IR H_2 lines observed on L1157 are combined with ISOCAM observations of 0–0 lines from S(1) to S(7) [5]. The latter clearly trace a gas at lower temperature that is spatially associated with the warmer gas probed by the v=1–3 transitions. An important consequence of this result is that the

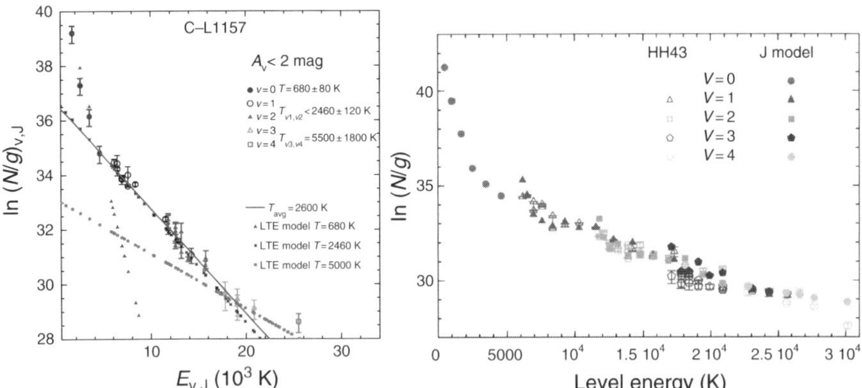

Fig. 7. Examples of H_2 Boltzmann diagrams deviating from a single temperature LTE gas. On the left, the diagram for L1157-C, where near-IR ro-vibrational lines obtained with NICS at TNG are combined with mid-IR pure rotational lines obtained with ISOCAM. Three different temperature regimes are evidenced by lines with different excitation energy (from [5]). On the right, the Boltzmann diagram of HH 43 obtained with SofI at NTT by [20] shows not only a temperature stratification but also deviations from thermodynamical equilibrium that have been fitted with a shock model at low pre-shock density (from [17])

H_2 column density derived from excitation diagrams employing near-IR lines only can largely underestimate the total column density and consequently the mass flux rate.

Deviations from a single temperature component may occur also when considering lines at high vibrational state ($v > 3$, see e.g. [20]), testifying that the molecular gas can reach $T > 4000$ K in the post shocked gas. It has been found that significant temperature stratifications are found more frequently in optically visible HH objects, where the molecular gas is spatially associated with atomic/ionic gas at higher excitation, while in jet knots showing up only in the IR the temperature rarely exceeds 3000 K and can be fairly well represented by a single-temperature slab of gas [21].

Another effect that has to be taken into account is the deviation from LTE conditions in low density jets. If the post-shocked gas is compressed at densities lower than $10^3 cm^{-3}$, the H_2 lines are not sufficiently thermalized, and this effect can be evidenced in the Boltzmann diagram by observing that the various vibrational manifold do not align on straight lines but present a curvature (see Fig. 7).

In principle, the Boltzmann diagrams can be used to constrain the type of shock giving rise to the H_2 emission. In practice, however, this is not very easy due to the large numbers of parameters involved in the models. In magnetically driven pure C-type shocks[2] the temperature remains fairly constant over most of the post-shock region, with values which depend on the shock velocity and range between 1000 and 3000 K [30]. Consequently, the Boltzmann diagram of this type of shock is characterized by a single temperature component. In curved C-type shocks, however, one can have the superpositions of several C-shocks at different strengths, resulting in a temperature stratification on the Boltzmann diagram. Low excitation temperatures ($\sim$300–1000 K) are predicted for the rotational lines and for 1–0 transitions, while moderate excitation (1500–2500 K) are generally predicted for the higher vibrational transitions (see e.g. [15]). In any event, pure C-type shocks do not predict the temperatures in excess of 3000 K that are sometime measured from the excitation diagrams. The H_2 spectrum of a pure J-type shock, on the other hand, strongly depends on the shock velocity. For high velocity J-type shocks, the gas is dissociated and ionized at the high temperatures attained at the shock front; H_2 reforms farther downstream, when temperature is decreased to values below $\sim$500 K. Such types of shocks, therefore, predict low H_2 excitation temperatures. In low velocity J-type shocks, on the other hand, H_2 is not dissociated and since the post-shock temperature profile significantly varies with the distance from the shock front, the corresponding excitation diagram displays a marked curvature structure, where also the high vibrational levels are excited [48].

---

[2] A full description of the difference between Continuous (C) and Jump (J) type shocks can be found in [27].

Further complications in the interpretation of Boltzmann diagrams derive from the presence of high velocity bow-shocks, where the velocity at the head of the bow is high enough to create a dissociative J-shock cup, while C-shocks are developed in the bow-wings, where the velocity is lower (e.g. [46, 47]). In addition, time-dependent shocks have been recently considered, with the development of models for non-steady-state conditions. Such models show that at early time shocks exhibit both C- and J- type characteristics and the presence of a large temperature stratification in the H_2 excitation diagram can be explained by this single shock. The time scales for a shock wave to attain steady-state depends on the density and for $n \sim 10^4 \text{cm}^{-3}$ is of the order of 10^4 years. It is indeed possible to see these time-dependent effects in young flows and measure the age of the flow, if the H_2 excitation diagram is sufficiently sampled over a large energy range [22, 32]. Deviations from steady-state can be also evidenced in Boltzmann diagrams by measuring an ortho/para ratio different from the equilibrium value of 3. In fact, if the gas temperature is low enough, the timescales to reach the equilibrium value are longer than the shock dynamical time. Such an effect has been indeed observed in the pure-rotational spectrum of some HH objects [33, 34, 53], where an ortho/para ratio of 1.2–1.6 has been measured (see Fig. 8).

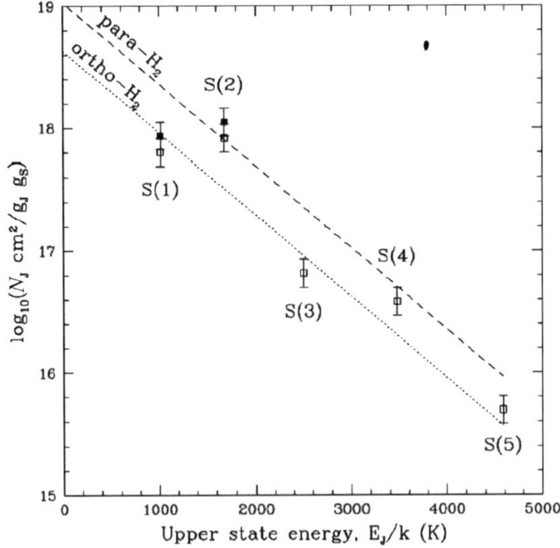

Fig. 8. Boltzmann plot of the H_2 pure rotational lines observed in HH 54 by ISO-SWS. Note how the ortho (J odd) and para (J even) lines display along two different lines in the diagram, where the ortho/para ratio is equal to the equilibrium value of 3. The observed displacement is consistent with an ortho/para ratio equal to 1.2 (from [33])

4 IR HAR Observations of Jet in Embedded Systems

As we have briefly described in Sect. 2, intrinsic and technical limitations set at present the maximum resolution that can be achieved from ground-based near-IR instrumentation on 8-m class telescopes to $\sim$0.2–0.5$''$, depending on the possibility of using AO systems, on the seeing conditions and on the brightness of the target. Nevertheless such limits permit the study of jets on spatial scales of the order of 100–200 AU in the nearby clouds, and several studies have been performed in the bast years to investigate the jet structure and physical properties in the region close to driving star, to get insights on the jet acceleration and collimation. Most of the works employing AO systems have been applied to the study of relatively evolved Classical T Tauri (CTT) systems and will be reviewed in the contribution by Dougados in this volume. Here, I will concentrate on the results obtained in the study of embedded young stars, the so-called Class I objects.

Many studies have been conducted in the past few years to investigate the structure and kinematics of the base of atomic jets in these systems and compare them to the more evolved CTT sources. This has been done through long-slit spectroscopy, with the slit positioned both along and across the jet and constructing the corresponding Position-Velocity (PV) diagrams illustrating radial velocity variations both as a function of the distance from the central source and in the jet transversal direction ([8], see also Fig. 9).

A detailed study of this kind has been done on the L1551-IRS5 jet by [42] with the IRCS instrument on Subaru. This is one of the first studies that recognized in a [Fe II] 1.64 μm PV diagram the presence of two velocity components, a narrow High Velocity component (HVC) at $V \sim -300$ kms^{-1} and a Low Velocity Component (LVC) at $V = -100$ kms^{-1} that is spatially confined within $\sim$400 AU from the IRS5 source. This velocity structure is typical of the Forbidden Emission Line (FEL) regions observed in optical studies of CTT stars (e.g. [25]) and indicates that the jets' kinematical structure is already defined at early evolutionary stages. Reference [40] constructed synthetic [Fe II] PV diagrams adopting the cold disk-wind model developed by [18] and compared them to the PV diagram observed in L1551 (Fig. 9). They found that the disk-wind model reproduces quite well the observed two velocity components but the LVC becomes weaker than the HVC too rapidly with respect to the observations, where the LVC dominates over the HVC further out, upto distances of $\sim$0.6$''$ from the star. Also, the predicted electron density is lower by one/two order of magnitude with respect to the values measured by [28] from [Fe II] line ratios. The same kind of problems have been found in the comparison of cold disk-wind models with optical observations of CTT stars (e.g. [31]). Additional heating at the jet base could improve the matching between the predicted and observed PV diagrams, enhancing the relative contribution of the LVC with respect to the HVC.

When more than one velocity resolved [Fe II] line is observed, one can study variations of the physical properties as a function of the jet

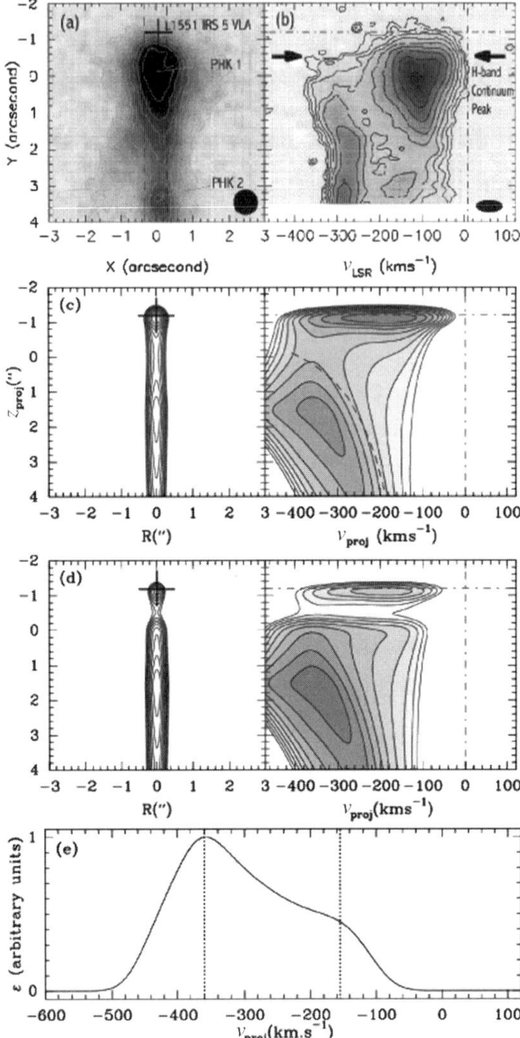

Fig. 9. (**a–b**) [Fe II] 1.644 μm emission maps (**left**) and position velocity diagrams (**right**) taken along the L1551 jet with Subaru by [42] (**c–d**) synthetic maps predicted for a disk-wind model by [40] convolved with the beam sizes of the L1551 observations without (c) and with (d) correction for extinction; (**e**) line profile extracted from the predicted PV diagram

velocity, to separately investigate the parameters pertaining to the different jet components. This kind of analysis has been performed by [19] on the HH 34 and HH 1 jets, using the [Fe II] 1.64 μm/1.60 μm ratio to derive the electron density and mass flux rate in different velocity bins. Like for L1551 IRS5, also in HH 34 the PV diagram at the jet base identifies an LV (at $V_{\mathrm{LSR}} \sim 0$ kms^{-1}) and an HV component (at ~ -200 kms^{-1}). The LVC observed close to the driving source has an electron density which is a factor of two higher than the density in the HVC (see Fig. 10): Most of the mass flux, however, is carried out by the HVC, which is the only one surviving at larger distances. This set constraints on the origin of the LVC: a high electron density seems in fact in contrast with scenarios in which the LVC is due to gas entrained by the HVC or to gas directly ejected in the external layers of a disk-wind (while the HVC should be ejected in the inner jet), since in both the cases one would expect both the ionization and the total density to decrease from high to low velocities.

Measurements of physical parameters at the base of IR jets in Class I sources have also been obtained by [51], through Subaru echelle spectroscopy. Electron densities of $\sim 10^{4}$cm^{-3} or greater are always found in the inner jet regions, while the H$_2$ line ratios indicate temperatures of 2–3 10^3 K with no evidence for temperature stratifications.

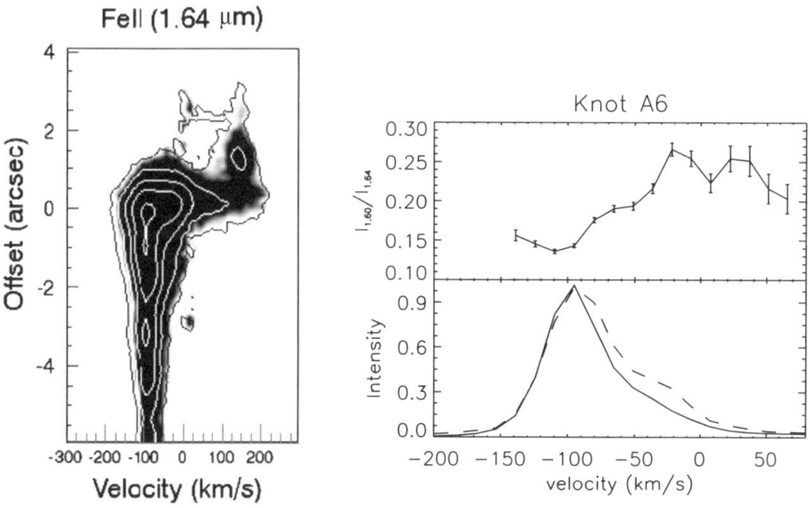

Fig. 10. (Left) continuum subtracted [Fe II] 1.64 μm PV diagram of the inner region of the HH 34 jet. Offsets are with respect to the IRS driving source; **(Right)** [Fe II] 1.64 μm/1.60 μm intensity ratio as a function of velocity (top) and profiles of the two lines (bottom) in the inner A6 knot located within $2''$ from the source. The plotted ratio is sensitive to density variations and indicates that the HVC gas (at $V_{\mathrm{LSR}} \sim -100$ kms^{-1}) has an electron density lower than the gas in the LVC (at $V_{\mathrm{LSR}} \sim -30$ kms^{-1}). From [19]

An interesting result found by [9] and [10] is that several Class I jets present high-velocity H_2 emission within a few hundreds of AU (see Fig. 11). Such small scale H_2 emission regions have been called 'molecular hydrogen emission-line regions' (MHELs) analogous to the atomic FELs regions observed in TT stars. In fact, like the FELs, both LVC and HVC are observed in these regions, with velocities of the order of 5–20 and 50–150 kms^{-1}, respectively. The origin of the MHELs is unclear, but the similarity of their kinematical signature with the ones observed in atomic jets suggests a common origin for the two. Such a tight relationship between MHELs and FELs has been demonstrated by [11], who showed that a FEL region, traced by [Fe II] lines, is often associated with the H_2 emission region: [Fe II] emission has usually higher velocities than the H_2 and a ratio between the HVC/LVC brightness is higher than the H_2 emission. The most likely interpretation is that the HV H_2 emission is excited in a layer between the atomic jet and the near-stationary and dense ambient medium. The H_2 LVC, however, could also be due to the cool molecular component of a disk-wind.

Adaptive optics assisted observations were obtained on one of this MHEL regions, namely the SVS13 jet, using NACO at the VLT [12]. The reached spatial resolution $<0.25''$, well sampled by a pixel scale of $0.027''$, was able to resolve the H_2 emission into several peaks along the jet, consistent with thermal expansion of packets of gas ejected during periods of increased accretion activity (see Fig. 11). Proper motion measurements obtained combining these

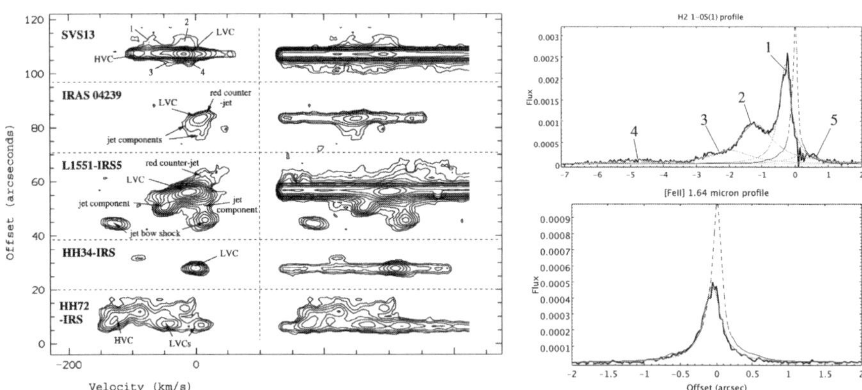

Fig. 11. (Left) H_2 position-velocity diagrams of the inner molecular jets associated with Class I objects obtained with CGS4 at UKIRT [9]. The left hand-panel, where the continuum emission from the central source has been removed, shows only the line emission associated with the Molecular Hydrogen Emission Line (MHEL) regions; **(Right)** H_2 2.12 μm and [Fe II] 1.64 μm intensity profiles along the MHEL region associated with SVS13, obtained with NACO at VLT [12]. The dashed line shows the profile through the (subtracted) continuum adjacent to each line. While several H_2 emission peaks have been resolved within $5''$ from the source, the [Fe II] emission is all confined to a region of only $0.5''$

data with Subaru data taken two years earlier indicate that the H_2 peaks possess a tangential velocity of $0.028''$/year. At variance, [Fe II] presents a barely resolved single peak which seems to be stationary. This evidence suggests that [Fe II] could be associated with a stationary collimation shock at the base of the jet.

This result also shows the potential of IR HAR observations to obtain proper motion measurements of pure H_2 jets, on data taken at relatively short intervals of time. Instrumentation performing IFU spectra-images appears particularly suited for studying the proper motion on emission knots very close to the driving stars, allowing one to obtain more precise continuum-subtracted line images. In general, the sub-arcsec resolutions that are obtained in IR imaging are suitable for getting proper motion measurements of H_2 jets when first epoch data of a good quality are available. Results obtained by different authors [5, 6] on large scale H_2 flows show that the total velocities of H_2 knots along the jet are remarkably high, up to $400\,\mathrm{kms}^{-1}$, much higher than the H_2 dissociation velocity of $\sim 50\,\mathrm{kms}^{-1}$. This indicates that in most cases the H_2 gas is traveling in a medium that has been already set into motion by precedent ejection events.

Interesting results have also been obtained from studies of the kinematics and excitation of IR jets in the transverse direction across the width of the jet. Reference [7], in particular, obtained high dispersion long-slit observations of the two small- scale H_2 jets from HH 26-IR and HH 72-IR with the slit perpendicular to the jet axis and in both jets detected an asymmetry of the measured radial velocity with respect to the jet axis that can be modeled by assuming that the jets possess a toroidal velocity due to rotation. These observations follow the detection of rotational signatures in jets obtained through HST observations of T Tauri micro-jets (e.g. [1]), showing that jets from younger protostars are also able to carry the excess of angular momentum away from the central source.

5 Conclusions

The progresses done in IR spectroscopy of jets during the last decade have been tremendous, thanks both to the access at sensitive instrumentation mounted on large telescopes and to the development of sophisticated models and tools for the data analysis and interpretation. The very next future will see a boost of new IR instrumentation specifically designed to improve the high-angular resolution. AO-assisted IFU spectrometers using laser guide stars are begining to be operational and are seeking to get 2D maps of excitation and kinematics at the jet base, allowing us a better comparison with the model for jet acceleration and heating. The VLTI/Amber interferometry which is able to get spectroscopic visibilities on mas angular resolution, is expected to give a major breakthrough in the understanding of the jet acceleration mechanisms as soon as it becomes fully operational. With a similar resolution, it will be possible in the next decade to obtain reconstructed interferometric images from planned facilities, like the VLTI/VSI instrument and Linc-Nirvana on LBT. Finally, the MIRI instrument on JWST will make it possible for us to observe in the 20 µm window with $\sim 1''$ of angular resolution, while PACS on the Herschel satellite will have a spatial resolution of $\sim 9''$ at 60 µm. These instruments will, therefore open, the mid- and far-IR spectral range to observations at the relevant spatial scales for the study of young stars flow.

When all these new facilities become will be fully operational it is expected that the study of jet physics at high angular resolution will move more and more into the IR regime.

References

1. Bacciotti, F., Ray, T.P., Mundt, R., Eislööffel, J., Solf, J.: Astrophys. J. **576**, 222 (2002)
2. Bally, J., Lane, A.P.: Astrophys. J. **257**, 612 (1982)
3. Bally, J., Reipurth, B., Davis, C.J.: Observations of jets and outflows from young stars. In: Reipurth, B., Jewitt, D., Keil, K. (eds.) Protostars and Planets V, pp. 215–230. University of Arizona Press, Tucson (2007)
4. Black, J.H., van Dishoeck, E.F.: Astrophys. J. **322**, 412 (1987)
5. Caratti o Garatti, A., Giannini, T., Nisini, B., Lorenzetti, D.: Astron. Astrophys. **449**, 1077–1088 (2006)
6. Chrysostomou, A., Hobson, J., Davis, C.J., Smith, M.D., Berndsen, A.: MNRAS **314**, 229 (2000)
7. Chrysostomou, A., Bacciotti, F., Nisini, B., Ray, T.P., Eislöffel, J., Davis, C.J., Takami, M.: The transport of angular momentum from YSO jets. In: Protostars and Planets V. Proceedings of the Conference, Hilton Waikoloa Village, Hawai'I, 24–28 October 2005. LPI Contribution No. 1286, p. 8156
8. Davis, C.J., Berndsen, A., Smith, M.D., Chrysostomou, A., Hobson, J.: MNRAS **314**, 241 (2000)
9. Davis, C.J., Ray, T.P., Desroches, L., Aspin, C.: MNRAS **326**, 524 (2001)
10. Davis, C.J., Stern, L., Ray, T.P., Chrysostomou, A.: Astron. Astrophys. **382**, 1021 (2002)
11. Davis, C.J., Whelan, E., Ray, T.P., Chrysostomou, A.: Astron. Astrophys. **397**, 693 (2003)
12. Davis, C.J., Nisini, B., Takami, M., Pyo, T.-S., Smith, M.D., Whelan, E., Ray, T.P., Chrysostomou, A.: Astrophys. J. **639**, 969 (2006)
13. Dionatos, O., Nisini, B., Giannini, T., Garcia Lopez, R., Davis, C., J., Smith, M.D.: (in preparation)
14. Draine, B.T.: arXiv:astro-ph/0304488 (2003)
15. Eislöffel, J., Smith, M.D., Davis, C.J.: Astron. Astrophys. **359**, 1147–1161 (2000)
16. Eislöffel, J.: Astron. Astrophys. **354**, 236 (2000)
17. Flower, D.R., Le Bourlot, J., Pineau des Forêts, G., Cabrit, S.: MNRAS **341**, 70 (2003)
18. Garcia, P.J.V., Ferreira, J., Cabrit, S., Binette, L.: Astron. Astrophys. **377**, 589 (2001)

19. Garcia Lopez, R., Nisini, B., Giannini, T., Eislöffel, J., Bacciotti, F.: Astron. Astrophys. (submitted)
20. Giannini, T., Nisini, B., Caratti o Garatti, A., Lorenzetti, D.: Astrophys. J. **570**, L33 (2002)
21. Giannini, T., McCoey, C., Caratti o Garatti, A., Nisini, B., Lorenzetti, D., Flower, D.R.: Astron. Astrophys. **419**, 999 (2004)
22. Giannini, T., McCoey, C., Nisini, B., Cabrit, S., Caratti o Garatti, A., Calzoletti, L., Flower, D.R.: Astron. Astrophys. **459**, 821 (2006)
23. Gredel, R.: Astron. Astrophys. **292**, 580 (1994)
24. Hartigan, P., Morse, J.A., Raymond, J.: Astrophys. J. **436**, 125 (1994)
25. Hirth, G.A., Mundt, R., Solf, J.: Astron. Astrophys. Suppl. **126**, 437 (1997)
26. Hollenbach, D., McKee, C.F.: Astrophys. J. **342**, 306 (1989)
27. Hollenbach, D.: The physics of molecular shocks in YSO outflows. In: Reipurth, B., Bertout, C. (eds.) Herbig–Haro Flows and the Birth of Stars, pp. 181–198. IAU Symposium No. 182, Kluwer Academic Publishers, Boston (1997)
28. Itoh, Y., et al.: PASJ **52**, 81 (2000)
29. Jones, A.P.: J. Geophys. Res. **105**, 10257 (2000)
30. Kaufman, M.J., Neufeld, D.A.: Astrophys. J. **456**, 611 (1996)
31. Lavalley-Fouquet, C., Cabrit, S., Dougados, C.: Astron. Astrophys. **356**, L41 (2000)
32. Le Bourlot, J., Pineau des Forêts, G., Flower, D.R., Cabrit, S.: MNRAS **332**, 985 (2002)
33. Neufeld, D.A., Melnick, G.J., Harwit, M.: Astrophys. J. **506**, L75 (1998)
34. Neufeld, D.A., et al.: Astrophys. J. **649**, 816 (2006)
35. Nisini, B., Caratti o Garatti, A., Giannini, T., Lorenzetti, D.: Astron. Astrophys. **393**, 1035 (2002)
36. Nisini, B.: Astrophys. Space Sci. **287**, 207 (2003)
37. Nisini, B., Bacciotti, F., Giannini, T., Massi, F., Eislöffel, J., Podio, L., Ray, T.P.: Astron. Astrophys. **441**, 159 (2005)
38. Nussbaumer, H., Storey, P.J.: Astron. Astrophys. **193**, 327 (1988)
39. Oliva, E., et al.: Astron. Astrophys. **369**, L5 (2001)
40. Pesenti, N., Dougados, C., Cabrit, S., O'Brien, D., Garcia, P., Ferreira, J.: Astron. Astrophys. **410**, 155 (2003)
41. Podio, L., Bacciotti, F., Nisini, B., Eislöffel, J., Massi, F., Giannini, T., Ray, T.P.: Astron. Astrophys. **456**, 189 (2006)
42. Pyo, T.-S., Hayashi, M., Kobayashi, N., Terada, H., Goto, M., Yamashita, T., Tokunaga, A.T., Itoh, Y.: Astrophys. J. **570**, 724 (2002)
43. Quinet, P., Le Dourneuf, M., Zeippen, C.J.: Astron. Astrophys. Suppl. **120**, 361 (1996)
44. Reipurth, B., Yu, K.C., Heathcote, S., Bally, J., Rodríguez, L.F.: Astron. J. **120**, 1449 (2000)
45. Reipurth, B., Bally, J.: Annu. Rev. Astron. Astrophys. **39**, 403 (2001)
46. Smith, M.D., Brand, P.W.J.L., Moorhouse, A.: MNRAS **248**, 451 (1991)
47. Smith, M.D., Brand, P.W.J.L., Moorhouse, A.: MNRAS **248**, 730 (1991)
48. Smith, M.D.: Astron. Astrophys. **296**, 789 (1995)
49. Smith, N., Hartigan, P.: Astrophys. J. **638**, 1045 (2006)
50. Stanke, T., McCaughrean, M.J., Zinnecker, H.: Astron. Astrophys. **392**, 239 (2002)
51. Takami, M., et al.: Astrophys. J. **641**, 357 (2006)

52. White, G.J., et al.: Astron. Astrophys. **364**, 741 (2000)
53. Wilgenbus, D., Cabrit, S., Pineau des Forêts, G., Flower, D.R.: Astron. Astrophys. **356**, 1010 (2000)
54. Wilking, B.A., Schwartz, R.D., Mundy, L.G., Schultz, A.S.B.: Astron. J. **99**, 344 (1990)

Observing YSO Jets with Adaptive Optics: Techniques and Main Results

C. Dougados

Laboratoire d'Astrophysique de Grenoble
Catherine.Dougados@obs.ujf-grenoble.fr

Abstract. Detailed studies of the central 100 AU's of young stellar jets are critical to constrain current ejection theories. They require however subarcsecond angular resolution, achievable in the optical/NIR domain either from space with HST or from the ground with adaptive optics (AO) technics. I review in this contribution this latter technique and its application to the study of young stellar jets. I start with a general description of AO observing and the combination of AO with spectroscopy, with special emphasis on integral field spectroscopy (IFS). I then summarize recent results obtained on T Tauri microjets. The prospects of future AO systems regarding jet studies is finally briefly reviewed.

C. Dougados: *Observing YSO Jets with Adaptive Optics*, Lect. Notes Phys. **742**, 105–121 (2008)
DOI 10.1007/978-3-540-68032-1_5

1 Introduction

The physical mechanism by which mass is ejected from young stars and collimated into jets remains a fundamentally open issue in star formation theory. The strong correlation between ejection and accretion found in pre-main sequence (PMS) stars, with a mass flux ratio as high as 0.1 [8, 20] favored accretion-driven magneto-hydrodynamic (MHD) wind models [1, 27]. However, it is not yet established whether the jet originates from the stellar surface [42], the magnetosphere/disk interface [43], or a wide range in disk radii ("disk winds", [16]) and whether it is launched mainly by magneto-centrifugal forces or by a strong thermal pressure gradient in an accretion-heated corona [18]. Distinguishing between these various scenarios is crucial not only for the jet phenomenon but also for models of exoplanet formation. Each scenario has distinct implications for the internal structure, angular momentum transfer, and heating/irradiation processes in the inner regions of protoplanetary disks.

The small-scale jets detected in the vicinity of the optically revealed classical T Tauri stars- (CTTSs) offer a unique opportunity to test current ejection theories [15]. They give access to the innermost regions of the wind (≤ 100 AU), where models predict that most of the collimation and acceleration processes occur and where the interaction with the ambient medium is minimized (see Fig. 3). In addition, the stellar and accretion disk properties are well characterized in these systems. Critical to the launching models are constraints on acceleration and collimation scales, kinematics (poloidal and toroidal velocities), jet densities, mass-loss rates, and ejection efficiencies (defined as the ratio of mass-loss to mass-accretion rates) on scales less than a few 100 AUs from the central source.

A fundamental difficulty in imaging a faint jet close to a bright CTTS is the contrast with the source itself. The line emission contrast can be improved either by decreasing the point spread function (PSF) or by increasing the spectral resolution. Obviously the optimum solution is a combination of both. The required angular resolution is on the order of $0.1''$ (which corresponds to 14 AU at the Taurus distance of 140 pc), i.e., well below the limit of $1''$ imposed by the atmospheric turbulence on ground-based observations. Such high angular resolution can be achieved either from space (with the Hubble Space Telescope (HST) for example) or from the ground on D>4 m class telescopes with adaptive optics correction. I summarize in this chapter the technique of Adaptive Optics (AO) observation and its application to the study of the launching process in jets from young stars. The chapter by Francesca Bacciotti, Mark Mc Caughrean and Tom Ray presents the results obtained from space with HST.

2 Observing with Adaptive Optics

2.1 General Remarks

Most ground-based 8–10 m class telescopes are now equipped with AO. This allows us in principle, to reach an angular resolution in the near-infrared domain, similar to the HST in the optical domain (i.e. $0'''05$, Table 1). In addition, ground-based telescopes, because of their larger collecting areas, offer an increased sensivity (by more than an order of magnitude with respect to HST), critical for detailed studies of faint extended emission. The principle of AO is detailed in the chapter by S. Esposito and E. Pinna. Here, I will briefly come back to a few issues critical for successful AO observations.

The level of turbulence is characterized by the Fried parameter r_0, the size scale of turbulent cells, and the correlation time t_0, related to r_0 and the speed of the turbulent layers. Both parameters increase with wavelength and are critical for AO performance. Most current astronomical AO systems are designed to provide close to diffraction-limited images in the near-infrared (1 to 2 μm) with a capability for partial correction in the visible. Because of the wavelength dependence of atmospheric turbulence, the level of correction strongly decreases with decreasing wavelength. Thus, only very partial correction can be achieved by current systems in the optical domain.

The image of a point source formed by an adaptively corrected telescope, called the PSF, is composed of a diffraction-limited core superimposed on an extended halo of typical size, the seeing disk (see Fig. 1). The strehl ratio, defined as the ratio between the peak intensity of the image divided by the peak intensity of a diffraction-limited image with the same total flux, quantifies the degree of correction achieved by the AO system (a strehl ratio of 1 corresponds to full correction). Because of inherent limitations such as the number of actuators used for the correction of the secondary mirror and the frequency of the correction loop, full correction is never achieved. Typical strehl ratios achieved by current AO systems are 50% in the near-infrared and a few percent in the optical domain where the correction is very partial. High contrast AO systems which will achieve strehl ratios of 80–90 are currently under study.

In the natural guide star mode, the choice of the reference star to analyze the wavefront perturbations induced by the turbulent layers of the atmosphere becomes critical. The wavefront analysis is usually performed in the red optical domain (R,I bands), in a wavelength range different from the scientific observations. A near-infrared wavefront sensor is also available on NAOS at the Very Large Telescope (VLT). The AO reference star must be within angular distance less than the isoplanatic angle ($\simeq 30''$ at 2 μm, $5''$ at 500 nm) of the scientific target so that the measured distortions on the incoming turbulent wavefront are correlated to the ones in the direction of the scientific target. The AO sensor reference star must also be bright enough to ensure good sig-

Table 1. Examples of diffraction limits at select wavelengths for Space-based versus Ground-based telescopes

λ	HST	$D = 8\,\mathrm{m}$
630 nm	$0''05$	$0''015$
2.2 μm	$0''.17$	$0''05$

nal to noise ratio on the wavefront sensor, and its intensity distribution must be dominated by a central peaked core of size typically $\leq 1''$. The limiting magnitude of the AO reference star will depend critically on the number of sub-apertures required to analyze the wavefront, i.e., the number of actuators on the secondary mirror. Current systems optimized for best performance in the near-infrared on 8 m class telescopes provide nominal correction (strehl ratios of 50%) for $R = 11$–12th magnitude, wavefront reference star and partial correction (strehl ratios of a few %) down to $R = 17$th magnitude. The scientific target itself can be used for the wavefront analysis, if it meets the above requirements. This on-axis analysis will give the best correction and is the preferred mode of correction.

Laser guide stars (LGS) are currently being developed on most 8–10 m telescopes (Keck, VLT, Gemini, Subaru). LGS mode still requires a natural star for tip–tilt correction. The requirements are similar but less stringent than in the natural guide star mode. For the VLT LGS, for example, the tip–tilt correction star must have optical magnitudes (V/R) in the range 11th–18th magnitude. A good estimate of the effective PSF at the time of the observations is also required for a proper interpretation of AO observations. Indeed, the PSF usually presents low-level structure that may be mistaken for an astronomical signal. If a good estimate of the observational PSF is available, deconvolution techniques may be applied to the observed image (see [44] for a discussion of the different algorithms). Because the level of correction is highly dependent on the turbulence properties, the PSF can vary a lot during typical on-source integrations. The PSF can either be reconstructed from the data recorded by the wavefront sensor or estimated through regular (typically every 15 minutes) observations of an unresolved star. The PSF reference star must be corrected at the same level as the science target. Therefore, it should have

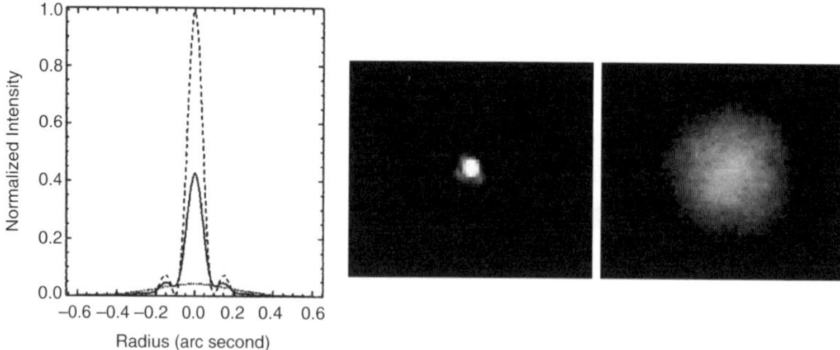

Fig. 1. (**Left**) Comparison between the diffraction-limited PSF *(dashed line)*, an AO-corrected PSF *(full line)* and the uncorrected PSF. PSFs were computed in the H band for a telescope of diameter $D = 3.60$ m and $r_0 = 18$ cm at $\lambda = 0.5\mu$m. The Strehl ratio achieved for the AO PSF is 40%. This figure is taken from [44]; (**Right**) Image of a point source in the K band after (Left) and before (Right) AO correction on the Subaru 8 m telescope. Image copyright of Subaru telescope

the same magnitude (as measured on the wavefront sensor) as the star used for the wavefront analysis. It should also be observed with a similar airmass (i.e., same path through the atmospheric turbulent layers) and not far in time from the scientific target.

Figure 1 illustrates the gain provided by adaptive correction over conventional (i.e., without AO) imaging. First, concentration of the light of the central star allows the separation of the stellar contribution from faint extended emission (such as the line emission from the jet) down to a distance of typically one PSF full width half maximum (FWHM). Second, the detectability of the faint jet emission increases dramatically with increasing angular resolution. With a seeing-limited angular resolution, the jet emission of intrinsic transverse width $<0''.2$–$0''.4$ (see below and Fig. 4) is spread over a larger area resulting in decreased surface brightness.

2.2 AO and Spectroscopy

In addition to direct imaging with filters, AO correction can be used in combination with a spectroscopic instrument. Spectral information combined with high angular resolution is critical to the study of young stellar jets as it allows to probe the kinematics in the acceleration region of the flow. The smaller PSFs achieved with AO correction allow the user to increase the spectral resolution and the signal to noise ratio on both the continuum (unresolved) and jet emissions (intrinsic FWHM $\simeq 0''.2$–$0''.3$). In addition, spectroscopic information allows an accurate subtraction of the central source photospheric absorption spectrum, critical to retrieve the intrinsic line emission profile close to the source. Moreover, in the near-infrared domain, jet [Fe II] emission lines can be blended with H_2 or HI lines.

Conventional long slit-spectroscopy in combination with AO correction is available, for example, on the NACO-VLT and IRCS-SUBARU instruments. In the case of complex extended objects, Fabry-Perot scanning or stepped long-slit spectroscopy can help retrieve the 2D spatial information. These two techniques however are very time consuming. In addition, because of the variable nature of the spatial PSF even after AO correction, it is highly desirable to get the 2D spatial and 1D spectral information in one single exposure. Integral Field Spectrographs (IFS) allow one to do just that. These instruments appear particularly suited for use in conjonction with AO correction. They provide an instantaneous estimate of the 2D spatial PSF during the observations, if for example an unresolved continuum source is included in the field of view. Their use with AO correction has been pioneered by OASIS at CFHT in the visible and 3D at Calar Alto for the near-infrared. Similar instruments are now in use or in development for most 8-m class telescopes, in particular, SINFONI-SPIFFI in the near-infrared at the VLT. I come back to a more detailed description of IFS in the next section. Spectral resolutions available range from typically 2000–3000 to 10000 (long-slit spectroscopy only), corresponding to velocity resolutions between 200 and 30 kms^{-1}.

The combination of spectroscopy with adaptive optics correction can suffer from a few spurious effects. Due to the variation of the turbulence properties with wavelength, the PSF changes across the spectrum, in particular, the strehl ratio increases with increasing wavelengths. Changes of 10% across the spectrum are observed in the near-infrared for typical spectral settings at a resolution of a few 1000. When the slit width is close to or larger than the PSF (which can easily be the case in the near-infrared at low spectral resolutions), the following spurious effects can be introduced:

- Artificial continuum slopes: The variation of enclosed energy in a finite size slit resulting from the wavelength dependency of the PSF can lead to changes in spectral slope [23].
- Poor telluric calibration: The change in PSF across the spectrum also induces a change in sky absorption line widths with wavelength. If in addition the adaptive correction for the telluric standard is different from the science object the subtraction of the atmospheric absorption lines becomes very difficult.
- Wavelength shifts: When the PSF is smaller than the slit width, a shift in the spectral direction is introduced if the source is not well centered in the slit (cf. [31]). This effect, known as the *slit effect*, can be of dramatic importance, if very accurate velocity calibrations are needed. The shift of the source emission with respect to the slit can be due to an intrinsically asymmetric brightness distribution or atmospheric differential refraction.
- Varying spectral resolutions: The spectrum has a complex spatial profile: It is the sum of a diffraction limited core superimposed on an extended halo of size the seeing disk. Different spectral resolutions can be achieved for these two components: If the PSF is smaller than the slit width, the spectral resolution of the diffraction limited core will be set by the PSF, while the spectral resolution of the extended halo will in general be set by the slit width.

Again these effects will be particularly important when the slit width is on the order or larger than the PSF. In general, spectral calibrations, especially spectro-photometric and velocity calibrations, will be less accurate with AO correction than without it.

2.3 Integral Field Spectroscopy

The main three different concepts proposed for integral field spectroscopy are illustrated in Fig. 2. In the image slicer IFS concept, the imaging field is cut into slices by a first set of mirrors. A second set of mirrors rearranges the slices into a line, which is then fed to the slit entrance of the spectrograph. The image plane can also be sampled by a lenslet array. The light from each lens is then either fed by a fiber to the entrance slit of the spectrograph (optical fiber bundles IFS) or imaged onto a pupil then dispersed by a grism (lenslet array IFS). In a fourth different concept, marginally used today, the image is spectrally sampled by a Fabry-Perot interferometer in a few discrete sets of wavelengths. Obtaining a significant spectral coverage requires scanning of the Fabry-Perot. Because of the competition between the spatial and spectral coverage on the detector, IFSs in general have lower spectral resolution/coverage than long-slit spectrographs.

The data reduction procedure for IFS data has become increasingly user-friendly. One of the major difficulties is the spectral flat-fielding, which measures the transmission of the slit/microlenses array/fiber plus grism. These spectral flat-fields are usually derived from a combination of sky and dome exposures. Flat-fielding the datacubes to better than a few percent is currently very difficult to achieve. For lenslet array IFSs, another crucial step in the reduction procedure is the extraction of the spectra and astrometric calibration. Since they sample spatially the PSF, IFSs combined with AO suffer in general much less than long-slit spectrographs from the spurious effects mentioned above. In particular a lenslet array type IFS (like OASIS at the CFHT) will not suffer at all from the slit effect.

The tools used to analyze the wealth of information provided by IFS data include channel maps reconstructed by integrating the emission over a given wavelength range, position–velocity maps provided by projection and summation of the 2D spatial information onto a single axis (mimicking long-slit spectroscopic data), with spectra at any spatial position in the field. IFS combined with AO provides an instantaneous wavelength-dependent PSF characteriza-

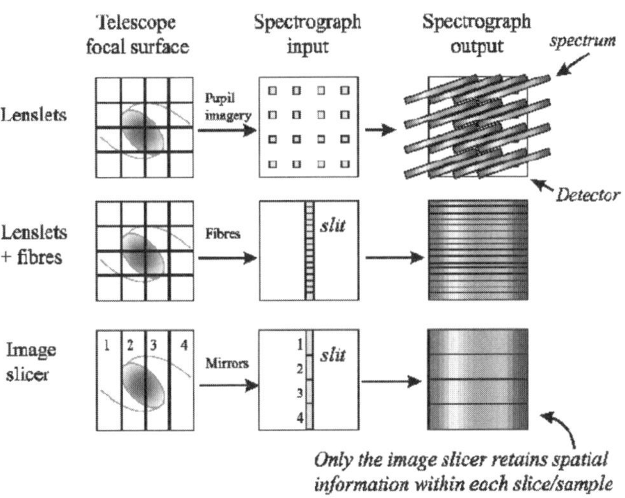

Fig. 2. Figure illustrating the different concepts proposed for integral field spectroscopy: lenslet arrays (**top**); optical fiber bundle (**middle**); and image slicers (**bottom**). See text for more details

tion. This PSF can be used for the deconvolution of reconstructed channel maps. In principle, a full tri-dimensional deconvolution of the datacube is also possible. This procedure is under development but not yet fully implemented.

IFS observations of protostellar jets started about 10 years ago with MPE-3D observations of the T Tau system by [22] and TIGER-CFHT observations of the DG Tau microjet by [28]. For more details on integral field spectrographs and their use in the context of star formation see the review by [19].

2.4 Optical and Near-infrared Jet Tracers

Spectra of atomic jets are characterized by strong forbidden emission lines of weakly ionized species. In the optical domain, the lines of [O i]6300Å, [S ii]6717,6731 Å, [N ii]6584 Å are strong and well-studied tracers of the atomic gas. A method proposed by [3] allows one, in particular, to derive estimates of jet excitation conditions (electron temperature T_e, electron density n_e, and ionization fraction x_e,) from the ratios of these lines. Their study allowed to (1) the jet emission to be probed on scales of 15–500 AU from the central source, (2) the kinematics of the jet to be probed at all velocities (from 0 to a few 100 km s^{-1}), and (3) an estimate of jet excitation conditions (n_e, T_e, x_e) and thus of nH and mass-loss rates to be deried. These are critical parameters for constraining ejection models.

The near-infrared (1–2.5 µm) spectrum of jets is dominated by the strong forbidden emission lines of [Fe ii] and transitions of molecular hydrogen (H_2). The iron lines provide diagnostics for ne (from the ratio of 1.533 to 1.644 lines), extinction (Av), Fe depletion and mass-loss rates (see [33, 34, 36]). The combination of optical and near-infrared [Fe ii] lines are a powerful plasma excitation conditions diagnostic tool [33, 36]. Other lines of interest in the near-infrared domain include the permitted atomic Hydrogen transitions, forbidden [S ii] lines at 1.03 µm, [C i] lines at 0.98 µm, and He sc i lines at 1.086 µm. These lines in combination with the main near-infrared [Fe ii] lines provide estimate of electronic temperature. More details on the properties of the near-infrared spectrum of jets from young stars are given in the contribution by Brunella Nisini.

2.5 AO Systems Used in Jet Studies

For a general review of AO observations in the context of star formation, see [32]. Pioneering results on pre-main sequence jets have been obtained by [22] on the T Tau system with the MPE 3D imaging spectrometer combined with the CHARM tip–tilt module mounted on the 3.5 m telescope in Calar Alto. The spatial sampling was 0″.52 and the spectral resolution ≃1000–2000 in the H and K bands. These observations revealed knots of molecular hydrogen emission in the immediate vicinity of the T Tau system. They were followed at much higher angular resolution (0.″ 15 over a field of view of 1″) by [26] using MPE 3D combined with the ALFA Adaptive Optics system [25]. This system was also used by [10] to study the close environment of the young stellar object LkH$_\alpha$225, an Herbig Ae/Be star, in the near-infrared.

A breakthrough in AO studies of jets was allowed by the adaptive optics bonnette PUE'O commissioned in 1996 at the 3.6 m Canada–France–Hawaii telescope (CFHT). The PUE'O bonnette consists of a secondary deformable mirror with 19 actuators combined with a shack-hartmann wavefront sensor sensitive in the R band. Under median seeing conditions, the system provides strehl ratios of 50% in the H and K bands, 5% in the optical domain (I band) for on-axis guide star magnitudes brighter than $R = 13.5$, and image FWHM ranging from $0''.1$ in the near-infrared to $0''.2$ in the R band. For a full characterization of the PUE'O system see [41]. In the context of jet studies, the PUE'O AO system has been used in the optical domain (R band) both in direct narrow-band imaging [13] and in combination with the IFS OASIS [14, 19, 29]. The OASIS instrument [5] is a lenslet array type IFS, providing a spectral resolution of $\simeq 3500$ over 2 spectral settings covering the main optical emission lines. The resulting FWHM after AO correction ranges between $0''.3$ and $0''.4$ (with a spatial sampling of $0''.11$ or $0''.16$). The OASIS instrument, located at the CFHT until 2002, is now installed at the William Herschel Telescope (WHT), Spain. It can be fed with the AO correction module NAOMI, which delivers corrected FWHM ranging from 0.4 (in the R band) to 0.2 (K band) for guide stars brighter than 11th magnitude (R band) under median seeing conditions ($0''.7$).

AO observations of jets are currently beginning to take full advantage of the combination of high sensitivity and high angular resolution provided by AO corrected 8 m class telescopes. The NACO VLT adaptive optics system was recently used in the long-slit mode on the embedded source SVS 13 by [11] providing a FWHM of $0''.25$. The results of these observations are presented in the contribution by Brunella Nisini. Striking results have also been obtained with the AO system of the SUBARU telescope (based on curvature wavefront sensing), combined with the near-infrared instrument IRCS in the long-slit mode. The combination achieved spatial resolutions (FWHM) of $0''.2$ and a spectral of resolution $R = 10000$ ($\Delta V = 30$ km s^{-1}). This allowed the detailed kinematic structure of the flow close to the source to be probed. The Subaru AO system has been recently significantly updated and near-diffraction limited images have been obtained in the K band (see Fig. 1).

3 Main Results on Jets

Only a handful of microjets from T Tauri stars have been studied so far at high angular resolution ($\simeq 10$), either from space with HST or from the ground with AO. This section is focused on the main results from AO observations of the atomic microjets from optically revealed young stars of ages of a few 10^6 yrs. Results obtained with AO on the younger, still embedded, Class I sources are presented in the contribution by Brunella Nisini. The results shown below illustrate the power of AO observations in the context of jet studies, but the general conclusions mentioned here derive from the analysis of all available high-angular resolution data sets.

3.1 Morphology and Collimation Scales

The pioneering work of [24] using long-slit spectroscopy with accurate central continuum subtraction has revealed spatial extensions of a few arcseconds in the forbidden emission line regions of a dozen CTTs. Conventional narrow-band imaging with the HST has provided the first images of jets from 2 CTTs [39], HH 30, and HL Tau. Observations with the PUE'O AO bonnette on the 4 m CFH telescope, used both in conventional narrow-band imaging and in combination with the OASIS IFS, has provided remarkable $0''.2$–$0''.4$ resolution images of the small-scale jets from DG Tau (Fig. 3), CW Tau, and RW Aur on scales $\leq 4''$[13, 28, 29]. Very high angular resolution images $(0''.1)$ reconstructed from stepped long-slit and slitless spectroscopy with STIS on the HST have also been obtained for the DG Tau, RW Aur [4, 45], CW Tau, UZ Tau E, and HN Tau microjets [21].

These observations reveal a complex morphology for the jet emission close to the central source (Fig. 3). The jet brightness distribution is dominated by knots with separations of order $1''$. A clear bow-shaped structure is detected in the DG Tau microjet outer knot, suggesting that shocks may play an important role in the line excitation process. T Tauri microjets thus appear as a scaled down version of the Herbig-Haro flows from younger sources, an indication that the same ejection mechanism is at work through the star formation phases.

Figure 4 shows the variation of jet width against projected distance from the source for a number of CTTs outflows, combining high-angular resolu-

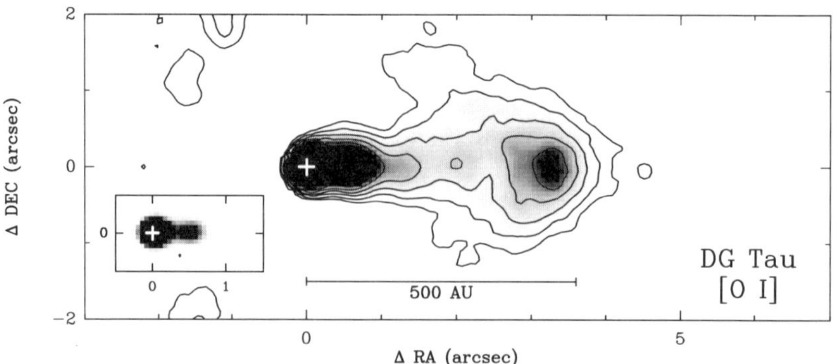

Fig. 3. [O I] image of the DG Tau A microjet obtained with PUE'O on the CFHT. PSF FWHM $= 0.''2$ after OA correction. After deconvolution final angular resolution of $0.1''$. Adapted from [14]

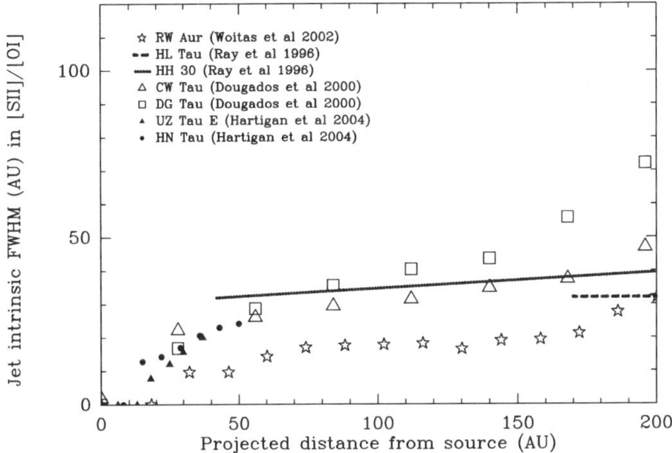

Fig. 4. Variation of jet widths against projected distance from the central source. Jet widths are measured from jet [O I]/[S II] emission intrinsic FWHM (corrected from the PSF FWHM), derived from space HST and ground-based AO observations. Figure adapted from [39]

tion space and ground-based observations (HL Tau, HH 30: [40]; DG Tau, CW Tau, and RW Aur: [13, 46]; HN Tau, UZ Tau E: [21]). At the highest spatial resolution currently achievable ($0''.1 = 14$ AU for the Taurus Auriga Star Formation Region) the jet innermost regions (below $\simeq 20$ AU) are unresolved. Jet widths are then seen to increase up to 20–30 AU at projected distances of around 50 AU from the source. Further out, jet widths increase slowly with distance with full opening angles $\leq 5°$. These observations show that jets achieve almost cylindrical collimation very early on, on scales $\simeq 50$ au. A few constraints available from younger flows show similar collimation properties. These observations put strong constraints on theories of jet launching and appear to rule out pure hydrodynamic models for jet focusing (see the chapter by Sylvie Cabrit in the first volume of this series).

3.2 Kinematics

Medium-resolution spectroscopic observations combined with AO of the inner regions of the DG Tau, RW Aur, HL Tauzcma, and SVS 13 microjets have been obtained with the integral field spectrograph OASIS(CFHT) in the optical domain [19, 28, 29] and/or in the near-infrared with the IRCS(SUBARU) [37, 38] and NACO(VLT) [11] instruments in the long-slit mode. HST/STIS long-slit observations have been also published on the DG Tau, RW Aur, TH 28 and LkHα321 microjets by [5, 6, 9, 45, 46].

Channel maps reconstructed from the IFS OASIS/CFHT observations (Fig. 5) clearly show an onion-like kinematical structure for the flow, with high velocity material more concentrated towards the jet axis than low velocity material [29]. This effect has been also detected in higher angular resolution HST observations [5]. Detailed kinematics along the jet axis have been investigated so far for 3 microjets: DG Tau, RW Aur and HL Tau. Figure 6 (right panel) shows a position–velocity diagram along the RW Aur jet axis in the near-infrared [Fe II] 1.644 μm emission line obtained by [38] with the IRCS/SUBARU spectrograph and AO correction (velocity resolution 30 kms^{-1}, angular resolution 0."15–0."2). These observations illustrate the fact that the acceleration scale of the HVC flow is below current resolution (< 30 AU). Terminal centroid flow velocities range from 150 to 350 kms^{-1}. Moderate variations along the jet axis of 10–20% of the average flow velocity are observed for the HVC on scales less than 3" from the source.

3.3 Excitation Conditions

High-angular resolution allows us to study the spatial variation of the jet plasma parameters and excitation conditions in the wind-launching regions, i.e., on spatial scales $\leq$ a few 100 AU. For example [29] and [14] compute the evolution of the optical forbidden line emission ratios ([S II]/[O I], [N II]/[O I], [S II]6716/6731) along the jet axis for the DG Tau and RW Aur microjets from AO+IFS data (OASIS/CFHT). In both cases, line ratios are best reproduced by Jump-type shock conditions, indicating that time variability plays a dominant role in the heating process.

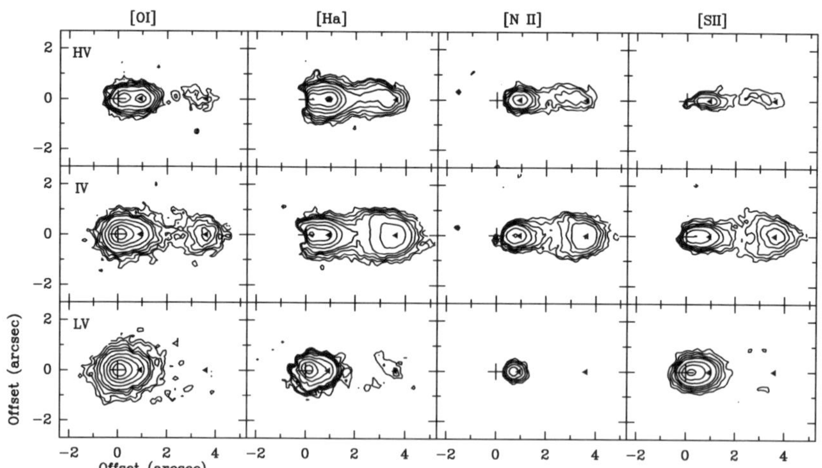

Fig. 5. Channel maps in three different velocity intervals (HVC:[−400,−250], IVC:[−250,−100], LVC:[−100,+10] km s^{-1}) of the optical emission lines of the DG Tau microjet reconstructed from IFS+AO data (OASIS/CFHT combined with AO correction). Angular resolution is 0."4, velocity resolution 100 km s^{-1}. Figure from [29]

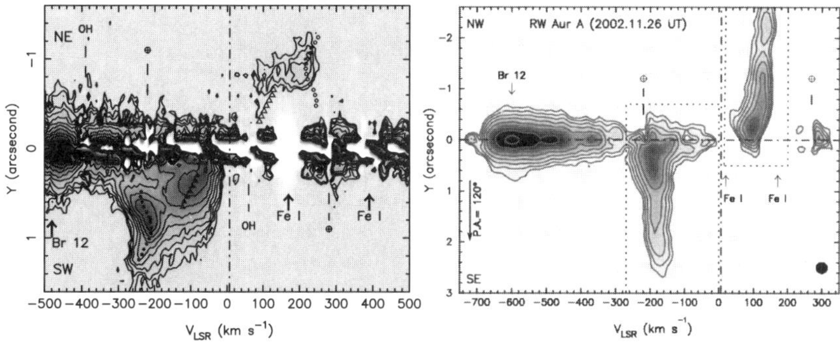

Fig. 6. Position–velocity diagram of the DG Tau (**Left**) and RW Aur (**Right**) microjets in the [Fe II] 1.644 μm obtained with IRCS/SUBARU (AO+long-slit spectrograph). Angular resolution is 0."2 angular resolution and velocity resolution 30 km s⁻¹. Figures taken from [37] and [38]

Using the method developed by [3, 14, 29] also derive the variation of plasma parameters (n_e, T_e, x_e, and total hydrogen density n_H) along the jet axis for the RW Aur and DG Tau microjets (Fig. 7). Similar studies have been conducted from STIS/HST data on these 2 microjets [2, 5, 46], and on the HH 30 [4] and Th 28 [3] microjets. In these 4 jets, strong gradients in excitation conditions are observed below 1″, illustrating the crucial importance of high-angular resolution (Fig. 7). Both the electronic and hydrogen densities and the excitation temperature strongly decrease with distance from the source. In contrast, the ionization fraction is seen to rapidly level off to values between 0.03 and 0.5. In DG Tau, the degree of excitation (x_e, T_e) clearly increases with flow velocity [5, 29]. The more accurate derivation of n_H and jet radius allows for a better estimate of mass-loss rates in these flows, a crucial parameter for launching models. Newly derived ejection to accretion rates are still uncertain by a factor 10 but typically range between 1 and 10%. See [7] for a detailed discussion of the remaining uncertainties in the derivation of mass-loss rates.

3.4 Time Variability and the Origin of Knots

Multi-epoch high-angular resolution imaging observations allow to probe the variability of the ejection process. Large proper motions, on the order of the jet flow velocity, have been inferred from the combination of space and ground-based AO observations for the small-scale knots in the DG Tau, RW Aur, and CW Tau jets [13, 21, 30, 38]. These properties, combined with the line ratios indicative of shock conditions and the morphology suggestive of bow-shock type structures as in the DG Tau case, suggest that knots are likely internal working surfaces produced by time variable ejection velocity and/or direction, as in the younger HH jets. Inferred variability timescales are ≃1–10 yrs. This variability timescale is also observed at the base of the younger,

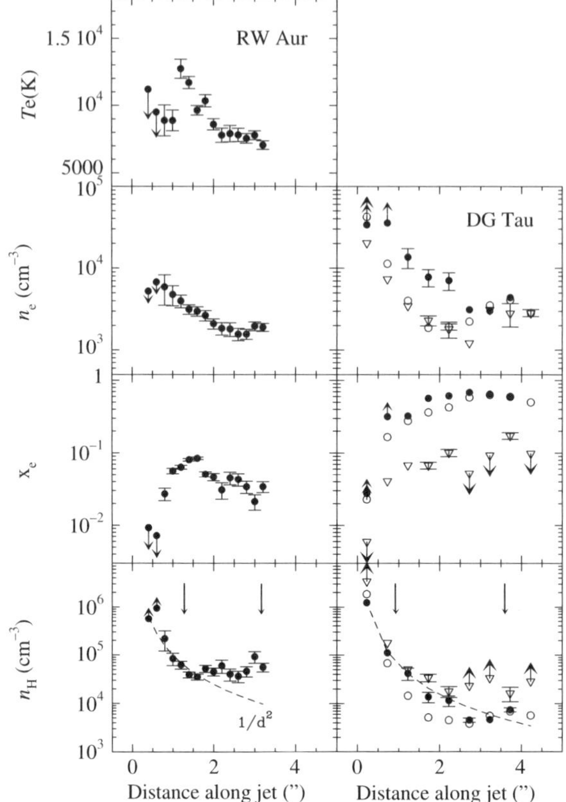

Fig. 7. Variation of plasma parameters along the jet axis derived from AO+IFS observations (OASIS at CFHT) for the DG Tau and RW Aur microjets. Strong gradients are observed in the inner arcsecond illustrating the need for high-angular resolution. Figure from [12]

embedded HH flows. Its origin may be related to instabilities in the disk or at the star-disk interface.

4 The Future of AO Regarding Jet Studies

Pioneering studies led in the past years have fully shown the potential of adaptive optics observations for jet studies. However, only a few objects have been studied so far in detail (only 3 for the variation of excitation conditions DG Tau, RW Aur, and HH 30, for example). A full exploitation of the current instrumentation for more systematic studies is required. In particular, it is essential to extend significantly the sample of sources to probe: a broader range in disk accretion rates (including Fu Ori type objects) and central stellar masses (going from HAeBe to brown dwarfs) and to investigate the influence

Table 2. A selection of IFS instruments

Instrument	Telescope	IFS-type	λ	Pixel	Spec. res.	Date operational
SINFONI	VLT	slicer	1.05–2.45 µm	25,100,250 mas	2000–4000	2004
OSIRIS	Keck-II	lenslet	0.98–2.4 µm	20–100 mas	3500	2005
NIFS	Gemini	slicer	0.95–2.4 µm	40×100 mas	5000	2005
MUSE	VLT	slicer	0.46–0.93 µm	25–200 mas	2000–4000	2010
NIRSPEC	JWST		0.6–5 µm	100 mas	3000	2013

of multiplicity. Proplyds, irradiated jets, and Herbig-Haro objects are also promising sources of studies at high-angular resolution.

One of the main limitations today to conducting these studies in a more systematic way is the limiting magnitude required for wavefront sensing. However, this limit will disappear in the coming years with the availability of Laser Guide Star mode on most large optical telescopes (Keck: in operation; VLT, Gemini: being commissioned, soon offered to the community; Subaru: first light late 2006). This guiding mode will significantly enlarge the sample of sources accessible (down to $R = 17$–18, which is the magnitude limit for the tip–tilt correction), opening interesting prospects for studies of jets around low-mass stars for example.

The near-infrared domain is currently the most suited to pursue studies of jet-launching regions, to take the best advantage of near-diffraction limited images (resolution of 30–50 mas) provided by current AO systems installed on most 8–10 m telescopes. An angular resolution of 50 mas (i.e., 7 AU at the Taurus distance) will help to probe with unprecedented detail the transverse structure of the microjets (width, velocity profile, and plasma parameters variation across the jet) in their launching regions, which is the most discriminant between the different ejection models. Such an angular resolution should, in particular, allow the angular velocity of the high velocity component of the flow to be constrained [36].

Within the past two years, many IFS systems, combined with AO, have been installed on most ground-based 8–10 m telescopes (the following webpage compiles the current list of IFS projects: http://www.aip.de/Euro3D/ instruments.html). Table 2 below lists some of the main instruments' characteristics.

This table also includes two longer term projects that will be of strong interest for the young jets community. First, the second generation VLT instrument MUSE, a very ambitious project. The ambition of MUSE is to deliver adaptive optics corrected images, combined with integral field spectroscopic capability in the optical domain over a large field of view (1 arc min^2). NIRSPEC is a near-infrared IFS on the next generation space telescope JWST.

Adaptive optics observations on 4 m class telescopes (such as the WHT) will remain critical to conducting complementary studies. In particular, the optical spectro-imager OASIS now at the WHT, combined with the NAOMI AO correction system, will allow pioneering studies to be conducted at the CFHT. The optical domain is still a choice domain to constrain the plasma parameters (such as the ionization fraction x_e). Multi-epoch spectro-imaging observations, for example, would allow the origin of the small-scale variability observed to be further constrained.

Also in a longer term, high-contrast AO systems, which plan to achieve strehl ratios $>80\%$ in the near-infrared domain, optimized for high contrast imaging close to a bright source, will be extremely powerful to probe jet properties even closer to the central source. The coming years will no doubt provide us with a wealth of new observations and constraints of the jet-launching region.

References

1. Anderson, J.M., Li, Z.-Y., Krasnopolsky, R., Blandford, R.: Astrophys. J. **590**, L107–L110 (2003)
2. Bacciotti, F.: Astrophys. Space Sci. **293**, 37–44 (2004)
3. Bacciotti, F., Eislöffel, J.: Astron. Astrophys. **342**, 717–735 (1999)
4. Bacciotti, F., Eislöffel, J., Ray, T.: Astron. Astrophys. **350**, 917–927 (1999)
5. Bacciotti, F., Mundt, R., Ray, T.P., Eislöffel, J., Solf, J., Camenzind, M.: Astrophys. J. **537**, L49–L52 (2000)
6. Bacciotti, F., Ray, T.P., Mundt, R., Eislöffel, J., Solf, J.: Astrophys. J. **576**, 222–231 (2002)
7. Cabrit S.: In Proceedings of Star Formation and the Physics of Young Stars. Bouvier J., Zahn J.P. (eds.), vol. 3, pp. 147–182. EDP Sciences, EAS Publications Series (2002)
8. Cabrit, S., Edwards, S., Strom, S.E., Strom, K.M.: Astrophys. J. **354**, 687–700 (1990)
9. Coffey, D., Bacciotti, F., Ray, T.P., Woitas, J., Eislöffel, J.: Astrophys. J. **604**, 758–765 (2004)
10. Davies, R.I., et al.: Astrophys. J. **552**, 692–698 (2001)
11. Davis, C.J., Nisini, B., Takami, M., Pyo, T.S., Smith, M.D., Whelan, E., Ray, T., Chrysostomou, A.: Astrophys. J. **639**, 969–974 (2006)
12. Dougados, C., Cabrit, S., Lavalley-Fouquet, C.: RMxAC, **13**, 43 (2002)
13. Dougados, C., Cabrit, S., Lavalley, C., Ménard, F.: Astron. Astrophys. **357**, L61–L64 (2000)
14. Dougados, C., Cabrit, S., Lopez-Martin, L., Garcia, P., O'Brien, D.: Astrophys. Space Sci. **287**, 135 (2003)
15. Eislöffel, J., Mundt, R., Ray, T.P., Rodríguez, L.F.: In Protostars and Planets IV. Mannings V., Boss A.P., Russell S.S. (eds.), p. 815. Univ. of Arizona Press, Tucson (2000)
16. Ferreira, J.: Astron. Astrophys. **319**, 340 (1997)
17. Garcia, P.J.V.: New Astron. Rev. **49**, 590–606 (2006)
18. Garcia, P.J.V., Cabrit, S., Ferreira, J., Binette, L.: Astron. Astrophys. **377**, 609–616 (2001)
19. Garcia, P.J.V., Thiébaut, E., Bacon, R.: Astron. Astrophys. **346**, 892 (1999)

20. Hartigan, P., Edwards, S., Gandhour, L.: Astrophys. J. **452**, 736–768 (1995)
21. Hartigan, P., Edwards, S., Pierson, R.: Astrophys. J. **609**, 261–276 (2004)
22. Herbst, T.M., Beckwith, S., Glindemann, A., et al.: Astron. J. **111**, 2403–2414 (1996)
23. Goto, M., et al.: Proceedings of the SPIE 4839, **1117** (2003)
24. Hirth, G. A., Mundt, R., Solf, J.: Astron. Astrophys. Suppl. **126**, 437–469 (1997)
25. Kasper, M.E., et al.: Exp. Astron. **10**, 49 (2000)
26. Kasper, M.E., Feldt, M., Herbst, T.M., et al.: Astrophys. J. **568**, 267–272 (2002)
27. Königl, A., Pudritz, R.: In Protostars and Planets IV. Mannings, V., Boss, A.P., Russell, S.S. (eds.), pp. 759–787. Univ. of Arizona Press, Tucson (2000)
28. Lavalley, C., Cabrit, S., Dougados, C., Ferruit, P., Bacon, R.: Astron. Astroph. **327**, 671 (1997)
29. Lavalley-Fouquet, C., Cabrit, S., Dougados, C.: Astron. Astrophys. **356**, L41–L44 (2000)
30. López-Martín, L., Cabrit, S., Dougados, C.: Astron. Astrophys. **405**, L1–L4 (2003)
31. Marconi, A., et al.: Astrophys. J. **586**, 868 (2003)
32. Ménard, F.: In Proceedings of the ESO Workshop Science with Adaptive Optics. Brandner, W., Kasper, M. (eds.), p. 163. Springer, Berlin (2005)
33. Nisini, B., Bacciotti, F., Giannini, T., Massi F., Eislöffel, J., Podio, L., Ray, T.P.: Astron. Astrophys. **441**, 159–170 (2005)
34. Nisini, B., Caratti o Garatti, A., Giannini, T., Lorenzetti, D.: Astron. Astrophys. **393**, 1035–1051 (2002)
35. Pesenti, N., Dougados, C., Cabrit, S., Ferreira, J., O'Brien, D., Garcia, P.: Astron. Astrophys. **416**, L9–L12 (2004)
36. Pesenti, N., Dougados, C., Cabrit, S., O'Brien, D., Garcia, P., Ferreira, J.: Astron. Astrophys. **410**, 155–164 (2003)
37. Pyo, T.-S., et al.: Astrophys. Space Sci. **287**, 21–24 (2003)
38. Pyo, T.-S., et al.: Astrophys. J., 649, 836 (2006)
39. Ray, T., Dougados, C., Bacciotti, F., Eislöeffel, J., Chrysostomou, A.: In Protostars and Planets V. Reipurth, B. Jewitt, D., Keil, K. (eds.), pp. 231–244. University of Arizona Press, Tucson (2007)
40. Ray, T.P., Mundt, R., Dyson, J.E., Falle, S.A.E.G., Raga, A.C.: Astrophys. J. **468**, L103 (1996)
41. Rigaut, F., et al.: PASP. **110**, 152–164 (1998)
42. Sauty, C., Tsinganos, K.: Astron. Astrophys. **287**, 893 (1994)
43. Shu, F.H., Najita, J., Ostriker, E.C., Shang, H.: Astrophys. J. **455**, L155 (1995)
44. Véran, J.P., Durand, D.: In Astronomical Data Analysis Software and Systems IX, ASP Conference Series. Manset, N., Veillet, C., Crabtree, D. (eds.), vol. 216, p. 345 (2000)
45. Woitas, J., Bacciotti, F., Ray, T.P., Marconi, A., Coffey, D., Eislöffel, J.: Astron. Astrophys. **432**, 149–160 (2005)
46. Woitas, J., Ray, T.P., Bacciotti, F., Davis, C.J., Eislöffel, J.: Astrophys. J. **580**, 336–342 (2002)

Spectro-astrometry: The Method, its Limitations, and Applications

E. Whelan[1] and P. Garcia[2,3]

[1] Dublin Institute for Advanced Studies, School of Cosmic Physics
ewhelan@cp.dias.ie
[2] Departamento de Engenharia Física, Faculdade de Engenharia, Universidade do Porto, Rua Dr. Roberto Frias, 4200-465 Porto, Portugal
[3] Centro de Astrofísica, Universidade do Porto, Rua das Estrelas, 4150-762 Porto, Portugal
pgarica@astro.up.pt

Abstract. Intermediate resolution spectroscopic observations provide a window into the immediate environment of young stars. In particular, such studies have yielded a lot of new information regarding outflow activity and binarity. Using spectroscopy, one can investigate the kinematics and excitation conditions of jets close to where they are launched and resolve close binaries. The primary constraint to the use of spectroscopy in this manner is the maximum spatial resolution achievable. Observers are all the time aiming to improve angular resolution and overcome the limiting effects of the Earth's atmosphere. This is where the technique of spectro-astrometry comes to the fore. The spectro-astrometric technique, which allows the user to recover sub-seeing spatial information from a simple spectrum has been applied to many interesting problems. For example, spectro-astrometry has been used to discover that brown dwarfs can drive outflows. In this review of spectro-astrometry, we shall discuss the technique beginning with a revision of angular resolution and astrometry. The limitations of the technique are described, and the uses to which it has been put to date are extensively discussed.

1 The Spectro-astrometric Technique

The modern astronomer's understanding of a wide range of astronomical phenomena is greatly limited by the need for better angular resolution. For example, we have not yet achieved the resolution required to obtain a comprehensive picture of how jets from young stars (the topic at the heart of this series of lectures) are launched. The complicated techniques of adaptive optics and interferometry (thoroughly described in this book) are designed to improve, and have immensely improved the resolution achievable. However, spectro-astrometry, which is now described, can offer the observer spatial information on comparable angular scales.

E. Whelan and P. Garcia: *Spectro-astrometry*, Lect. Notes Phys. **742**, 123–149 (2008)
DOI 10.1007/978-3-540-68032-1_6 © Springer-Verlag Berlin Heidelberg 2008

Spectro-astrometry simply translates as a combination of spectroscopy and astrometry. It is a straight-forward technique and its application to a simple long-slit or echelle spectrum will allow the user to recover sub-seeing spatial information. The concept in its present form was first applied to the study of binarity in young stars [1], and since then it has been used effectively to not only probe binary stars but also protostellar jets. In this chapter, we shall give a complete introduction to spectro-astrometry and its limitations. The important applications of this technique shall also be discussed. As the confusion between angular resolution and astrometry is one of the common sources of error in the interpretation of spectro-astrometry, we shall begin with a discussion of both these topics before dealing with the technique itself.

1.1 Angular Resolution

Consider a point source located at a great distance and ignore atmospheric effects for simplicity. Any ideal telescope will produce an angular image of the point source that is not point-like anymore but is instead "spread" over a finite angular width and is surrounded by rings. The source image angular diameter (or "spreading") decreases with the diameter of the telescope. This is the well-known phenomenon of diffraction.

A fundamental lower bound to the angular size of the image can be estimated by recalling Heisenberg's uncertainty relations for the momentum of the incoming photon,

$$\Delta x \cdot \Delta p_x \geq \frac{\hbar}{2} \tag{1}$$

For a photon arriving at the telescope there is an uncertainty in the position, which is the diameter of the telescope $\Delta x = D$. On the other hand, the projection of the photon's momentum in the mirror plane is $p_x = h \sin \theta / \lambda$, where θ is the angle of the photon momentum vector ($\boldsymbol{p}$) with the mirror normal. The limit to the uncertainty in the angle theta ($\Delta \theta$) of the incoming photon might be derived by assuming that it dominates the momentum uncertainty $\Delta p_x \simeq h \Delta \theta / \lambda$. Substituting these expressions in (1) we obtain,

$$\Delta \theta \geq \frac{\lambda}{4 \pi D} \tag{2}$$

The limit so derived is in qualitative accordance with the observation of the increasing angular size of the image of the point source, with a decreasing telescope diameter.

To quantify the shape of the "spreading" of a point source, the diffraction phenomenon should be treated taking into account that the astronomical sources are in practice located at infinity and telescope systems are designed such that Fraunhoffer diffraction is a good approximation to the observed intensity. The amplitude of the electric field of the radiation, with wave number

k, from a point source, passing through an aperture A, in the image plane ZY at a distance R from the aperture is given by the Fraunhoffer expression

$$E(Y, Z) = \int_{-\infty}^{+\infty} \int_{-\infty}^{+\infty} A(y, z) \exp\left(ik\frac{Yy + Zz}{R}\right) dydz, \qquad (3)$$

where the integration takes place in the plane zy where the aperture is located. For convenience, the integration is now unbounded and $A(y, z)$ defines the aperture, being 1 inside the aperture and 0 outside it.[1] Introducing the variables $K_y = kY/R$ and $K_z = kZ/R$, it becomes clear that (3) states that the electric field at the image plane is the fourier transform of the aperture. The image of the point source is, therefore, the power spectrum of the fourier transform of the aperture. This image is normally called the point spread function – PSF.

For the case of a circular aperture, the PSF is an analytical function and is presented in Fig. 1. The full width at half maximum of the PSF is λ/D. Considering a 10 m telescope operating at 1 µm the width of the PSF is 10^{-7} rad i.e., 0.021″. This number is a striking factor of ~ 25 smaller than the actual PSF of a normal telescope operating on the ground. This is due to the degradation in angular resolution caused by atmospheric turbulence, and this is dealt with in detail in the chapter by Esposito.

Classical Resolution Criteria

In astronomy, the most simple object where imaging is of use is a binary star. However, any image of a binary star is only helpful if the stars can be separated. This is the basis of simple two-point resolution criteria. The most

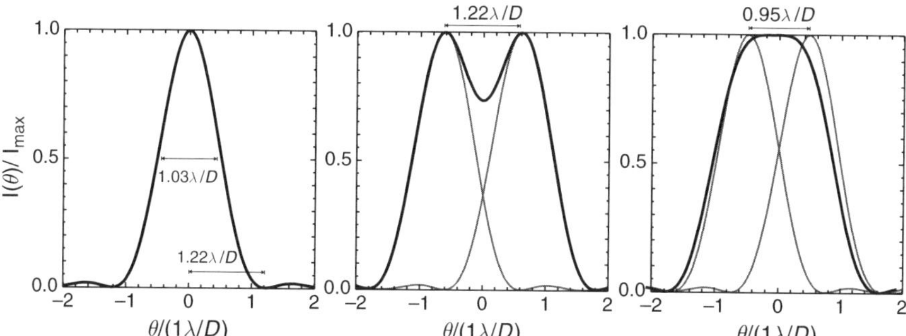

Fig. 1. Resolution criteria. (**Left**) A cut through the center of an image of a point source. (**Center**) the Rayleigh resolution criteria; (**Right**) the Sparrow resolution criteria

---

[1] A can be a complex or real function for more general apertures such as the ones using in apodization, choronography, or nulling.

famous example is the Rayleigh criterion, which when applied to the imaging case states that two stars are resolved if they are separated by the angular distance between the maximum of the PSF and its first minimum (located at $1.22\,\lambda/D$). A second more interesting criterion is that of Sparrow [31]. It states that the resolution limit is the separation at which the second derivative of the intensity at the mid-point of the two sources vanishes. The "valley" between the two maxima disappears at $0.95\,\lambda/D$. As can be seen in Fig. 1, the above two classical resolution criteria are not really limits, and it is immediately seen that a binary star is the image.

Resolution Depends on Signal to Noise

An important influence on the angular resolution of an image is the signal to noise ratio, as pointed out by [16]. Indeed, if the signal to noise ratio is infinite, any binary could be inferred through analysis of the width of the central peak. However, as the signal decreases, background noise becomes important compared to the photon noise of the object. Figure 2 illustrates the effect of sampling the noise in an image of a binary separated by the Sparrow criteria. At a high signal to noise, it is clear that if the object is a binary we could recover the separation, position angle, and flux ratio. On the other hand, at low signal to noise, it becomes impossible to recover the binary separation in the Sparrow limit. Of course, if the separation was larger, it would be possible to measure it even at the low signal to noise of the figure.

Super-resolution

A different approach to understanding the effect on a stellar signal as it passes through a telescope is to think of the telescope as a bandpass filter. Using this line of thinking, a telescope is a low bandpass filter, filtering out high

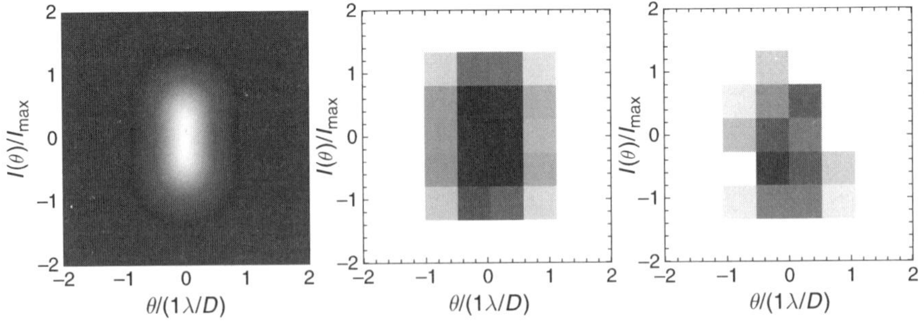

Fig. 2. The effect of noise on resolution. (**Left**) image of a binary separated by the Sparrow criteria of $0.95\lambda/D$; (**Center**) the same image sampled with $0.5\lambda/D$ pixels, total number of photons is 10^4, Poisson noise; (**Right**) total number of photons is 10^2, Poisson noise

frequency information. However, in certain conditions, normally associated with the absence of noise and perfect knowledge of the telescope transfer function, it becomes possible to completely reconstruct the object [16]. In practice, using inverse problem techniques (deconvolution), one can obtain information several factors below the telescope diffraction limit and then determine from the data both the object and the transfer function [35]. In summary, due to diffraction, a telescope filters the high frequency content of an *unknown* object image. Using several techniques, it might be possible to recover part of the information of the *unknown* image into an observed image. It is also possible (dependent on the signal to noise of the observation) for this observed image to have an angular resolution higher than the classical limits, or to be *super-resolved*.

1.2 Astrometry

Astrometry deals with locating the central position of a *known* unresolved source, primarily a star. This problem is indeed a classic problem of astronomy, since the first days of positional astronomy to the modern astrometric techniques used by planet hunters. It has, therefore, received a great deal of attention. For further discussion see [41].

The fact that we know that the source is unresolved is a qualitative difference with respect to effects of diffraction on an *unknown* object image. As we have seen above, an unresolved source produces an image of the telescope PSF. In astrometry, the interest is not on this image but only on the spatial center of the image. There is, therefore, a dramatic contrast between astrometry and "deconvolution" in terms of final outcomes of the data mining process. In a deconvolution, an *unknown* image is recovered from a bandpass-limited system (via diffraction or atmospheric seeing). The unknown image can have hundreds of pixels and, therefore, the information to recover is of the order of hundreds of intensity measurements. In contrast, in astrometry the full image information is used to recover only a few measurements, the center of the star and possibly its width and intensity. The precision with which the star center is determined cannot be confused with the angular resolution of the image. These are very different quantities. *This mistake is common in newcomers to spectro-astrometry.*

Several techniques have been developed in order to recover the center of a stellar image. These include (a) computing the first moment of an image, (b) fitting a symmetric Gaussian function, and (c) median centering. Reference [32] compared the three methods and found Gaussian fitting to be the most robust to bias. The considered biases included background, image truncation, and image undersampling. The typical astrometric accuracies are of the order of 1% of a pixel. Achieving a better accuracy is hampered by the detector itself. Detector fabrication limitations create positional inhomogeneities in the detector pixels, which act to limit the astrometric accuarcy [28].

1.3 Spectro-astrometry

Spectro-astrometry is the astrometric measurement and analysis of spectrally dispersed images. To our knowledge, the first ideas of spectro-astrometry were recorded in the proceedings of the IAU Colloquium n.62, "Current techniques in double and multiple star research", held at Flagstaff, AR, during May 19–21, 1981. In this study, the positions of objects in different filters are compared to obtain information on the nature of the objects. Interesting papers that followed include those by [7, 13]. Beckers (1982) discussed how in the context of speckle interferometry (see Sect. 1.4) spatial information at much smaller scales than those achieved by the seeing-limited refractometer can be obtained. Again, the angular position at two different colors/wavelengths is combined. Indeed, it was Beckers who coined the term differential speckle interferometry. Reference [13] explained how a refractometer can be used to detect positional shifts with wavelength in a close binary star [7].

It is a fair statement that following these papers spectro-astrometry and the analagous technique of differential phase were born. Spectro-astrometry is based on seeing-limited observations currently done with spectrographs and differential phase on interferometric observations achieved done with optical/NIR interferometers. Differential phase is briefly discussed in Sect. 1.4.

We shall begin with assuming (for simplicity) that we have an ideal instrument that generates spectrally dispersed images of a star without atmospheric effects. If the instrument is an integral field spectrometer, the resulting data are a cuboid of 2D angular dimensions versus wavelength (Fig. 3). If we apply astrometric positional fitting (Gaussian fitting) to each of the images (of fixed wavelength) and analyze the positional evolution with wavelength, we are doing spectro-astrometry (refer to [18]). For a long-slit or echelle spectrum the spatial profile is extracted at each pixel along the dispersion direction (see Fig. 4). Gaussian fitting is used to measure the centroid of the spatial profiles

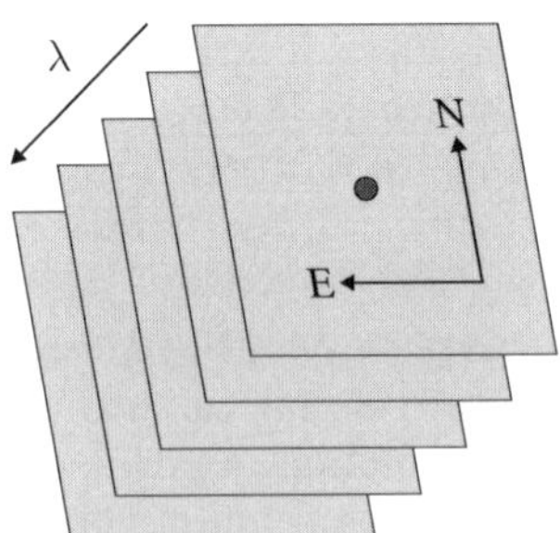

Fig. 3. A cuboid of 2D angular dimensions versus wavelength. Spectro-astrometry is applying astrometric positional fitting in each of the images. When using long-slit or echelle data, each pixel along the dispersion direction corresponds to a fixed wavelength. In this case, spectro-astrometry is done by extracting the spatial profile at each pixel and using astrometric fitting (usually Gaussian fitting) to measure its centroid

for the continuum and then pure emission line region as a function of wavelength. In Fig. 5, the case of a long-slit spectrum is presented in several steps (again atmospheric effects are ignored).

The top row of Fig. 5 illustrates the observation of a single star. The star is initially a point source (T1) but when imaged by the telescope it is diffracted and the typical ring pattern appears (T2). When passing through the slit (T3), diffraction takes place, and the dispersed/diffracted Spectrum is observed in the detector. The process is now recorded on the two-dimensional detector as a function of wavelength and position along the slit (T4). The positional information perpendicular to the slit was (in a first approximation) lost. A cut at a fixed wavelength in the detector shows a profile of the diffracted image (T5). This profile can be fitted, e.g., with a Gaussian, and the resulting Gaussian centre versus wavelength can be plotted and analyzed (T6). In practice, the result is a curve caused by optical distortion. In the presence of atmosphere, differential refraction would disperse the stellar spectra (before entering the slit) in the paralactic direction. If the slit was aligned parallel to the paralactic direction, then a slope would be present on the Gaussian center versus position curve (T6).

In the bottom row of Fig. 5, a binary star with different spectra is considered. For simplicity, suppose that one of the stars spectra is featureless, with an emission line, the other stellar spectra having only one absorption

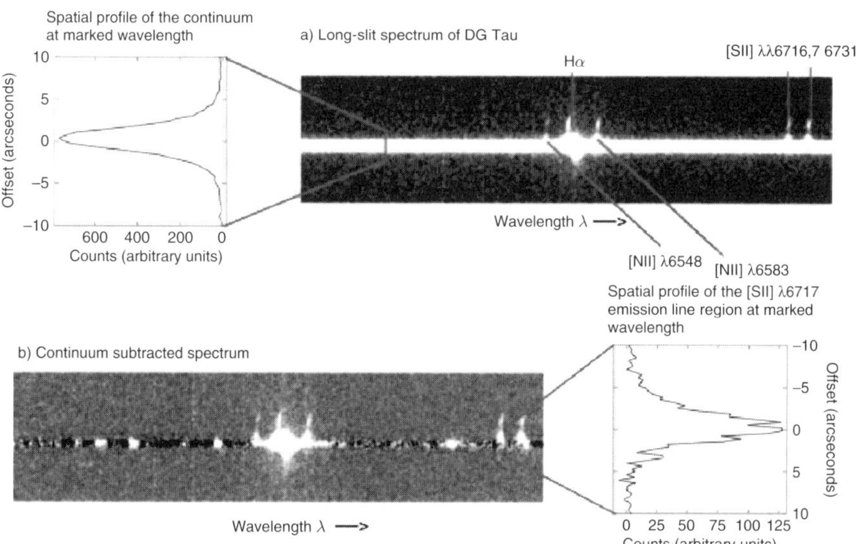

Fig. 4. Example of applying spectro-astrometry to the long-slit spectrum of a typical classical T Tauri star DG Tau. The extraction of a 1D spatial profile from, first, the continuum emission and then from an emission line in the continuum subtracted spectrum is illustrated. Here the continuum has been stubracted using IRAF fitting routines

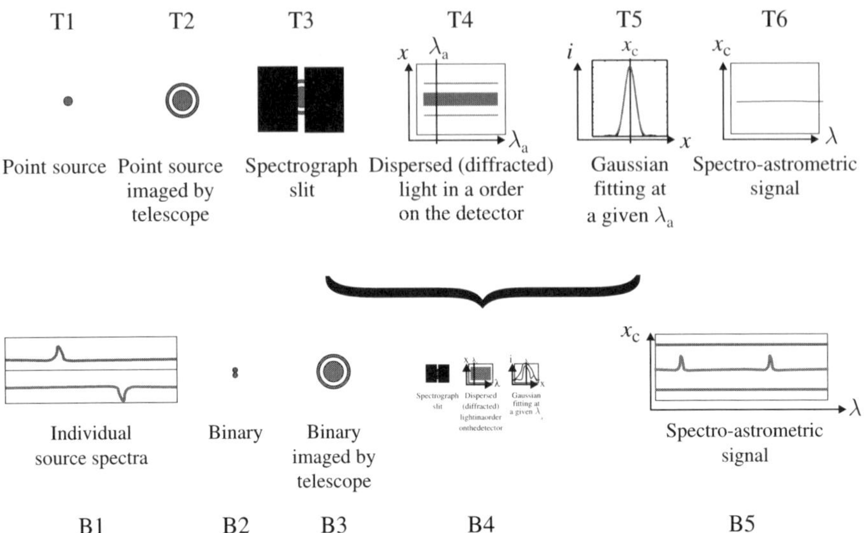

Fig. 5. Spectro-astrometry applied to a long-slit spectrum of a single star (**top**) and a binary star (**bottom**). See text for full description

line (B1). The binary is very close (B2), such that it is unresolved by the telescope. The diffracted image thus formed (B3) has no spatial information that would reveal the binary nature of the object. After passing through the slit and imaged in the detector and fitted (B4), the resultant spectro-astrometric signal is presented (B5). Several observations can be made (B5):

1. The signal is in between the two stellar positions. This is expected from a barycenter.
2. The amplitude of the spectro-astrometric signal is smaller than the separation of the binary. Only in a very special case, where one of the components would have negligible flux with respect to the other, and vice-versa, should the amplitude be equal to the separation.
3. In the spectral region of the emission line of the top component, the spectro-astrometric signal shifts towards the top, and in the spectral region of the absorption line of the bottom component the signal shifts to the bottom.

The interesting feature of the results we have just obtained is that we know that small-scale structure is present in the object and that this small structure could not be found by simply inspecting the diffracted image B3. The difficulty is now in interpreting and quantifying the spectro-astrometric signal in terms of actual object properties. Some of the more interesting results obtained to date with spectro-astrometry are described in Sect. 3.

Spectro-astrometric Accuracy and Continuum Subtraction

In our above description, for simplicity we have ignored atmospheric effects. However, in reality the width of the PSF of a point source is set by the atmospheric seeing. Spectro-astrometry is measuring the centroid of a flux distribution as a function of wavelength, and when estimating the accuracy to which we can do this, we must also consider the signal to noise of our observation. In summary, while the angular width of the flux distribution for a point source, is set by the seeing, the centroid of the gaussian fit is measured to an accuracy given by

$$\sigma_{\mathrm{centroid}} = \frac{\mathrm{seeing(mas)}}{2.3548\sqrt{N_{\mathrm{p}}}},\tag{4}$$

where N_{p} is the number of detected photons. Here, the seeing is the FWHM of the distribution. We are assuming that the only source of noise is photon noise and that the distribution is better than Nyquist sampled. Also, as any variation in the seeing with wavelength is small, a fixed average seeing can be used when calculating the errors. Note that in general the total wavelength range under consideration is short. Assuming a fixed average seeing means that the precision in the measurements is dominated by N_{p}. A value of $N_{\mathrm{p}} = 10^{6}$ and a seeing of 1 arcsecond results in an accuracy of less than 1 milliarcsecond. Such conditions are not difficult to achieve.

Any spectro-astrometric analysis of a spectrum (whether long-slit, integral field, or echelle) will be contaminated due to, firstly, misalignment effects of the optics. As a result, the spectro-astrometric trace may be curved or tilted; however, this curvature is easily fitted and thus removed. A more serious consideration is the contamination of the emission line regions by the continuum emission. If one is using spectro-astrometry to study the origin of certain emission line regions (for example a line region displaced from the source position can be deduced to be formed in an outflow), the first step is to accurately map the source position by finding the centroid of the continuum emission. The position of a emission line region can be displaced from the source position; however the the extent of the offset is weighted by the ratio of the intensities of the continuum and line regions. The factor by which the measured shifts are weighted by the continuum is given by the following (see [33]):

$$I_{\lambda(\mathrm{line})}/[I_{\lambda(\mathrm{line})} + I_{\lambda(\mathrm{cont})}]\tag{5}$$

Simply, contamination by the continuum will tend to drag the line position back towards the source, and therefore it is important that this continuum contamination be removed. This process is illustrated in Fig. 4. In addition, Fig. 6 shows the spectro-astrometric analysis of the [OI]λ6300 forbidden line and Hα permitted line in the spectrum of the classical T Tauri star (CTTS) V536 Aql, before and after continuum subtraction. The [OI]λ6300 line originates in the jet driven by V536 Aql, and the resultant shift while still visible

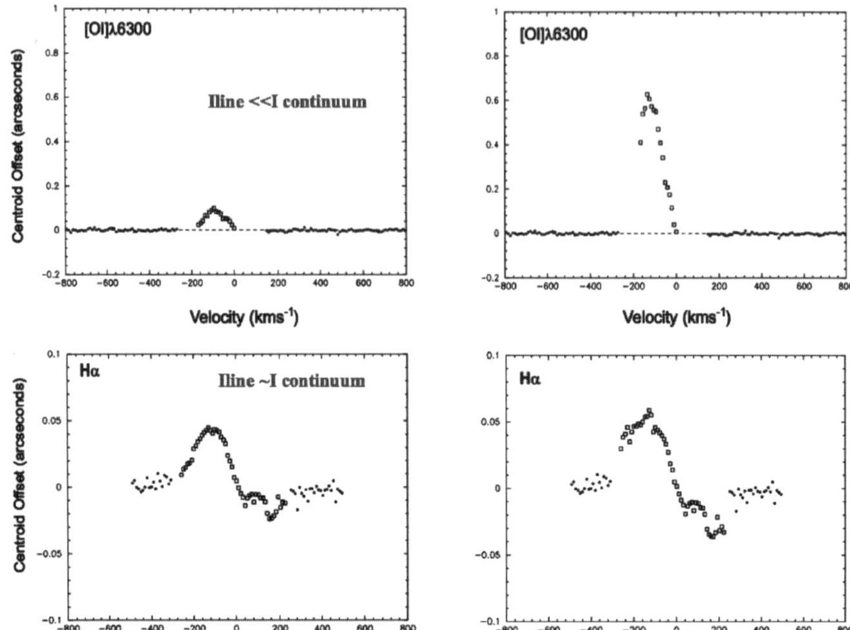

Fig. 6. Spectro-astrometric measurements of the [OI]λ6300 forbidden line and the Hα permitted line taken from the continuum contaminated (**left**) and continuum subtratced (**right**) spectrum of the T Tauri star V536 Aql. Note how contamination by the continuum emission will tend to drag the position of the line region back toward the source. This effect is much smaller in Hα, as the intensity of the Hα emission is comparable to the continuum intensity. The ratio of the continuum contaminated shifts to those measured from the continuum subtracted spectrum is $I_{\lambda(\text{line})}/[I_{\lambda(\text{line})} + I_{\lambda(\text{cont})}]$, from [33]

before the continuum subtraction is much reduced. If the continuum emission is much stronger than the pure line emission, it is possible that a real displacement in the line region will remain hidden until the continuum contamination is removed. This is especially true where spectro-astrometry has been applied to the study of permitted emission line regions in CTTSs (see Sect. 3). Part of the Hα emission (the line-wings) also originates in the jet and is thus displaced. However, in this case the Hα line is strong compared to the continuum emission, and so the removal of the contamination results in only a small increase in offset.

There have been two approaches to the problem of continuum contamination. First, the continuum can simply be removed to reveal the pure emission line regions. This is shown in Fig. 4. This method has been adopted by [15, 43, 44, 46]. Second, the true line shifts are recovered by calculating the factor by which the offsets are weighted by the continuum and dividing the raw positional displacement by this number. This factor is given above and this method is used by [33, 34]. Both approaches yield similar results.

1.4 Spectro-astrometry and Interferometry

As explained by Neal Jackson in this book, an interferometer measures the complex spatial coherence of an astronomical source, at an angular frequency determined by the separation of the telescopes (B), and the wavelength of the observations (λ). See also [39] for a useful review of optical interferometry. This spatial coherence of the fringes generated by the interference of the light from the telescopes is encoded in the observed fringes. The spatial coherence is a normalized quantity and a complex number, and in astronomy it is commonly referred to as the visibility. The visibility or spatial coherence is related to the object-normalized brightness distribution by a Fourier transform and is the value of the Fourier transform at a position in the frequency domain. This position is determined by the relative positions of the telescopes. We can denote the visibility by $\tilde{V} = V \exp 2\pi i\phi$, where V is the visibility amplitude (varies between 0 and 1) and ϕ is the visibility phase (or simply the phase).

Because of the Fourier transform link between the visibility and the object brightness distribution, some insights can be gained on the visibility by recalling some Fourier transform properties. In particular, the Fourier transform of a center-symmetric function is real, and therefore the phase is either $\phi = 0$ or $\phi = 2\pi$. Hence, the Fourier phase stores a deviation from center-symmetry.

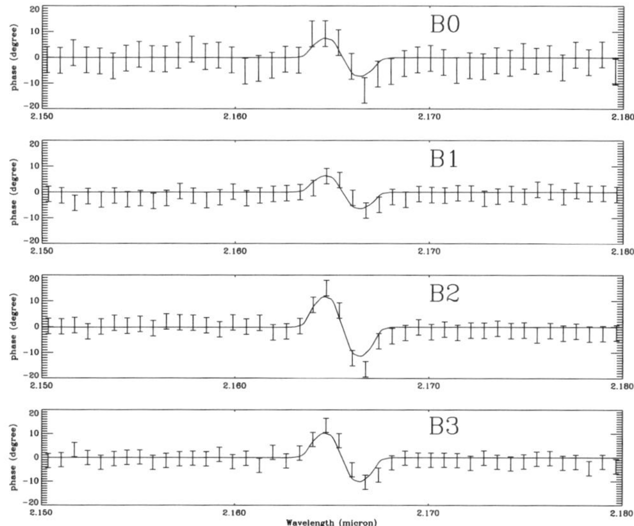

Fig. 7. Differential phase of the visibility across the Brγ line of the Be star α Arae (from [37]). The data were obtained with the AMBER instrument. The fit is for a for a keplerian disk

The most simple deviation from center-symmetry is a shift. An angular shift (θ) of the brightness distribution translates in to a visibility phase shift of $\phi = B\theta/\lambda$. Therefore, an astrometric signature is stored in the visibility phase and not in the visibility amplitude.

In interferometry *differential phase*, the evolution of phase with wavelength, $\phi(\lambda)$ is analogous to spectro-astrometry as outlined above. This differential phase can be measured in the detector by dispersing the fringes with a spectrograph, which is what is done, for example, by the AMBER instrument (see the chapter by Malbet). The initial suggestion of differential phase by [7] was made in the context of speckle interferometry. Signal-to-noise analyses were published [10, 11, 27], and successful applications were initially achieved with optical interferometry e.g. [40]. State-of-the art results using differential phase are presently achieved with AMBER in the NIR. For example, several AMBER papers were published, where the differential phase technique was applied to model Novae [12], the η Car Luminous Blue Variable [14], Be stars [36, 37], and a Wolf Rayet/O star binary system [38]. Figure 7 shows the result of applying the differential phase technique to the Be star A Arae.

2 Limitations

As the use of the spectro-astrometric technique becomes increasingly popular, it is clear that instrumental effects can affect results. Instrumental effects such as misalignment of the spectrum with CCD columns, departure of the of the CCD pixels from a regular grid, and imperfect flat-fielding can introduce artifacts into a spectrum, which result in false spectro-astrometric signatures [6]. Especially important is the slit-effect which we describe below. Reference [6] demonstrate how artificial bipolar structures can be introduced when the PSF of point source suffers distortion in a relatively wide slit.

Figure 8 from [26] illustrates the path of two light rays from different positions across the slit. The red ray crosses the center of the slit, then passes through a collimator, and is diffracted by the grating toward the camera and then the detector. The ray is diffracted by angle θ_0 which is connected to the incident angle ϕ. A located at an angle δ of the central ray will be focused by the telescope on the slit, at a distance y of the slit center. This ray will reach the diffraction grating at a different incident angle ($\phi - \psi$) and will be diffracted at a different angle (θ_1) from the ray that crossed the slit center. In practice, at the detector, rays diffracted at angles θ_0 and θ_1 will be registered in different "wavelength pixels". However, these two rays do have the same wavelength.

The slit-effect bias then consists of the spurious wavelength or velocity shift of the points of the object that are not located at the center of the slit. A point located above the center of the slit will be wavelength-biased in an opposite direction from a point located below the center of the slit. Therefore, if the star is centered on the slit, there is no final bias in the central velocity of a photospheric line. The line (if originally spectrally unresolved)

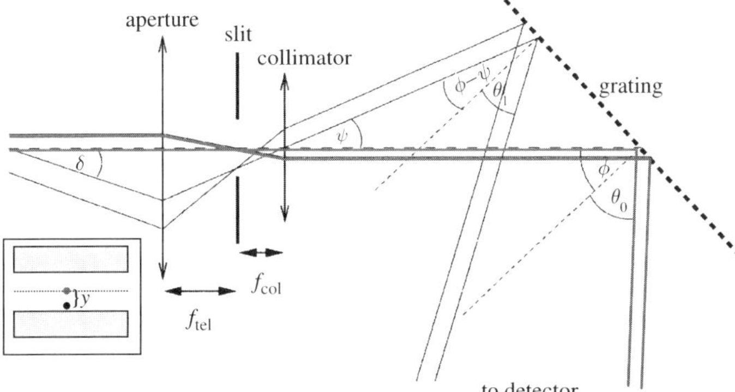

Fig. 8. Illustration of the slit-effect taken from Maciejewski & Binney (2001). The two dots in the lower left are the positions of two points in the slit. The red ray crosses through the center of the slit. The black ray crosses near the lower edge of the slit

will be artificially broadened into a width $\Delta\lambda$ which is the connected to the spectral resolution of the spectrograph $R = \lambda/\Delta\lambda$. Spectrographs are designed such that the width of unresolved spectral lines are Nyquist sampled and the width of the slit for operating spectral resolution matches the seeing. We can, therefore, estimate the typical magnitude of the slit-effect. A seeing of $1''$ corresponds to λ/R Å$/''$.

We have seen so far how different regions of the slit are registered at different wavelengths at the detector, in spite of having the same wavelength. Hence, if the slit is unevely illuminated, or if the centering of the source is inaccurate, a false spectro-astrometric signature is introduced. Reference [6] computed the spectro-astrometric signature caused by the slit-effect for the case of a binary star.

Protecting Against False Spectro-astrometric Signatures

It is vital to elimiate artifacts from any spectro-astrometric analysis. This is especially important where the measured offsets are at the milliarcsecond scale, for example, as measured for RU Lupi in [33]. Methods for overcoming the slit-effect include (a) accurately calibrating the image quality at the slit, (b) ensuring that reference photospheric lines are present in the spectra, (c) the use of spectrographs that provide high spectral resolution, and most importantly (d) using a slit width which is narrow with respect to the seeing.

In addition, by taking antiparallel slit positions, it is possible to check for the presence of artifacts. In binary studies, it is common to take data at position angles of $0°$ and $180°$. Real spectro-astrometric signatures are inverted as we go from $0°$ to $180°$ but artifacts are not (see [34]). In studies of protostellar jets, it has been more common to indirectly elimate any artifacts. In most cases, the slit is placed along the jet axis (refer to Sect. 3), and all signatures measured to date have been shown to agree with the known tracers of outflows of the forbidden emission lines. In spectro-astrometric studies of brown dwarf outflows (again refer to Sect. 3), the slit was placed at 0°. For these studies, artifacts were elimated by analyzing lines where no offset was expected. If a false signature existed, it would be measured in all lines. Lines like HeI 6678 and Li I6708 were shown to be confined to the source position, and therefore the offsets meaured in the forbidden emission lines were certainly real.

3 Astronomical Applications

Since the beginning of the 1980s, there have been several documented approaches to the problem of extracting sub-seeing information from a flux distribution. These techniques were, over time, much refined to the current spectro-astrometric technique, as described above. To date, spectro-astrometry has proven a useful tool in the study of young binaries and protostellar jets primarily, and here its application to these phenomena is described.

Reference [1] coined the term spectro-astrometry when describing its use in finding young binaries. With this technique, it is possible to detect binaries with separations down to $\sim$10 mas [2, 3, 34]. If an unresolved star is actually a close binary, then its position, measured as a function of wavelength, is the centroid position of the two stars (refer again to Fig. 5). In the very likely scenario that the spectra of the two stars differ significantly (e.g. one is a strong Hα emitter), this centroid position will change with wavelength as the relative contributions of the two spectra vary. Commonly, one star will be an accretor and thus emit strongly in Hα. The centroid position will shift toward the accretor, at the Hα line, indicating the binary nature of the source. In addition, by taking spectra at orthogonal slit positions the binary position angle can be estimated.

Figure 9, taken from [2], shows the result of the spectro-astrometric analysis of the pre-main sequence binary KK Oph. Note the positional displacement detected at the Hα line. The double-peaked nature of the displacement is due to the fact that both components are Hα emitters, hence the observed Hα line profile is a combination of two different line profiles with differing line widths. In the wings of the observed line, the broader component will dominate, while at the center the narrower component will be more important. In KK Oph, the line center is displaced to the east of the continuum position and the wings to the west. The binary status of KK Oph was known prior to this work, and its position angle had been estimated from optical speckle observations. Reference [2] demonstrates the power of the spectro-astrometric technique, as the binary position angle of KK Oph (and several other sources) is seen to agree closely with the accepted value. In Fig. 10, the straight line through the spectro-astrometric data is the binary position angle from previous data.

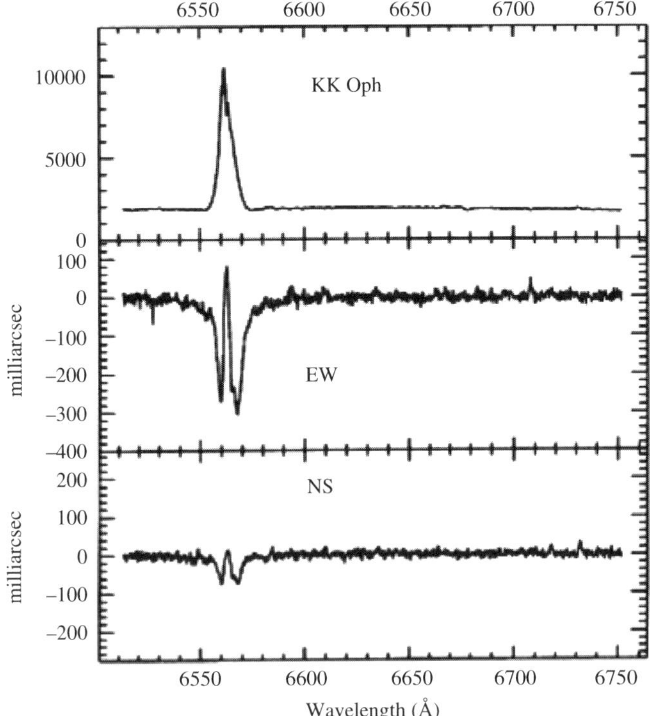

Fig. 9. Position spectra of the binary star KK Oph taken from [2]. Spectra were taken at orthogonal position angles in order to make an estimate of the binary position angle. A large displacement is measured at the Hα line. The displacement is double-peaked as both stars are Hα emitters. The signal from the star with the broader Hα line dominates at the line-wings, while the narrower component is more important at the line peak

Since the applicability of spectro-astrometry to the study of binary stars was demonstrated by [2] much work has followed. For example, [34], presented spectro-astrometric observations for 28 pre-main sequence stars. Many of the sources included in this study were already known to have binary companions; however, several new binaries were found. Overall results demonstrated that spectro-astrometry was capable of recovering binary separations down to $\sim$10 mas. Some of the results from this study are shown in Fig. 11. Garcia et al (1999) used spectro-astrometry to analyse OASIS integral field spectrograph observations of the ZCMa binary. The spectro-imaging data allowed the authors to do 2D-spectro-astrometry. This study was important as the authors successfully recovered the spectrum of each component of the binary in spite of a spatial sampling of 0.$''$ 11 similar to the system separation of 0.$''$ 1. Also, see [3, 4] for further information on the use of spectro-astrometry in the study of pre-main sequence binaries.

A spectro-astrometric technique was first applied to the study of bipolar outflows from CTTSs by [29], and altogether between 1984 and 1998 the

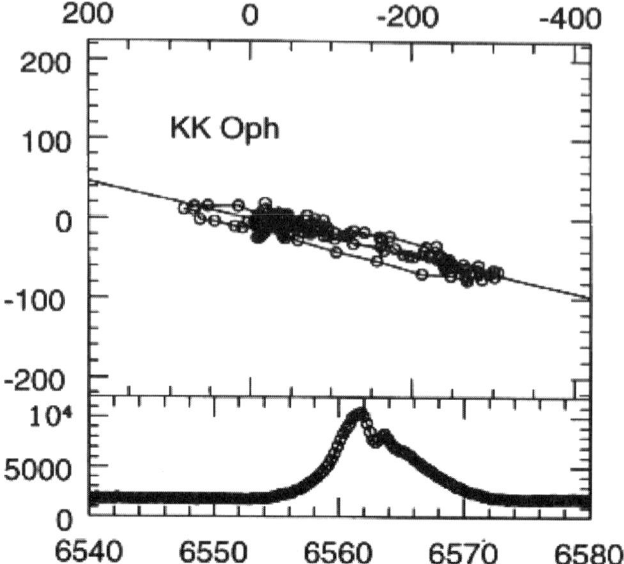

Fig. 10. The Hα centroid position as a function of wavelength for the orthogonal spectra is plotted. The x, y positions trace the position angle of the binary, and the solid line through the spectro-astrometric data is the binary position angle, estimated from an older study. This figure is also taken from [2], and it confirms that spectro-astrometry has an important role to play in the study of binary stars

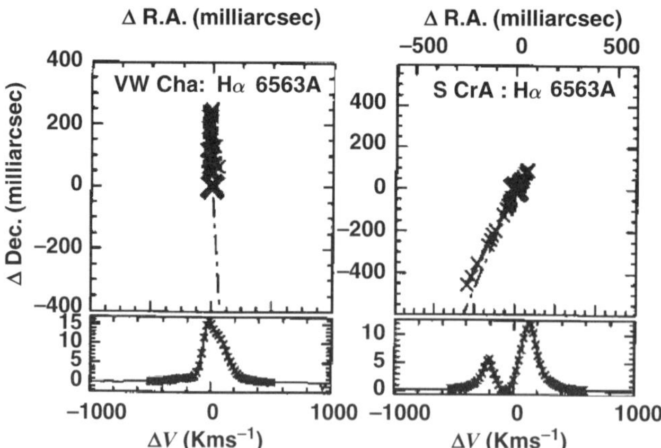

Fig. 11. A sample of the spectro-astrometric results presented in [34]. Again, the centroid position (extracted at orthogonal position angle) is plotted as a function of velocity. Results show the source to be a binary and again allow the postion angle of the binary to be estimated

spatial characteristics of the spectra of 12 CTTSs were investigated [20, 21, 29, 30]. The primary goal of this work was to probe the spatio-kinematic properties of the forbidden emission line regions (FELRs) of CTTSs as a means of distinguishing between the different models describing their formation. By spatially resolving the FEL components, the authors were able to confirm the unique origins of the different velocity components, thus verifying the models of [24, 25]. FELs, e.g. [OI]$\lambda\lambda$6300, 6363, are important coolants of the shocks that result when a super-sonic jet interacts with its ambient medium. Thus they are excellent tracers of jets. In particular, the approach taken by the authors listed above was to study the FELRs of CTTSs with jets using long-slit spectroscopy. The long-slit technique minimized any contrast problems with the source and allowed the FELRs to be easily isolated. In many cases, the slit was placed along the jet (jet position angle was known from imaging), and, therefore, the FELRs of the resulting spectrum provided an excellent snap shot of the jet. This is illustrated in Fig. 12. In the article by Pat Hartigan in this book, the importance of FELs to the study of protostellar jets and their use as diagnostics of the physical conditions in the jets are outlined.

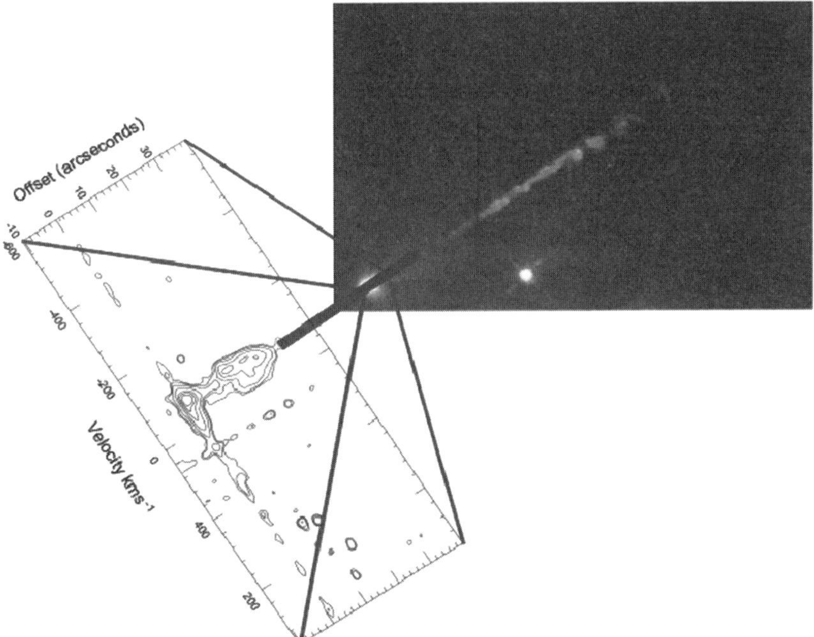

Fig. 12. HST Wide-Field Planetary Camera II image of the collimated outflow from the Class I YSO HH 34 and the corresponding United Kingdom Infrared Telescope (UKIRT) [FeII] 1.644 µm spectrum, taken from [15]. The observation was obtained by aligning the instrument slit to the previously known axis of the outflow. In the spectrum, one can clearly make out strong emission from the source, a weak continuum and emission from the closest knot

Through the work discussed above it became possible to probe outflows from CTTSs to within ~1000 AU of their central engines, with long-slit spectra of their FELRs. With the arrival of HST and the advent of adaptive optics techniques the spatial resolution that could be achieved was further greatly increased (see Chapters by Dougados and Bacciotti), and thus the FELRs of CTTSs were mapped on much smaller spatial scales. This work allowed the morphology, kinematics, and physical conditions of a sample of CTTS jets to be studied at an unprecendented angular resolution. In CTTSs, the critical density at which we see forbidden emission occurs far enough from the driving source for the FELRs to be well extended (see Fig. 13). Thus, the high angular resolution techniques now available to us ensure that we can easily and directly resolve spatially the FELRs. Hence, the use of spectro-astrometry in studying FELRs (optical and near-infrared) is limited. Notable results that came from analyzing CTTS FELRs with spectro-astrometry (refer to 14; 15; 42) are that high velocity FE is displaced further from the driving source than low velocity emission (underlining their separate origins) and that when mapping FELRs an initial displacement as a function of velocity is seen in all cases. However, this behaviour is easily seen in the HST/STIS spectra presented by [5] and the AO assisted integral field spectra by [17]. Note also that 14, 15 among were the first to study (and apply spectro-astrometry to) the near-infrared FELRs of CTTS. When it comes to the study of jets/outflows from young stars, where spectro-astrometry has proven its use, is in the spatial analysis of permitted emission lines (PELs). Spectro-astrometry has been sucessfully used to dis-entangle the outflow component to the PELRs of CTTSs.

Considerable effort has been made over the last 20–30 years to model the PEL regions (e.g. Hα, Paβ) commonly found in the spectra of CTTS. To date, the principal models have involved magnetically driven winds [9, 19] and, more recently, magnetospheric accretion [8]. While infall models based on magnetospheric accretion have produced the most promising results, some features of

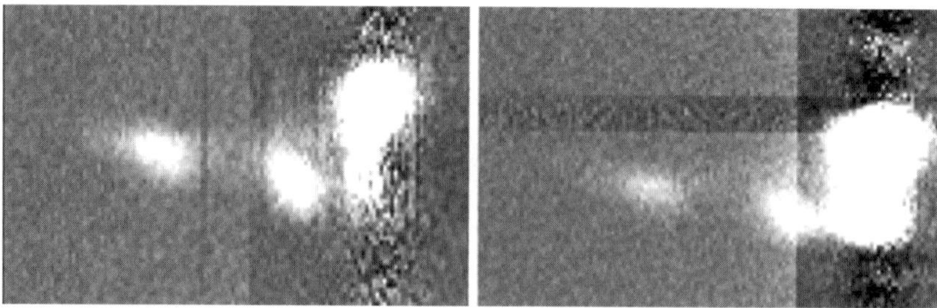

Fig. 13. Here the [SII]λ6731 and [OI]λ6300 emission line regions in the spectrum of DG Tau are presented. The jet driven by DG Tau is very obvious at these wavelengths. This figure is given as a comparison to Fig. 14. Due to the higher critical density of the [OI] line the jet is brighter in [OI] close to the source and the farthest knot along the jet is brightest in [SII]

the PELs cannot be reproduced, e.g., the large FWHM of the lines and their extensive, symmetrical line wings [43]. These features are normally associated with outflow activity, hence they suggested that outflows also contribute to the PELRs of CTTSs. This has recently been confirmed through the use of spectro-astrometry [33, 43]. A spectro-astrometric analysis of the PELRs of a sample of CTTSs has shown the emission making up the extensive line-wings also to be tracing the outflow. Simply put, the bulk of the Hα emission say was confined to the source position, while the wing emission was displaced in the same direction as the jet. Important to this work is a comparison between the analyses of the PELs and that of the FELs tracing the jet. Figures 13 and 14 illustrate this comparison. The FELs of DG Tau are very obviously spatially extended and therefore formed in a jet. However, there is no evidence that the wing emission of the Paβ line region is spatially extended. This extension is only recovered through spectro-astrometric analysis.

In Figs. 15, 16, 17, and 18, the spectro-astrometric plots of the Paβ, Hα, [OI]λ6300 and [SII]λ6731 ELRs found in the spectra of DG Tau and V536 Aql

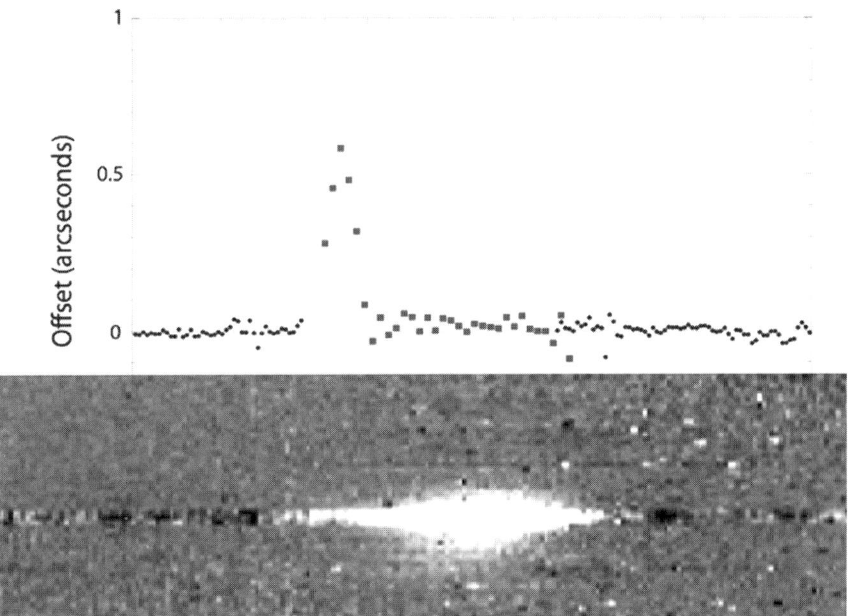

Fig. 14. The Paschen beta emission line region in the spectrum of DG Tau and the corresponding spectro-astrometric measurement. This figure beautifully illustrates the power of the spectro-astrometric technique, especially when compared to Fig. 13. The blue-shifted wing of the Paβ lines is formed in the DG Tau jet. However, this is only revealed using spectro-astrometry. Unlike for the FELs, there is no extension in the Paβ line region. This is simply because the displacement in the Paβ line is much less than the seeing of the observation. The results shown here were presented in [43]

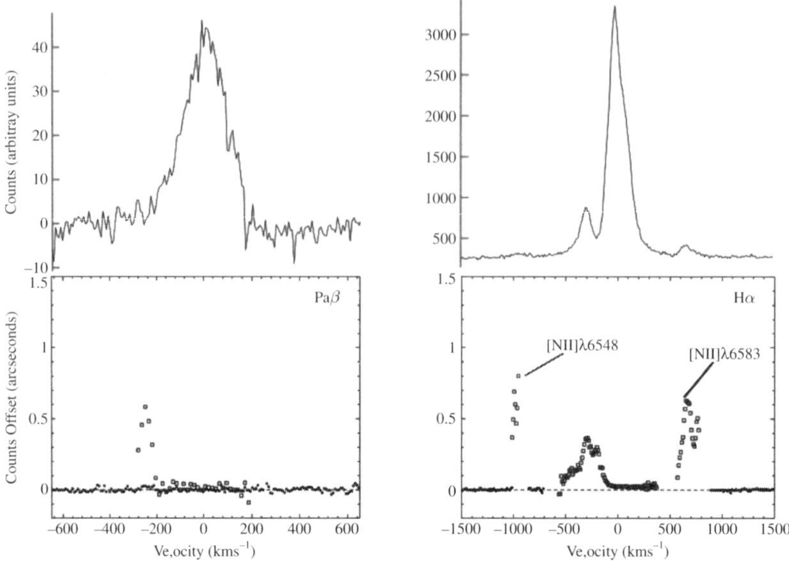

Fig. 15. Spectro-astrometric analysis of DG Tau adapted from [43]. Here the results of the analysis of the PELs are shown. The important result here is the displacement measured in the Paβ line. The offsets measured for both Paβ and Hα agree with the FELs. This is an important comparison

are shown. These figures are adapted from [43]. For both these sources, the PEL wings are found to be displaced in the same direction as the known jets (studied through their FELRs). This displacement is also measured within the velocity range of the jets. For example, the blue-shifted jet driven by DG Tau has a radial velocity of $\sim$200 kms^{-1}. The blue wing of DG Tau Paβ lines is displaced to $\sim$0.6 arcseonds at a velocity of ~ -250 km/s. This corresponds to what we see in [OI]λ6300 and [SII]λ6731. The DG Tau jet was already known to be bright in Hα.

Reference [33] also investigated the optical spectra of RU Lupi with spectro-astrometry. The Hα emission line is symmetric with no obvious outflow component. However, Gaussian fitting of the spatial profile revealed the wing emission to be displaced in the direction of the blue- and red-shifted jet. What was interesting about the results of the RU Lupi and V536 Aql analysis was that the PEL wings traced the blue- and red-shifted jets, while only the blue-shifted jets are seen in the FELs. FELRs are commonly only blue-shifted, and this is explained through the obscuration effect of the circumstellar disk. Reference [33] explained the RU Lupi result using an idea based on circumstellar disk dust holes. As FE is quenched close to a typical CTTS (densities are too high), the FELRs occur further from the central engine than the PELs. Hence, a PELR originating in a red-shifted jet could be seen through a gap in the circumstellar disk while a red-shifted FELR would

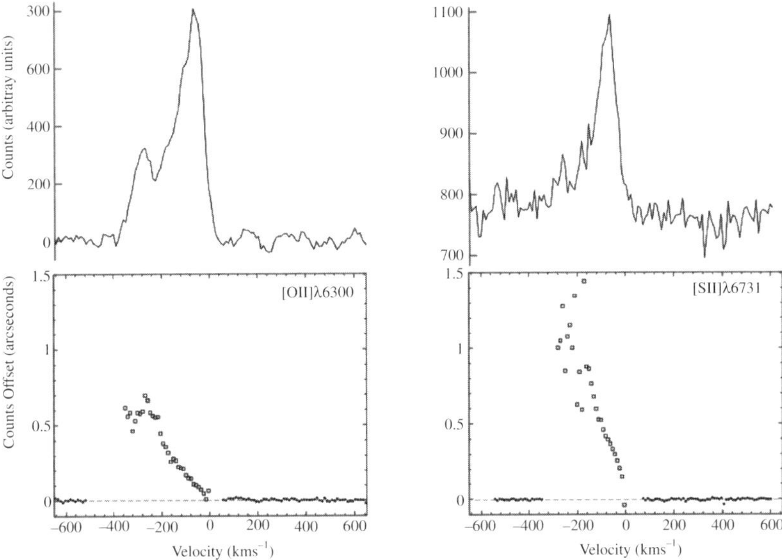

Fig. 16. Spectro-astrometric analysis of DG Tau adapted from [43]. Here the results of the analysis of the FELs is shown. The spectro-astrometric analysis of the FELs is important as a comparison to what is seen in Paβ

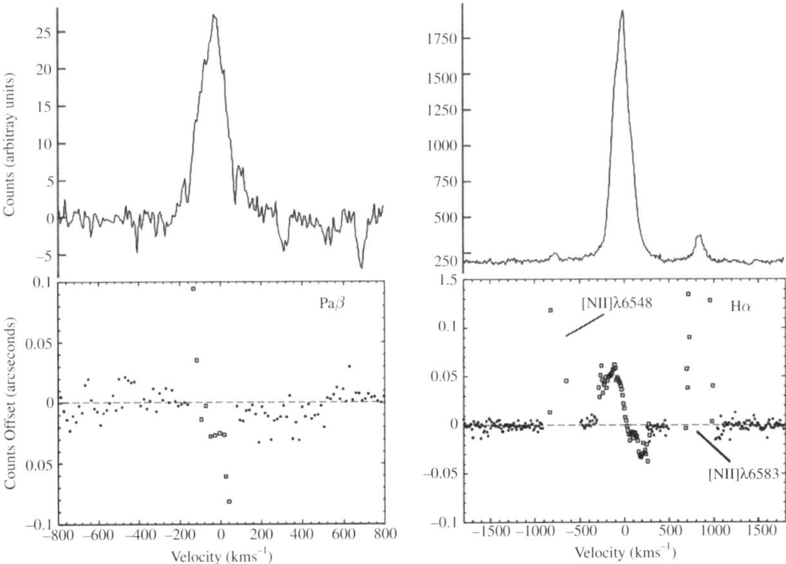

Fig. 17. Spectro-astrometric analysis of V536 Aql adapted from [43]. Offsets are measured in both the Paβ and Hα lines, demonstrating that the line emission originates in an outflow. These displacements are not directly visibile in the V536 Aql but are only recovered through spectro-astrometry. Note that we are seeing both the blue- and red-shifted jets

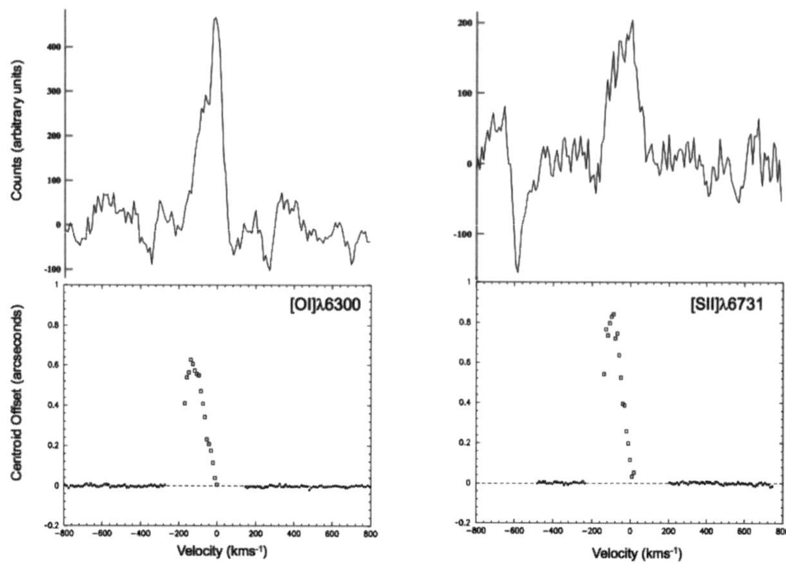

Fig. 18. Spectro-astrometric analysis of V536 Aql adapted from [43]. Again, the spatial extension in the FELs is probed as a comaprison to what is measured in the PELs, shown above. Here, it is important to note that only the blue-shifted jet is detected while both the blue and red jets are seen in Paβ and Hα. This is explained by the presence of a dust hole close to the star which allows the red-shifted jet to be seen. At this distance from the star the forbidden emission is quenched due to high density

still be obscured by the disk. Reference [43] use the same idea to explain the bipolar offsets measured in the V536 Aql Paβ and Hα lines.

Circumstellar disk dust holes are commonly studied through the near-infrared spectral energy distributions (SED) of the protostar. The star is modeled by a simple black body, and the infrared excess seen in the SED is the disk. Often, dips are observed in the infrared excess that correspond to a lack of disk material (primarily dust) at a certain temperature (or distance from the star). The SEDs of both RU Lupi and V536 Aql show such a dip in their infrared excesses. Assuming a radial temperature dependence for a circumstellar disk, the distance at which we estimate the dust hole to exist corresponds well in both cases to the scale of the jet estimated from the spectro-astrometric results.

Lastly, on the subject of outflows, I will discuss the use of spectro-astromety in the detection of outflows driven by brown dwarfs (BDs). The optical spectrum of young, newly forming BDs is strikingly similar to the spectrum of a typical CTTSs [22, 23]. In particular, BDs exhibit strong Hα emission. Through the study of the Hα ELRs of BDs, many BDs were identified as accretors and shown to have a CTT-like accretion phase. As outflow activity in young stars is intrinsically linked to accretion activity, a reasonable next question was, "Can accreting BDs also drive outflows?" The only hint that brown dwarfs could power outflows/jets was the discovery of a number of brown dwarfs with FELRs. However, the FELRs of brown dwarfs were faint and confined to the source, i.e., not spatially extended. Hence, their originating in an outflow could not be confirmed. This is where spectro-astrometry comes into use. To date, two BDs with T Tauri-like outflows have been discovered and this was achieved purely through the use of spectro-astrometry [44, 45, 46].

Figure 19 is a continuum subtracted position velocity (PV) diagram of the [OI]λ6300 line region in the spectrum of the BD 2MASS1207-3932. 2MASS1207 is a 24 M_{JUP} BD known to have a near edge-on circumstellar

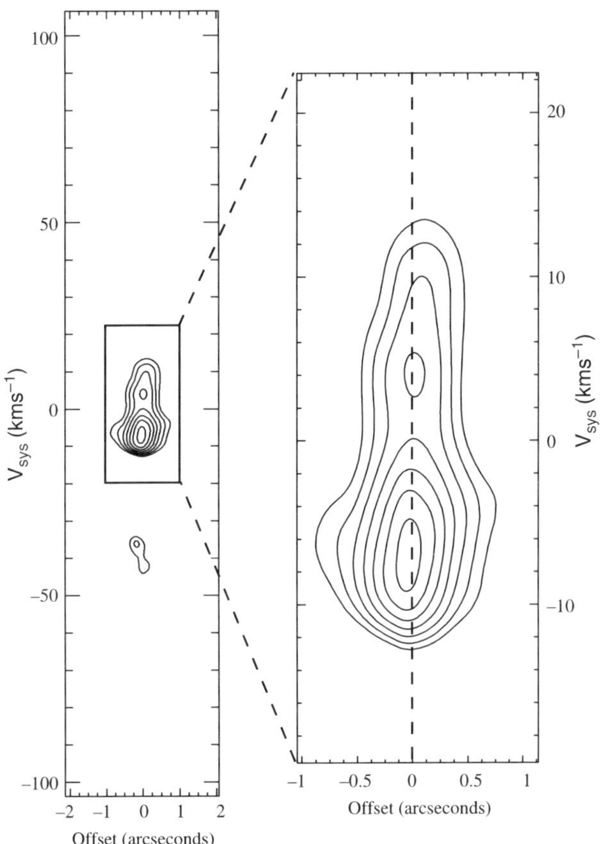

Fig. 19. Continuum subtracted position velocity diagram of the [OI]λ6300 line in 2MASS1207 taken from [46]. Here, the spectrum was smoothed in both the spectral and spatial directions, using an elliptical Gaussian function of FWHM 0.12 Angstroms $\times$ 0".22. The contours begin at 3 times the r.m.s noise and increase in intervals of the r.m.s noise. The zero spatial offset line is measured from an accurate mapping of the continuum position using spectro-astrometry. The red- and blue-shifted components to the line are obvious, and an opposing offset in these components is suggested

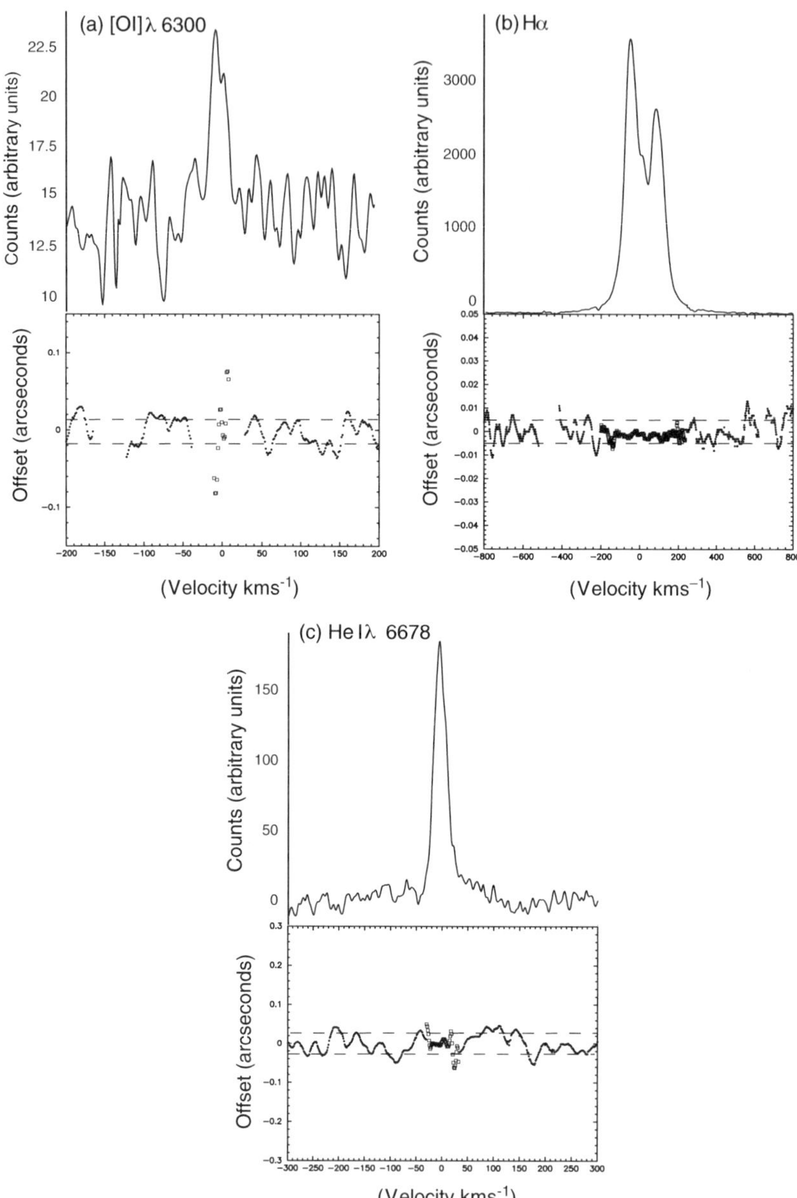

Fig. 20. Offset velocity diagrams in the vicinity of the [OI]λ6300, Hα, and He I lines. This figure is taken from [46]. The line profiles are an average of three pixel rows in the spatial direction centered on the continuum. Note that the continuum has not been subtracted. The green dashed lines delineate the ±1 σ error envelope for the centroid position of the continuum. The line and continuum centroid errors are the same in all cases. The bipolar offset in the [OI]λ6300 reveals the presence of an outflow. No offsets are measured in the Hα and He I lines as expected, ruling out the possibility of spectro-astrometric artifacts

disk. The recent discovery that it is driving a bipolar outflow makes it the lowest mass galactic object with an associated outflow. As is clear from Fig. 19, the [OI]λ6300 line has two velocity components at blue-and red-shifted radial velocities of -8 and $+4$ kms^{-1}, respectively. While there is a hint from the PV diagram that the different components are extended in opposing directions, this is by no means convincing. The relative displacement between the blue and red emissions is only recovered through spectro-astrometric analysis (see Fig. 20). 2MASS1207 was not the first BD discovered to have an outflow. Reference [44] used spectro-astrometry to study the FELRs of the 65 M$_{\mathrm{JUP}}$ BD ρ-Oph 102. Results showed the FELRs to be extended to $\sim$0".1 at a blue-shifted radial velocity of ~ -45 kms^{-1}.

4 Concluding Remarks

The simplicity and effectiveness of the spectro-astrometric technique means that it has enormous potential in the study of phenomena associated with star formation. Our discussion of some of the discoveries made so far in Sect. 3 underlines this statement, and further studies are planned to build on those described above. For example, we plan to use spectro-astrometry to continue to probe outflow activity in BDs, and spectra of 5 more BDs is under analysis at present. From an observational point of view, other ways in which we can develop this work is to analyze 3D data taken with an integral field spectrometer, i.e., do 2D spectro-astrometry and combine spectro-astrometry with adaptive optics. It is also critical that we grow our understanding of spectro-astrometric artifacts and how to elimiate them. Work is underway to build on the results of [6]. The different ways in which instrumental effects cause false spectro-astrometric signatures need to modeled, and we also need to apply spectro-astrometry to data obtained with many different instruments.

References

1. Bailey, J.A.: Spectroastrometry: A new approach to astronomy on small spatial scales, SPIE **3355**, 932B (1998a) 161
2. Bailey, J.: Detection of pre-main-sequence binaries using spectroastrometry. Mnth **301**, (1998b) 161
3. Baines, D., Oudmaijer, R.D., Porter, J.M., Pozzo, M.: On the binarity of Herbig Ae/Be stars. MNRAS **367**, 737 (2006)
4. Baines, D., Oudmaijer, R.D., Mora, A., Eiroa, C., Porter, J.M., et al.: The pre-main-sequence binary HK Ori: Spectro-astrometry and EXPORT data. MNRAS **353**, 697 (2004)
5. Bacciotti, F., Mundt, R., Ray, T.P., Eislöffel, J., Solf, J., Camenzind, M.: Hubble space telescope STIS spectroscopy of the optical outflow from DG Tauri: Structure and kinematics on subarcsecond scales. Astrophys. J. Lett. **537**, L49 (2000)
6. Brannigan, E., Takami, M., Chrysostomou, A., Bailey, J.: On the detection of artefacts in spectro-astrometry. MNRAS **367**, 315 (2006)
7. Beckers, J.M.: Differential speckle interferometry. Opt. Acta **29**, 361 (1982)

8. Calvet, N., Hartmann, L.: Balmer line profiles for infalling T Tauri envelopes. Astrophys. J. **386**, 239 (1992)
9. Calvet, N., Hartmann, L., Hewett, R.: Winds from T Tauri stars. II- Balmer line profiles for inner disk winds. Astrophys. J. **386**, 229 (1992)
10. Chelli, A., Petrov, R.G.: Model fitting and error analysis for differential interferometry. I. General formalism. Astron. Astrophys. **109**, 389 (1995)
11. Chelli, A., Petrov, R.G.: Model fitting and error analysis for differential interferometry. II. Application to rotating stars and binary systems. Astron. Astrophys. **109**, 401 (1995)
12. Chesneau, O., et al.: AMBER/VLTI interferometric observations of the recurrent Nova RS Ophiuchii 5.5 days after outburst. Astron. Astrophys. **464**, 119 (2007)
13. Christy, J.W., Wellnitz, D.D., Currie, D.G.: Detection of double stars with the two-color refractometer. Lowell Observatory Bull. **9**, 28 (1983)
14. Davis, C.J., Ray, T.P., Desroches, L., Aspin, C.: Near-infrared echelle spectroscopy of Class I protostars: molecular hydrogen emission-line (MHEL) regions revealed. MNRAS **326**, 524 (2001)
15. Davis, C.J., Whelan, E., Ray, T.P., Chrysostomou, A.: Near-IR echelle spectroscopy of Class I protostars: Mapping forbidden emission-line (FEL) regions in [FeII]. Astron. Astrophys. **397**, 693 (2003)
16. den Dekker, A.J., van den Bos, A.: Resolution: A survey. J. Opt. Soc. Am. A **14**, 547 (1997)
17. Dougados, C., Cabrit, S., Lavalley, C., Ménard, F.: T Tauri stars microjets resolved by adaptive optics. Astron. Astrophys. **357**, L61 (2000)
18. Garcia, P.J.V., Thiebaut, E, Bacon, R. "Spatially Resolved Spectroscopy of Z Canis Major is Components" Astron. Astrophysics, 346, p. 892–896, (1999).
19. Hartmann, L.W., Kenyon, S.J.: Optical veiling, disk accretion, and the evolution of T Tauri stars. Astrophys. J. **349**, 190
20. Hirth, G.A., Mundt, R., Solf, J.: Jet flows and disk winds from T Tauri stars: the case of CW Tau. Astron. Astrophys. **285**, 929 (1994)
21. Hirth, G.A., Mundt, R., Solf, J.: Spatial and kinematic properties of the forbidden emission line region of T Tauri stars. Astron. Astrophys. **126**, 437 (1997)
22. Jayawardhana, R., Mohanty, S., Basri, G.: Probing disk accretion in young brown dwarfs. Astrophys. J. Lett. **578**, L141 (2002)
23. Jayawardhana, R., Mohanty, S., Basri, G.: Evidence for a T Tauri phase in young brown dwarfs. Astrophys. J. **592**, 282 (2003)
24. Kwan, J., Tademaru, E.: Disk winds from T Tauri stars. Astrophys. J. **454**, 382 (1995)
25. Kwan, J.: Warm disk coronae in classical T Tauri stars. Astrophys. J. **489**, 284 (1997)
26. Maciejewski, W, Binney, J, "Kinematics from spectroscopy with a wide slit: detecting black holes in galaxy centers", MNRAS, **323**, p. 831–838, (2001)
27. Petrov, R., Roddier, F., Aime, C.: Signal-to-noise ratio in differential speckle interferometry. J. Opt. Soc. Am. A **3**, 634 (1986)
28. Shaklan, S.B., Pravdo, S.H.: Astrometric calibration of array detectors. SPIE **2198**, 1227, (1994)
29. Solf, J.: High-resolution observations of bipolar mass flow from the symbiotic star HM Sagittae. Astron. Astrophys. **139**, 296 (1984)
30. Solf, J., Boehm, K.H.: High-resolution long-slit spectral imaging of the mass outflows in the immediate vicinity of DG Tauri. Astrophys. J. Lett. **410**, L31 (1993)

31. Sparrow, C.M.: On spectroscopic resolving power. Astrophys. J. **44**, 76 (1916)
32. Stone, R.C.: A comparison of digital centering algorithms. Astrophys. J. **97**, 1227, 1989
33. Takami, M., Bailey, J., Gledhill, T.M., Chrysostomou, A., Hough, J.H.: Circumstellar structure of RU lupi down to AU scales. MNRAS **323**, 177 (2001)
34. Takami, M., Bailey, J., Chrysostomou, A.: A spectro-astrometric study of southern pre-main sequence stars. Binaries, outflows, and disc structure down to AU scales. Astron. Astrophys. **397**, 675 (2003)
35. Thiébaut, E.: Introduction to image reconstruction and inverse problems. NATO ASIB Proc. 198: Opt. Astrophys. **397** (2005)
36. Meilland, A., et al.: An asymmetry detected in the disk of ? Canis Majoris with AMBER/VLTI. Astron. Astrophys. **464**, 73 (2007a)
37. Meilland, A., et al.: First direct detection of a Keplerian rotating disk around the Be star ? Arae using AMBER/VLTI. Astron. Astrophys. **464**, 59 (2007b)
38. Millour, F., et al.: Direct constraint on the distance of ?2 Velorum from AMBER/VLTI observations. Astron. Astrophys. **464**, 107 (2007)
39. Monnier, J.D.: Optical interferometry in astronomy. Rep. Prog. Phys. **66**, 789 (2003)
40. Vakili, F., Mourard, D., Bonneau, D., Morand, F., Stee, P.: Subtle structures in the wind of P Cygni. Astron. Astrophys. **323**, 183 (1997)
41. van Altena, W., Stavinschi, M.: The future of astrometric education. Astrometry in the Age of the Next Generation of Large Telescopes **338**, 311 (2005)
42. Weigelt, G., et al.: Near-infrared interferometry of ? Carinae with spectral resolutions of 1 500 and 12 000 using AMBER/VLTI. Astron. Astrophys. **464**, 87 (2007)
43. Whelan, E.T., Ray, T.P., Davis, C.J.: Paschen beta emission as a tracer of outflow activity from T-Tauri stars, as compared to optical forbidden emission. Astron. Astrophys. **417**, 247 (2004)
44. Whelan, E.T., Ray, T.P., Bacciotti, F., Natta, A., Testi, L., Randich, S.: A resolved outflow of matter from a brown dwarf. Nature **435**, 652 (2005)
45. Whelan, E.T., Ray, T.P., Bacciotti, F., Jayawardhana, R.: Probing outflow activity in very low mass stars and brown dwarfs. New Astron. Rev. **49**, 582 (2006)
46. Whelan, E.T., Ray, T.P., Randich, S., Bacciotti, F., Jayawardhana, R., Testi, L., Natta, A., Mohanty, S.: Discovery of a bipolar outflow from 2MASSW J1207334-393254, a 24 MJup Brown Dwarf. Astrophys. J. Lett. **659**, L45 (2007)

Observations of YSO Jets from Space: HST and Beyond

F. Bacciotti[1], M. McCaughrean[2], and T. Ray[3]

[1] INAF - Osservatorio Astrofisico di Arcetri
  `fran@arcetri.astro.it`
[2] School of Physics, University of Exeter
  `mjm@astro.ex.ac.uk`
[3] School of Cosmic Physics, Dublin Institute for Advanced Studies
  `tr@cp.dias.ie`

Abstract. Many important aspects of star formation, including the generation and propagation of stellar jets, are best studied with high angular resolution and wide wavelength coverage. These are primarily achieved with observations from Space. In this chapter, we describe the state-of-the-art of Space Astronomy in the context of the study of stellar outflows.

After a discussion of the basic principles, we illustrate the main facilities operating in Space at the time of writing. In particular we describe the capabilities of the Hubble Space Telescope (HST), that has provided a giant leap in our understanding

F. Bacciotti et al.: *Observations of YSO Jets from Space*, Lect. Notes Phys. **742**, 151–171 (2008)
DOI 10.1007/978-3-540-68032-1_7   © Springer-Verlag Berlin Heidelberg 2008

of newly forming systems: the major advances provided in the field of stellar outflows by this instrument are illustrated.

The final part of the chapter is dedicated to the next generation space observatories, and in particular to the James Webb Space Telescope (JWST), that is expected to provide new insights in the star formation process with observations in the infrared wavelength range.

1 Introduction

The formation of stars is accompanied by complex phenomena as, for example, the collapse of the parent magnetized molecular core, mass accretion from the circumstellar disk, and the generation of outflows. Although scientists have explained with success the global process in terms of basic physical ingredients (gravity, opacity to radiation, magnetic fields, rotation, etc., see, e.g. [25] and [19]), there are aspects of the standard theory that still pose problems. For example, "Can the mechanism envisaged for the birth of low mass stars i.e., solar-like stars, be generalized to stars of all masses or even sub-stellar objects? What is the exact nature of the accretion/ejection engine?" "How are planets generated, and what is the influence of energetic radiation from young stars on the early development of planets?"

To answer these and other questions, one needs to obtain from the observations a good knowledge of the physical conditions of the immediate environment of young stars. Most of the interesting phenomena occur within 100–200 AU from the Young Stellar Object (YSO). Thus even for the nearest stellar nurseries, at distances of 120–140 parsecs from the Earth, one needs to use sub-arcsecond resolution to explore this region. In addition, to properly mount an investigation of the physics of the regions of interest, information must be collected from tracers at a wide variety of wavelengths. Similarly, high angular resolution at different wavelengths is required to examine the structure and physical conditions of the outflows at larger distances from the star, a study that gives information on the past accretion/ejection history of the system, as well as on the properties of the interstellar medium in the parent nebula.

Diffraction-limited resolution and full wavelength coverage is primarily achieved with observations from outside the Earth's atmosphere, using telescopes mounted on orbiting spacecraft. Although considerable progress has been made recently in ground-based astronomy with the development of Adaptive Optics (see the chapter by C. Dougados in this book), the observations "of space from space" have been, and will remain, a unique tool for the exploration of the Universe. Instruments like the Hubble Space Telescope, in fact, have not only gifted us with images of breathtaking beauty but also provided a giant leap in our understanding of many celestial phenomena, including the formation of stars. Moreover, new-generation telescopes being planned at the time of writing are expected to provide yet unexplored opportunities.

In this chapter, we will describe the state-of-the-art of Space Astronomy in the context of the study of stellar outflows. We first discuss the basic principles

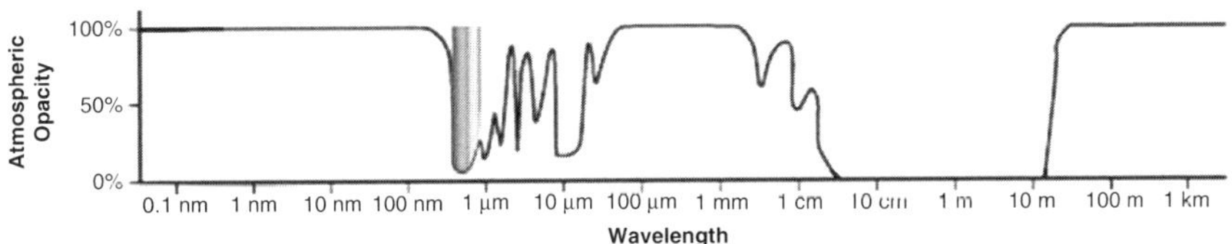

Fig. 1. Atmospheric opacity in the various wavelength bands

guiding space observations in Sect. 2, listing the main facilities useful for star formation studies that are operating at the time of writing. The capabilities of the Hubble Space Telescope are then illustrated (Sect. 3), together with a brief review of the major advances in the field provided by observations with this instrument (Sect. 4). The final part of the contribution (Sect. 5) is then dedicated to the development of new facilities, in particular, the James Webb Space Telescope.

2 Space Astronomy Basics

Although celestial objects can emit at all wavelengths of the electromagnetic spectrum, up to a few decades ago astronomical observations were greatly limited by the poor transparency of our atmosphere, illustrated in Fig. 1. In particular, the opacity reaches 100% at wavelengths shorter than about 3000 Å, i.e., for gamma-rays, X-rays and Far- and Mid-Ultraviolet (UV), as well as in the Far-Infrared. Observations from space are, therefore, required if one is to work at these wavelengths. Space observations also have the advantage of overcoming the "seeing" degradation of the spatial resolution caused by the Earth's turbulent atmosphere.

The observation in bands other than the visible, and the development of space observatories has followed the overall progresses in technology. A good review of the advances achieved in the second half of the last century can be found in [20]. For example, radio-astronomy started entering mainstream astronomy in the late forties, immediately after the development of radar during the 2nd World War. Also note that Karl Jansky did his work in the 1930s. Then, the skies began to be investigated at Infrared (IR) and millimeter wavelengths with the launch of the first high-altitude satellites, between 1960 and 1970, an epoch that also saw the infancy of X-ray astronomy. Almost all the electromagnetic spectrum, except for a small portion of the Far Infrared (FIR), was then accessible by 1980. Finally, by 1990, the advent of the ISO satellite (see below) opened the last window.

The collecting area of telescopes has also increased with time. There is however a dependency on wavelength, as the minimum telescope's diameter D must be proportional to $\lambda^{1/3}$. So, for the telescopes for Gamma- and X-ray astronomy, diameters smaller than 1 m are sufficient, while present-day optical and infrared telescopes vary between 1 and 8 meters. The antennas for radio astronomy are large paraboloids with diameters of tens of meters, up to the giant size of the antennas located in Bonn (100 m) and Arecibo ($\sim$200 m).

A large collecting area also means good spatial resolution, since in diffraction-limited conditions the minimum angular distance that can be resolved by the telescope is $\theta = \lambda / D$ at the wavelength λ. In the nineties, one striking advancement in spatial resolution in the visible was provided by the Hubble Space Telescope (HST). The sub-arcsecond angular resolution of HST went well below the limit imposed for ground-based observations ($\sim$0.''6 in good conditions) by atmospheric turbulence. Interferometry follows a different concept (see the chapter by N. Jackson in this volume) and can reach impressively high spatial resolution (of the order of one milliarcsecond) for progressively larger separations ('baselines') of the receivers. This may be at the expense however of "resolving-out" larger scale structure. In general, high angular resolution has been achieved first for long radio wavelengths and in the visible, but new instrumentation is catching up in the UV and X-ray bands, as well as in the IR. In the IR, in particular, this has been accompanied by the impressive growth in the size of detector arrays, that following the so-called Moore's Law reached a size of 4096$\times$4096 pixels in 2005 and are getting larger still.

Aside from access to important spectral bands and higher angular resolution, one should also mention the greater sensitivity and stability of measurements from space. The sensitivity is increased with respect to ground-based observations because of the substantial reduction of background signal in space. This is especially important when one has to do imaging experiments in background-limited conditions. Stability in time is also important when one has to subtract the instrument PSF in high contrast situations when repeating observations of the same target. An example of such a situation would be a study of the optical variability of a faint X-ray binary in a globular cluster.

The primary disadvantage of observatories in space is their complex design and associated cost. First of all space telescopes have generally a smaller collecting area than that reachable with telecopes constructed for ground-based astronomy. This in turn implies smaller sensitivity but also a limit in angular resolution. Since the diffraction limit is inversely proportional to the observing wavelength, one sees that ground-based AO systems and interferometry are catching up with the resolution of space observations in the infrared. As for the instrumentation, this is generally less developed than state-of-the-art ground-based equivalents. The instruments are also usually less flexible and are upgraded much less frequently, if at all. Also the detectors lag significantly and the field of view (FOV) is much smaller. Last, but not least, space projects take a long time to implement and is a huge financial effort by the funding agencies.

Despite the difficulties on both the technical and budgetary sides, it is universally recognized that space astronomy gives astronomers a unique opportunity to deliver top-quality results. In recent years, a number of space

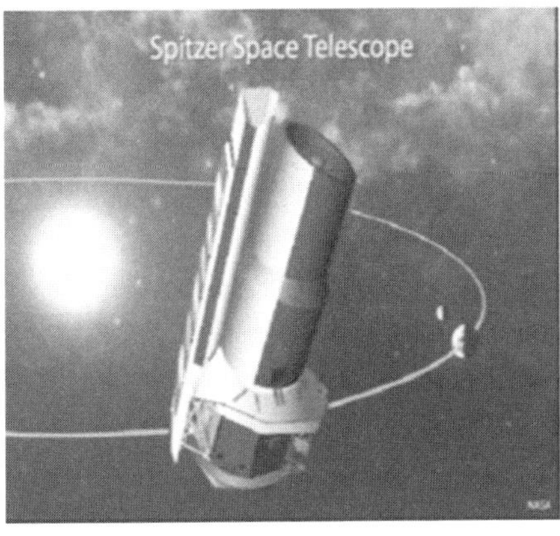

Fig. 2. (**Left**) the Spitzer Space Telescope; (**Right**) the Chandra X-ray Observatory

facilities have been launched, among which we list in the following that have contributed/contribute the most to star formation studies:

– the *Infrared Space Observatory (ISO)*,
  for near- mid-, and far-IR imaging and spectroscopy, realized by the European Space Agency (ESA), launched in 1994 and operational until 1998 (www.iso.vilspa.esa.es);
– the *Hubble Space Telescope (HST)*,
  for UV, optical and near-IR imaging and spectroscopy, resulting from a collaboration between NASA and ESA, launched in 1990 and still operational (www.stsci.edu/hst), see Fig. 3;
– the *CHANDRA X-ray Observatory*,
  that yields (relatively) high angular resolution X-ray imaging, as well as low resolution spectroscopy, launched by NASA in 1998 (chandra.harvard. edu), see Fig. 2;
– the *SPITZER Space Telescope*,
  launched by NASA in 2003, working at near- mid-, and far-IR wavelengths for imaging and spectroscopy (www.spitzer.caltech.edu), see Fig. 2.

Among these, pride of place is taken by the **Hubble Space Telescope**. Combining imaging and spectroscopy with high angular resolution, HST has provided cornerstone results in all fields of astronomy (from planetary science to the early epoch Universe). The benefit to astronomy and science in general from HST is of primary importance, beyond the scientific achievements (visit the Hubble Heritage Project at heritage.stsci.edu). This unique instrument is described in more detail in the next section, where we also illustrate some examples of the forefront science that can be done through its use in the field of star formation.

Fig. 3. The Hubble Space Telescope is a long-lived space observatory, resulting from a collaborative project by NASA and ESA

3 The Hubble Space Telescope

The HST is a long-lived space-based observatory working at a wide range of wavelengths (from the Far-UV to the Near-IR), resulting from the joint effort of the National Aeronautics and Space Administration (NASA) and the European Space Agency (ESA). Launched in April 1990 and greatly extended in its scientific powers (through new instrumentation installed during four servicing missions with the NASA Space Shuttle), the HST has returned data of unique scientific value.

The HST primary mirror has a diameter of 2.4 meters and is a reflecting telescope orbiting Earth every 97 minutes at a relatively low altitude ($\sim$600 kilometers). Hubble is not large by ground-based standards, but in its 16 years of operation it has achieved extraordinary results thanks to the escape from the distorting effects of the Earth's atmosphere. The flight of the telescope is controlled by the Goddard Space Flight Center, while the scientific operations are handled by the Space Telescope Science Institute (STScI). Both institutes are in Maryland (USA). The first years since the launch of HST in 1990 were indeed momentous, with the discovery of unwanted spherical aberration in the primary. The Shuttle Endeavor mission of December 1993 restored the full functionality of HST instruments with carefully designed corrective mirrors. The instruments installed in subsequent missions all have built-in corrective optics.

The instruments currently mounted on the HST are designed to perform imaging and spectroscopy in a wide wavelength range, at diffraction-limited spatial resolution (e.g., 0.″1 in the optical and about 0.″06 in the NUV). These are

- the *Wide Field Planetary Camera 2 (WFPC2)* provides high resolution images of astronomical objects over a relatively wide field of view and a broad range of wavelengths (1150 to 11000 Å), through 48 different filters. The detector is composed of four CCDs exposed simultaneously, three wide field CCDs arranged in a 'L' shape the long side of which projects to 2.′5, with a spatial scale of 0.″1/pixel, plus one smaller 'planetary' CCD with a smaller FOV (35″ × 35″) and a projected pixel size of of 0.″0455/pixel. Besides the unique science results, WFPC2, the "workhorse" of HST, has produced most of the spectacular images in the visible range known to the large public;

- the *Near Infrared Camera and Multi Object Spectrometer (NICMOS)*, which at present is the only Infrared instrument on HST. Its three 256 × 256 pixel cameras provide high angular resolution images from 0.8 μm down to 2.5 μm, the longest usable wavelength given HST's warm optics. The cameras are equipped with a set of different filters, polarimeters, and coronagraphic masks. NICMOS has delivered data extremely useful to the study of circumstellar disks and embedded jet sources;

- the *Space Telescope Imaging Spectrograph (STIS)*, designed to perform long-slit spectroscopy at moderate/high spectral resolution, as well as narrowband imaging in the ultraviolet and visible ranges (1150–11500 Å) through a variety of different gratings, filters, and detectors. STIS is currently not working (see below) and kept in "safe" mode. During its operations, however, STIS provided scientists with data of paramount importance for the understanding of the physics of line-emitting clouds like stellar jets;

- the *Advanced Camera for Surveys (ACS)* integrates the capabilities of WFPC2 toward larger sensitivity and shorter wavelengths. ACS works through three channels: a Wide Field Channel (WFC), covering the range from 3700 to 11000 Å, with a FOV of 202″× 202″ and a plate-scale of 0.″05/pixel; a High Resolution Channel (HRC), with FOV = 26″× 29″, sensitive to wavelengths from 2000 to 11000 Å and having a plate-scale of 0.027 arcsec/pixel; and a Solar Blind Channel (SBC), spanning the range from 1150 to 1700 Å, with a plate-scale of 0.″032/pixel and a FOV of 31″× 35″;

- the *Fine Guidance Sensors (FGS)* assures the correct pointing of the telescope on the requested targets but can also be used as a scientific instrument. In this mode, the FGS offers accurate relative astrometry of stars and high angular resolution. In "Position" mode, it can measure the relative positions of objects in its 69 arcmin^2 FOV with a precision of 1 milliarcsec, while multi-epoch observations measure proper motions with an accuracy down to 0.3 mas/yr or less. In "Transfer" mode it is used to measure the angular size of extended objects or the separation of close binary systems.

Unfortunately, a failure in the power supply stopped the operations of STIS in August 2004 after more than seven years, while ACS is not functioning (except for the Solar Blind Channel) since January 2007, due to an electrical problem. Both these instruments have essentially met their expected lifespan, but the suspension of their operations is felt as a great loss by the community. Repair of STIS will be attempted during a servicing mission planned for late 2008 (SM4).

In addition, during SM4 the astronauts will install two new instruments, the *Cosmic Origins Spectrograph (COS)* and the *Wide Field Camera 3 (WFC3)*. The spectrograph COS will perform high sensitivity, moderate- and low-resolution spectroscopy in the 1150–3200 Å wavelength range, enhancing the capabilities of HST/STIS at ultraviolet wavelengths, especially for faint sources. WFC3 is a camera equipped with state-of-the-art detectors and optics, which will provide wide-field imaging with continuous spectral coverage from the UV into the IR (see Fig. 4), in two independent channels. The UVIS channel will be sensitive at UV and optical wavelengths, approximately from 2000 to 10000 Å, with a field of view of 2.7×2.7 arcmin and a plate scale of 0.04 arcsec/pix. The IR channel will work between 0.85 μm and 1.7 μm, and will have have a scale 0.13 arcsec/pix and a field of view of 2.1×2.3 arcmin. WFC3 will be positioned in the center of the focal plane of the telescope.

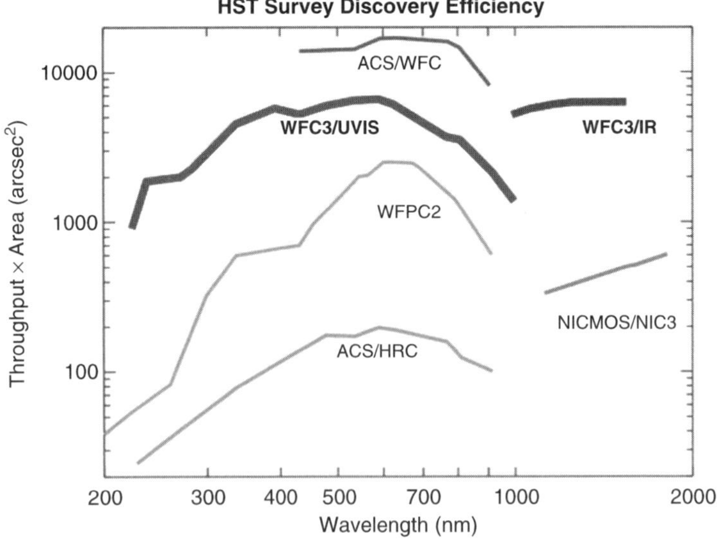

Fig. 4. Efficiency and wavelength coverage of the current HST cameras (WFPC2, ACS) and the WFC3, the new camera to be installed during servicing mission 4

Refurbished with these new instruments and with several sub-systems during SM4, HST is expected to function until at least 2013, working at the peak of its capabilities. In any case, the HST mission has fully demonstrated the importance of performing observations at high angular resolution. HST has returned data of unique scientific value that have led to big advancements and unexpected discoveries. In the next section, we will focus on the studies of stellar jets and accretion disks pursued with HST, illustrating some examples of the forefront science that was done using it. Before concluding this section, we would like to stress that the large datasets delivered by HST have been investigated only partially. The full exploitation of the data stored in the HST archive will take many years, and the science return from the planned archive exploration projects is expected to be quite important.

4 Jets from Young Stars: Clues from the HST

In stellar outflows, the gas flows at high supersonic velocities. As a consequence, instabilities in the flow generate shocks of various effective strengths that give rise to a large variety of emission lines (see the chapter by P. Hartigan in this book). These lines trace very different regions in terms of density, temperature, and excitation conditions that often have a size not larger than a few tens of AUs. To resolve such regions, even for the nearest star formation sites, one needs sub-arcsecond resolution. At the same time, a wide wavelength coverage is necessary to measure lines that have an important role in the diagnostics of the physical conditions of the gas. The best way to fulfill these requirements is to observe from space, and in this section we describe some of the many outstanding results obtained in this field with HST.

4.1 Imaging

Valuable insights have come from high angular resolution imaging conducted from space with the HST. The complexity revealed by HST is well illustrated by the WFPC2 images of the HH 46/47 and HH 111 jets shown in Fig. 5 [14, 24] . HST images have allowed us to discover many structures inside and around outflows, such as excavated cavities, large bow shocks, multiple flows from close binaries, etc. HST images have also allowed us to clarify the nature of the bright knots, at least for some of the flows. In Fig. 5, for example, the bright condensations clearly present an "arc" shape. These "mini working surfaces" are thought to arise from pulsations in the mass ejection process at the jet base. The mechanism that generates such pulsations, however, has not yet been identified.

As illustrated in Fig. 6, pairs of images taken at different epochs have enabled astronomers to identify the proper motion of the condensations in

160   F. Bacciotti et al.

direction and velocity [16]. This information, combined with the velocities along the line of sight (derived from Doppler shifts), have allowed us to obtain the total velocity in 3D space and of course the inclination of the jet with respect to the plane of the sky. Combination images taken a number of years apart have been put together to make "movies" (see Pat Hartigan's page at sparky.rice.edu/hartigan/movies.html) where the outflow is seen to propagate away from the source and new knots form close to the star (e.g. [7, 18]).

A target that has revealed interesting properties is the HH 30 jet in Taurus (Fig. 7, left panel). This beautiful bipolar jet is seen associated with two cusp-shaped reflection nebulae [8, 21]. The dark lane between them identifies a circumstellar disk seen edge-on, and the adjacent bright nebulae are illuminated by the light from the obscured star. The blue-and red-shifted jets emerge in directions almost exactly perpendicular to the disk plane and propagate at large distances. This system has often been taken as an illustration for the paradigm of the formation of an isolated low mass star. Other more recent HST observations, however, identified variable asymmetries in the illumination of the reflection nebula, which were tentatively interpreted as the possible presence of a compact binary system in the disk center [26].

Another outcome of the observations of jets with the WFPC2 is the delineation of shock pairs [14]. Images, taken in different excitation tracers, revealed the presence, predicted by jet propagation theories, of two adjacent

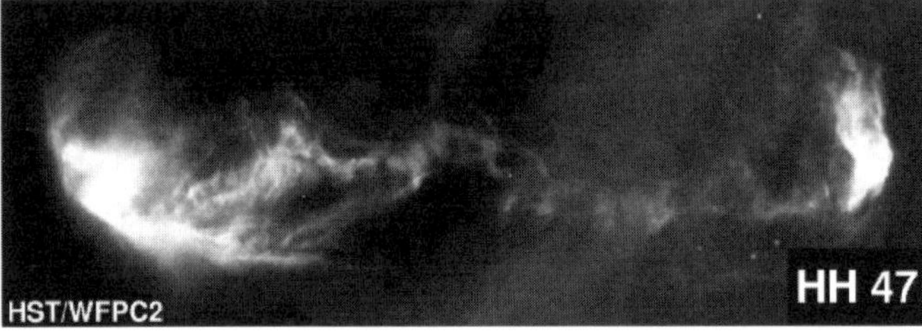

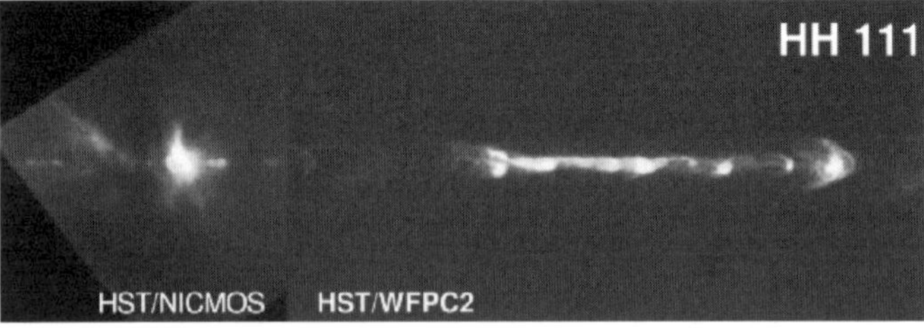

Fig. 5. HST images of the HH 46/47 and HH 111 jets. The bright 'knots' along the beam trace the position of internal working surfaces. In the NICMOS segment of the HH 111 image, the presence of the circumstellar (possibly circum-binary) disk is revealed as a dark lane. In both cases, one lobe of the jet is brighter because it is coming out of the dark cloud from which the star formed (adapted from [14, 24])

shock fronts defining a working surface. These are the so-called Mach disk, where the fast material coming from the star is decelerated and the "bow-shock", where the slower external medium ahead of the working surface is accelerated. These data have provided a powerful benchmark for radiative shock models and large-scale numerical simulations.

One should also not forget that NICMOS and WFPC2 have allowed us to identify, and "see" in silhouette, for the first time, many circumstellar disks , thus definitely validating the theories that predicted their existence around young stars ([6], see Fig. 8). Many of these sub-arcsecond disks were identi-

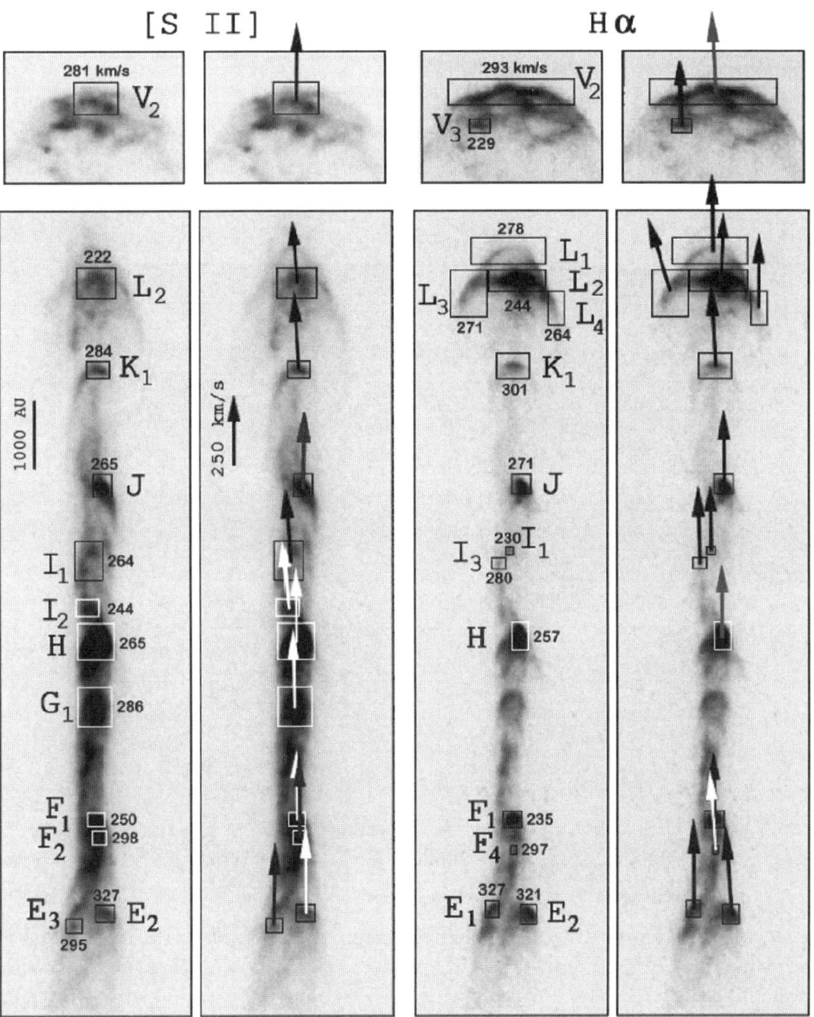

Fig. 6. Proper motions of the condensations along the HH 111 jets derived from HST/WFPC2 images taken at different epochs (from [16])

162   F. Bacciotti et al.

fied in the Orion Nebula, which is ionized by the stars of the Trapezium. This incidentally led to the discovery of so-called irradiated jets, which shine in recombination lines independent of the internal shock fronts. Beautiful new examples of such jets have also been found in the Carina nebula in recent observations performed with the ACS (Stapelfeldt et al., 2007 in prep.). Irradiated jets are photoionized, and this opens the interesting prospect of being able to measure directly the mass loss rate by just determining the electron density, which is relatively easy (see the discussion in P. Hartigan's chapter).

4.2 Spectroscopy

Another big chapter in the study of stellar jets opened when high angular resolution was first combined with long-slit spectroscopy, thanks to the Space Telescope Imaging Spectrograph (STIS). In this case, we have been literally flooded with information that has allowed us to shed light on the physical properties of the flow in critical regions such as the initial jet channel or the post-shock cooling zones (see the review by [22]). In this way, we have

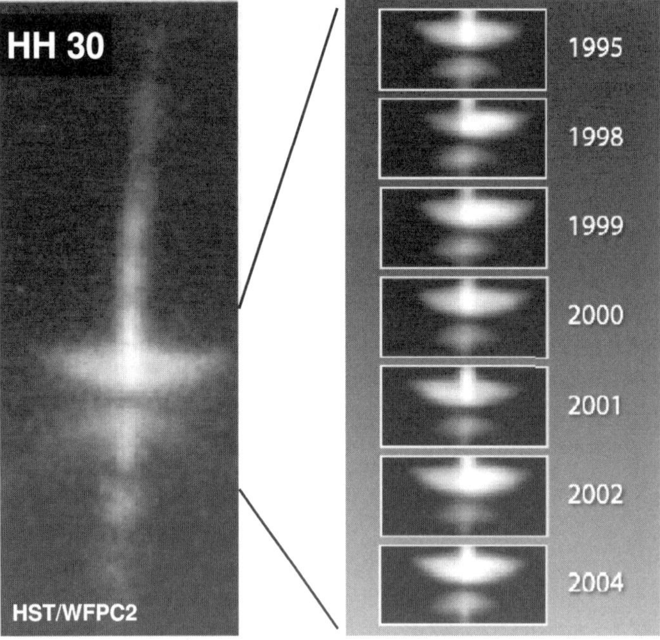

Fig. 7. (**Left**) Composite HST/WFPC2 image of the bipolar HH30 jet taken with several narrow-band filters. The jets emanate from the source in a direction perpendicular to the dark lane representing the circumstellar disk (adapted from [21]); (**Right**) Images of the system taken with the HST cameras WFPC2 and NICMOS at different epochs, identifying variability in the illumination of the reflection nebula around the source ([26])

been able to constrain observationally the variation of physical quantities in different velocity intervals with sub-arcsecond resolution, and, for the first time, in the direction perpendicular to the jet axis.

For example, the jets from the T Tauri stars DG Tau and RW Aur have been observed with STIS with multiple exposures of a 0.″1 wide slit, stepping the slit position across the flow every 0.″07. 3-D datacubes (2-D spatial, 1-D in velocity) have been built in a way similar to what would have been obtained with an Integral Field Unit (IFU), to study in detail the region of interest ([3, 27]). One projection of the datacubes gives 2-D images of the jets in different velocity intervals, the so-called channel maps in the first 100–200 AU from the star. The DG Tau flow, in particular, shows an onion-like kinematic structure, with a faster, denser, and more excited flow concentrated toward the axis, as predicted by magneto-hydrodynamic models (see Fig. 9, left panel).

The maps of the line ratios constructed with STIS can then give information about the gas physics. 1- and 2-D maps of the various quantities of interest in the different velocity channels have been obtained by [2, 4]. The electron density maps, for example, confirm that n_e is higher closer to the star, the axis, and at higher velocity. At the jet base, one then typically finds an ionization fraction of $0.01 < x_e < 0.4$, and total number densities up to 10^6 cm^{-3}. In the same region, electron temperatures range from 8×10^3 to 2×10^4 K. A similar study has been performed on slitless spectroscopy data obtained with STIS for the base of several T Tauri jet [17] and for the HH 30 jet

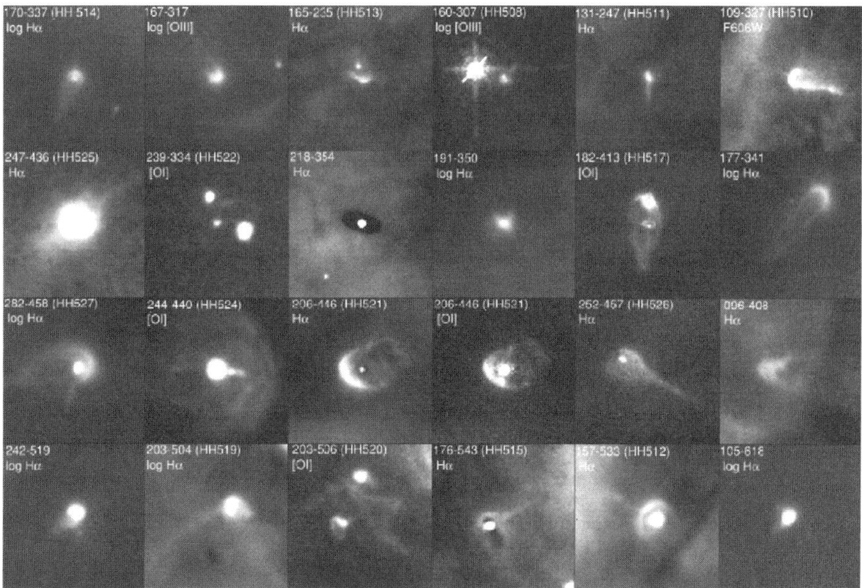

Fig. 8. Disks seen in silhouette, microjets and associated ionization fronts imaged with HST/WFPC2 in the photoionized portion of the Orion Nebula (from [6])

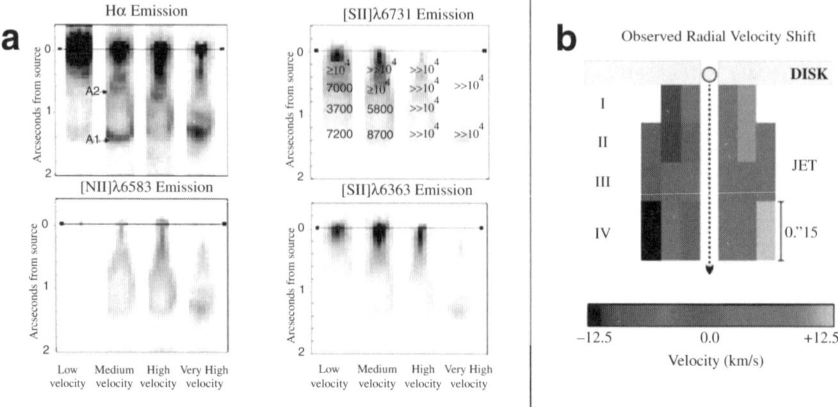

Fig. 9. (a) Channel maps in different velocity intervals of the jet from DG Tauri, in various emission lines, and at sub-arcsecond resolution. These images, reconstructed from HST/STIS spectra, reveal the onion-skin structure of the jet at its base, in agreement with the indications of magneto-hydrodynamic models. The numbers superimposing the [SII] image indicate the derived values of the electron density (figure taken from [3]); **(b)** Map of the velocity differences across the jet from DG Tauri derived from HST/STIS spectra. The velocity shift at the jet borders can be interpreted as rotation of the flow around its symmetry axis (adapted from [5])

[18]. Work is in progress to compare the obtained values with the indications from magneto-hydrodynamic models.

Another exciting study that has become possible, thanks to the high angular resolution of HST, is the search for signatures of *jet rotation* around the symmetry axis. According to the magneto-centrifugal scenario a key element of the formation of stars is the rotation of the system ([12, 19, 25]). Since, in this framework, the jet is believed to carry away the excess angular momentum, some trace of rotation should be observable in the outflow immediately above the accelerating region. Indeed, possible rotation signatures have been detected with HST/STIS in the flows from about ten T Tauri stars, starting from the DG Tau jet (see Fig. 9, right panel), and then using different observation techniques and emission lines (e.g. [5, 9, 10, 28]). Systematic shifts in the radial velocity of about 6 to 25 ± 5 kms^{-1} are found in positions of the jet displaced symmetrically at 20–30 AU from the axis and at 50–60 AU from the disk. These values are compatible with models for jet launching. Generally, the derived sense of rotation is the same in the various elements of the system (bipolar lobes, disks) and between different datasets. Only one discrepancy has been found at the time of writing, between the sense of rotation of the RW Aur disk and its jets. Studies are underway to better understand this case, as well as to extend the sample of jets and disks in which rotation signatures are detected.

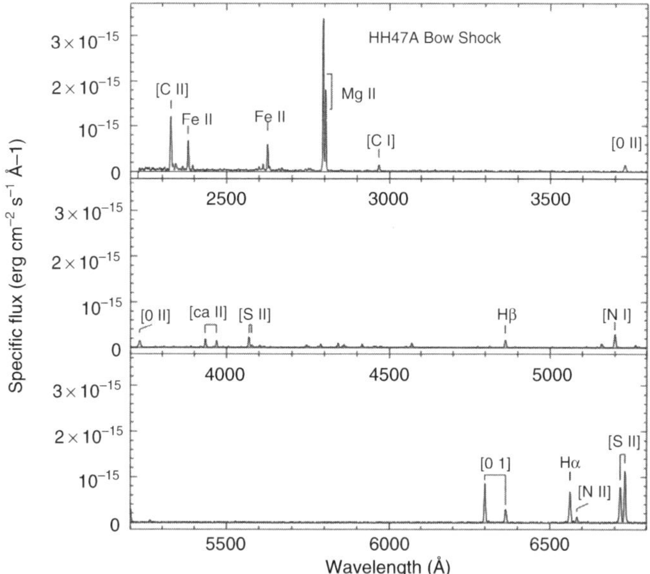

Fig. 10. Near UV-optical spectrum of the HH 47 bow-shock. The Mg doublet at 2800 Å is stronger than the classically observed optical lines (from [15])

A definitive confirmation that the observed velocity shifts are indeed rotation will only come from a larger statistical sample and an understanding of rogue cases, although for the time being this is the simplest explanation for the observed shifts. If rotation is indeed the cause, the above measurements will constitute the first global validation of the models proposed for the jet launching. In a rotation scenario, from the combination of measured toroidal and poloidal velocities, one can determine the *footpoint radius* of the wind, i.e., the location in the disk from where the observed portion of the rotating wind is launched. The observations are consistent with the footpoint being located between 0.5 and 2 AU from the star ([1, 9, 28]). This finding appears to support the disk-wind models, which consider an extended region in the disk for the origin of the wind, although X-winds may be present inside, closer to the symmetry axis. It has also been verified that a consistent fraction of the excess disk angular momentum present in the inner portion of the disk (more than 70%) can be carried away by the jet and that the magnetic field at the observed locations is mostly toroidal, as indicated by the models.

We conclude this section mentioning that the spectrographs on board HST have also allowed us to detect UV lines from the jets that arise in conditions of high excitation, i.e., close to the axis of the flow, at the jet base, or immediately downstream of the terminal strong shocks. Indeed, [15] demonstrated that lines like the MgII doublet at 2800 Å are actually stronger than the more classical optical lines used to identify Herbig-Haro jets, namely, Hα, [OI]λ6302, the [SII] doublet at 6730 Å , and [NII]λ6585 (see Fig. 10). These measurements, together with the jet spectra around the Lyman α line taken with STIS by [11], indicate that this is a very interesting field for exploration, as soon as new UV facilities become available.

5 Future Space Facilities

At the time of writing, a number of projects are under development for second generation space facilities. The most important in the context of star formation are

- the *Herschel Space Observatory*,
  for imaging and spectroscopy in the Mid-, and Far-Infrared (60–670 µm range), with a primary mirror 3.5 m, which will focus light onto three instruments (HIFI, SPIRE, and PACS). It is an ESA project with participation from NASA and will be launched in 2008 (herschel.esac.esa.int/overview.shtml);
- the *James Webb Space Telescope (JWST)*,
  for near-, mid-, and far-IR imaging and spectroscopy, seen in many respects as the successor of HST, which will provide high sensitivity imaging and spectroscopy in the red optical band, as well as in the near and mid-Infrared. JWST should be launched toward 2013, and it is a project developed as a collaboration between NASA, ESA, and the Canadian Space Agency (CSA) (www.jwst.nasa.gov);
- the *Darwin/TPF-I interferometer*,
  an ambitious project to install an interferometer in space operating at mid-infrared wavelengths, to find habitable planets. The project is under study by NASA and ESA, envisaged to be operational not before 2020 (sci.esa.int/science-e/www/area/index.cfm?fareaid=28 and planetquest.jpl.nasa.gov/TPF-I);
- the *XEUS telescope*,
  for X-ray imaging and deep spectroscopy, developed by ESA and expected in 2025 (www.rssd.esa.int/XEUS).

Although a comparative study of the capabilities offered by these instruments would be extremely interesting, we do not have the space here to discuss such a topic in full. We suggest that the interested reader explore the corresponding web sites for more information. Nevertheless, here we give a more detailed description of **JWST**.

5.1 The James Webb Space Telescope

The James Webb Space Telescope (JWST) is scheduled for launch in 2013 as an essential successor to the Hubble Space Telescope. It was originally known as the Next Generation Space Telescope but was renamed by NASA, with the

Fig. 11. The James Webb Space Telescope. The spacecraft will be positioned at the Lagrange 2 point so that the sunshield will not only guard it from solar radiation but also from infrared radiation from the Earth to Moon system. Star and planet formation is one of the four major themes that will direct the JWST science goals

agreeement of ESA and the CSA, to honor the man who led the famous Apollo program and NASA itself in its fledgling days. The telescope will have a deployable 6.6 m mirror consisting of 18 hexagonal beryllium segments (Fig. 11). It will be diffraction-limited at 2 μm, and its total light-gathering area will be 4.5 m^2, i.e., some 7 times greater than HST. Unlike HST, however, JWST will operate primarily in the infrared (0.6–28 microns), and thus not only will its suite of instruments be very different but it cannot be placed in near-Earth orbit as the telescope has to be kept cooled (<50 K). With this idea in mind, JWST will be positioned at the Lagrange 2 (L2) point. It will, therefore, be along the line joining the Sun to the Earth, opposite the Sun's direction; this is advantageous because not only will the sunshields protect the telescope from the infrared radiation of the Sun but also from radiation of the Earth and Moon. JWST is expected to operate for at least 5 years but the aim is to extend its life to 10 years. The telescope will have four main science goals, including understanding "The Birth of Stars and Protoplanetary Systems" and will thus provide a major boost to star formation studies.

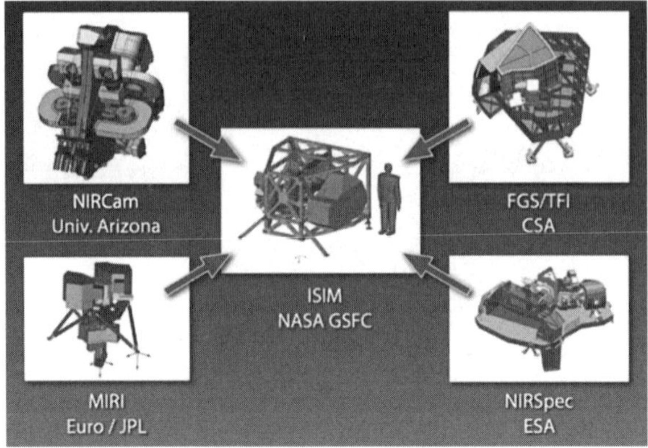

Fig. 12. Intrumentation on-board the JWST, with corresponding agencies

On board will be four science instruments: NIRCam, a 0.6–5 micron camera, NIRSpec, a 1–5 micron multi-object spectrometer, MIRI (Mid-Infrared Instrument), a 5–28 micron imager (and coronagraph) with spectrometer, and finally FGS/TFI, a 1–5 micron fine guidance sensor and tunable filter imager (see Fig. 12). While NIRCam and NIRSpec will be passively cooled to 30 K, MIRI will, in addition, be actively cooled to 7 K. The instruments are being built by a set of research organizations and space agencies: NIRCam by the University of Arizona, NIRSpec by the European Space Agency, MIRI by a European consortium and the Jet Propulsion Lab (JPL) and FGS/TFI by the Canadian Space Agency (CSA). All four instruments will be mounted on the NASA Goddard-supplied Instrument Science Integration Module (ISIM). The ISIM is responsible for acquisition of the JWST science data, including the fine guidance data for telescope-pointing control and wavefront sensing data for in-flight adjustment of the telescope optics. We will briefly review the capabilities of each instrument.

NIRCam is a dual channel near-infrared camera (0.6–2.4/2.4–5.0 microns) using a dichroic to divide the incoming beam along its long wavelength and short wavelength arms. Both arms have focal plane detectors consisting of 8 (2k×2k) and 2 (2k×2k) HgCdTe arrays for the short and long wavelength channels, respectively. Each channnel has a Field of View (FOV) of 2.1×4.2 arcminutes. Data from NIRCam will be used by the ISIM for wavefront sensing and is thus pivotal to the operation of the telescope. NIRCam, with its exceptional sensitivity, will be particularly useful in deep field imaging of distant galaxies, improving substantially with respect to the capabilities of HST.

NIRSpec is a Multi-object spectrograph, capable of making spectra of more than one hundred objects simultaneously, with configurable slit lengths and a FOV of 3×3 arcminutes. It has two spectral resolution modes, one operating with a prism at $R = 100$ between 0.6 and 5 µm, and the second at $R = 1000$ with 3 grating settings between 1 and 5 µm. The multiple slits are defined by a micro-shutter array activated by a magnet, which is controlled electronically and sweeps over the array. This device is supplied by NASA, while the instrument itself is being built by European industry and managed by ESA. In the $R = 100$ mode, NIRSpec will be used to do deep multi-colour imaging of galaxies. The $R = 1000$ mode, on the other hand, will be particularly useful in star formation studies, through the measurements of emission lines. In this way, one will be able to investigate the star formation rate, the stellar ages, the elemental abundances, and the physical conditions in the parent cloud.

MIRI presents great technological and scientific opportunities. The sensitive spectroscopy will enable the study of the properties of materials forming around ne born stars, like circumstellar disks, in unprecedented detail, and the coronagraphs will allow one to image directly massive planets. MIRI is a joint project from JPL, who are developing it, and several European institutes, and it has both imaging and spectral capabilities. As an imager, the instrument has a wide wavelength coverage, from 5 to 27 µm, with a FOV of 1.3×1.7 arcminutes. It is diffraction limited at 5 µm, with a plate scale of $0.''1$ per pixel. As a spectrometer, it works as an Integral Field Unit, combining a resolving power of $R = 1000\text{–}3000$ with high angular resolution ($0.''2$/pixel), on a FOV of either $3'' \times 3''$ or $7'' \times 7''$.

Finally, the FGS is a very broadband guide camera used for both "guide star" acquisition and fine pointing. It operates over a wavelength range of 1 to 5 micrometers, with a field of view sufficient to provide a 95% probability of acquiring a guide star for any valid pointing direction. The associated Tunable Filter Camera (TFI) is a wide-field, narrow-band camera that provides imagery over a wavelength range of 1.6 to 4.9 µm (with a gap between 2.6 and 3.1

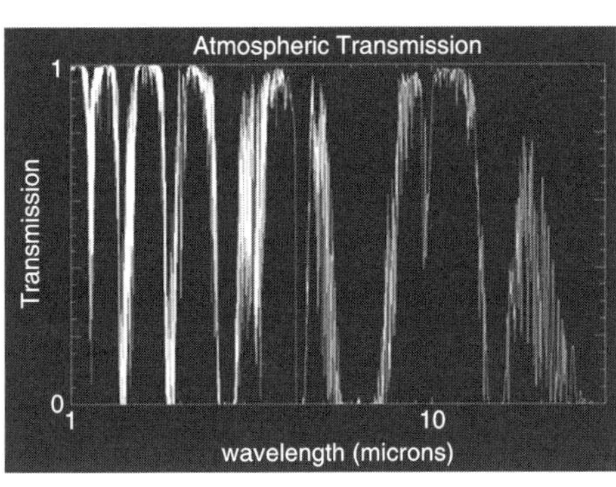

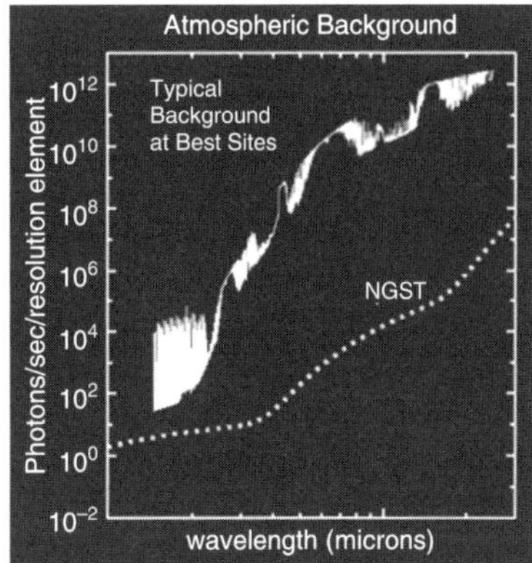

Fig. 13. Advantages of JWST: accessibility to infrared wavelength bands blocked by the atmosphere (**left**) and reduced sky background (**right**)

µm) via Fabry-Perot etalons, which are configured to illuminate the detector array with a single order of interference, at a user-selected wavelength.

The instruments on board JWST are expected to open a new era of exploration of the formation of young stars. These instruments are designed to fully exploit the advantages of JWST. These are, basically, the accessibility of wavelength windows almost totally forbidden from the ground, the high angular resolution achievable from outside the atmosphere (in the IR domain (0.″08 versus the 0.″6 obtainable on Earth in the best conditions), the extraordinary sensitivity of the optics and of the detector arrays, and, last but not least, the reduced sky background, which can be many orders of magnitude smaller than that at the best observing sites (see Fig. 13). For more details on the JWST project, see the web site www.jwst.nasa.gov and the review by [13].

References

1. Anderson, J. M., Li, Z.-Y., Krasnopolsky, R., Blandford, R.: Astrophys. J. **590**, L107–L110 (2003)
2. Bacciotti, F., Eislöffel, J., Ray, T.: Astron. Astrophys. **350**, 917–927 (1999)
3. Bacciotti, F., Mundt, R., Ray, T.P., Eislöffel, J., Solf, J., Camenzind, M.: Astrophys. J. **537**, L49–L52 (2000)
4. Bacciotti, F.: Rev. Mex. Astr. Ap. **13**, 8 (2002)
5. Bacciotti, F., Ray, T.P., Mundt, R., Eislöffel, J., Solf, J.: Astrophys. J. **576**, 222–231 (2002)
6. Bally, J., O'Dell, C.R., McCaughrean, M.J.: Astron. J. **119**, 2919 (2000)
7. Bally, J., Heathcote, S., Reipurth, B., Morse, J., Hartigan, P., Schwartz, R.: Astron. J. **123**, 2627, (2002)
8. Burrows, C., et al.: Astroph. J. **473**, 437 (1996)
9. Coffey, D., Bacciotti, F., Ray, T.P., Woitas, J., Eislöffel, J.: Astrophys. J. **604**, 758–765 (2004)
10. Coffey, D., Bacciotti, F., Ray, T.P., Eislöffel, J., Woitas, J.: Astrophys. J. **663**, 350 (2007)
11. Devine, D., et al.: Astroph. J. **542**, L115 (2000)
12. Ferreira, J.: Astron. Astrophys. **319**, 340 (1997)
13. Gardner, J.: Space Sci. Rev. **123**, 485 (2006), downloadable at uk.arxiv. org/abs/astro-ph/0606175
14. Heathcote, S., Morse, J., Hartigan, P., Reipurth, B., Schwartz, R.D., Bally, J., Stone, J.: Astron. J. **112**, 1141 (1996)
15. Hartigan, P., Morse, J., Tumlinson, J., Raymond, J., Heathcote, S.: Astrophys. J. **512**, 901 (1999)
16. Hartigan, P., Morse, J., Reipurth, B., Heathcote, S., Bally, J.: Astrophys. J. **559**, L157 (2001)
17. Hartigan, P., Edwards, S., Pierson, R.: Astrophys. J. **609**, 261–276 (2004)
18. Hartigan, P., Morse, J.: Astrophys. J. **660**, 426 (2007)
19. Königl, A., Pudritz, R.: Protostars and Planets IV. In: Mannings, V., Boss, A.P., Russell, S.S. (eds.), pp. 759–787. Univ. of Arizona Press, Tucson (2000)
20. Lena, P., Lebrun, F., Mignard, F.: Observational Astrophysics, 2nd edn. (Springer Verlag, 1998) Astron. Astrophys. **356**, L41–L44 (2000)

21. Ray, T.P., Mundt, R., Dyson, J.E., Falle, S.A.E.G., Raga, A.C.: Astrophys. J. **468**, L103 (1996)
22. Ray, T., Dougados, C., Bacciotti, F., Eislöffel, J., Chrysostomou, A.: invited in chapter to Protostars and Planets V. Reipurth, B., Jewitt, D., Keil, K. (eds.), 951, p. 231. University of Arizona Press, Tucson (2007)
23. Ray, T., Dougados, C., Bacciotti, F., Eislöeffel, J., Chrysostomou, A.: Protostars and Planets V. In: Reipurth, B., Jewitt, D., Keil, K. (eds.), pp. 231–244. University of Arizona Press, Tucson (2007)
24. Reipurth, B., Nature **340**, 42 (1989)
25. Shu, F.H., Najita, J., Ostriker, E.C., Shang, H.: Astrophys. J. **455**, L155 (1995)
26. Watson, A.M., Stapelfeldt, K.R.: Astron. J. **133**, 845 (2007)
27. Woitas, J., Ray, T.P., Bacciotti, F., Davis, C.J., and Eislöffel, J.: Astrophys. J. **580**, 336–342 (2002)
28. Woitas, J., Bacciotti, F., Ray, T.P., Marconi, A., Coffey, D., Eislöffel, J.: Astron. Astrophys. **432**, 149–160 (2005)

Observations and Models of X-ray Emission from Jets of Infant Stars

S. Orlando[1] and F. Favata[2]

[1] INAF - Osservatorio Astronomico di Palermo, Piazza del Parlamento 1, 90134 Palermo, Italy
orlando@astropa.inaf.it
[2] Astrophysics Div., Space Science Dep. of ESA, ESTEC, Postbus 299, NL–2200 AG Noordwijk, The Netherlands
Fabio.Favata@rssd.esa.int

Abstract. In the light of recent observations, this lecture revisits the main lines of evidence indicating that high energy phenomena, emitting in the X-ray band, take place in protostellar jets, possibly down to the jet acceleration site. For this new class of X-ray sources, we review the emission properties and the variability in the X-ray band. We discuss the possible impact of X-ray emission originating from jets on the environment of young stellar objects. We also show the importance of theoretical modeling in the interpretation of X-ray data and in the investigation of the nature of X-ray emission from jets. Finally, we discuss the intrinsic limitations in studying these X-ray objects and the future prospects.

S. Orlando and F. Favata: *Observation and Modeling of X-ray Emission from Jets of Infant Stars*, Lect. Notes Phys. **742**, 173–190 (2008)
DOI 10.1007/978-3-540-68032-1_8 © Springer-Verlag Berlin Heidelberg 2008

1 Introduction

In this contribution, we review the recent discovery of X-ray emission from jets of infant stars and describe their emission properties in the X-ray band. We also discuss the importance of studying the X-ray emission from protostellar jets, the state-of-the-art of this research field and the future prospects.

Herbig-Haro (HH) objects are known to be luminous condensations of gas in star-forming regions. In the past, many models have been proposed to explain their nature: stellar winds clashing with nebular material, dense pockets of interstellar gas excited by shocks from outflows, etc. Now the so-called jet-induced shock model is well established, in which material streams out of young stellar objects (YSOs) and collides with the surrounding interstellar medium.

In the last 50 years, HH objects have been observed at different wavelength bands, from radio to infrared to optical. The nature of this emission provides evidence for highly excited material. A clear prediction of the jet-induced shock model is that the most energetic HH objects should emit X-rays, although they had not been detected until a few years ago [6, 21]. In fact, one can derive useful relations between the physical parameters of interest (like the plasma temperature and the shock velocity) of the post-shock region as, for example:

$$T_{\mathrm{psh}} = \frac{\gamma - 1}{(\gamma + 1)^2} \left(\frac{m v_{\mathrm{sh}}^2}{k_{\mathrm{B}}} \right),$$

where T_{psh} is the post-shock temperature, γ is the ratio of specific heats, v_{sh} is the shock front speed, m is the mean particle mass, and k_{B} is the Boltzmann constant (see [28]). Assuming a typical velocity $v_{\mathrm{sh}} \approx 500 \ \mathrm{kms}^{-1}$ (as measured in most HH objects; see, for instance, [9] in the case of HH 154), the expected post-shock temperature is a few million degrees, thus leading to X-ray emission. Raga et al. [24] derived a simple analytic model that predicts X-ray emission originating from protostellar jets with the observed characteristics.

However, although the temperature of the post-shock plasma in high-excitation HH objects is expected to reach values of the order of few million degrees, X-rays had not been detected because past instruments (before the era of *Chandra* [19] and XMM-*Newton* [14], two world-class X-ray observatories in operation) lacked the necessary spatial resolution and sensitivity.

The idea that HH objects should emit X-rays is not new. In 1981, Pravdo and collaborators [20] reported the possible discovery of X-ray emission from HH 1, analyzing the data collected with the *Einstein* X-ray observatory. They interpreted the X-ray emission as due to strong stellar winds interacting with the interstellar medium.

The present contribution is structured as follows. In Sect. 2, we discuss the possible effects of X-ray emission from protostellar jets on YSOs environment and the importance of studying the emission properties of jets in the X-ray band; in Sect. 3, we illustrate the observational methodologies in the X-ray band and the evidence of X-rays from protostellar jets derived from the analysis of X-ray data collected with *Chandra* and XMM-*Newton*; in Sect. 4, we discuss the importance of hydrodynamic and magnetohydrodynamic modeling in the interpretation of the X-ray data and in the investigation of the nature of X-ray emission from jets; in Sect. 5, we discuss, on one hand, the intrinsic limitations in studying this new class of X-ray objects and, on the other hand, the importance of this study and the future prospects.

2 Impact of X-rays on the Environment of YSOs

In addition to the intrinsic interest, understanding the nature of X-ray emission from protostellar jets is important in the context of the physics of stars and planet formation. X-rays (and more in general ionizing radiation) affect many aspects of the environment of YSOs and, in particular, the physics and chemistry of the accretion disk and its planet-forming environment. The ionization state of the accretion disk around YSOs determines its coupling to the ambient and protostellar magnetic field, and thus, for example, influences its turbulent transport. In turn, this will affect the accretion rate and the formation of structures in the disk and, therefore, the formation of planets. Also, X-rays can act as catalysts of chemical reactions in the disk's ice and dust grains, significantly affecting its chemistry and mineralogy.

Recently, Natta et al. [18] derived the mass accretion rate from the luminosity of the near-infrared hydrogen recombination lines (Paβ or Brγ) for 45 class II objects and upper limits for 39. They found a correlation between the mass accretion rate, $\dot{M}_{\rm acc}$, and the mass of the star, M_* (see Fig. 1). In particular, $\dot{M}_{\rm acc}$ increases sharply with M_*. They also found a large spread (roughly two orders of magnitudes) of $\dot{M}_{\rm acc}$ for any given value of M_* that is likely a lower limit to the true dispersion because of the upper limits found for 39 objects (see empty symbols in Fig. 1).

The observed behavior of $\dot{M}_{\rm acc}$ does not have an obvious explanation and is difficult to understand in terms of disk physics only. The correlation may be due to a dependence of the disk physics on the properties of the central star. A possible cause may be the effect of the X-ray emission from the central star on the disk ionization and angular momentum transfer (see also [17]). The large spread of values of $\dot{M}_{\rm acc}$ for any M_* may be a side product of the X-ray emission from the central star. In this context, therefore, X-ray emission from jets may play an important role.

YSOs are strong X-ray sources, expected to irradiate the disk from the center (see Fig. 2) so that stellar X-rays will illuminate the disk with grazing incidence, concentrating their effects in the central region of the disk. On the other hand, protostellar jets are located above the disk so that they would illuminate the disk with near normal incidence, maximizing their effects even in the outer disk regions shielded from the stellar X-rays. As extensively

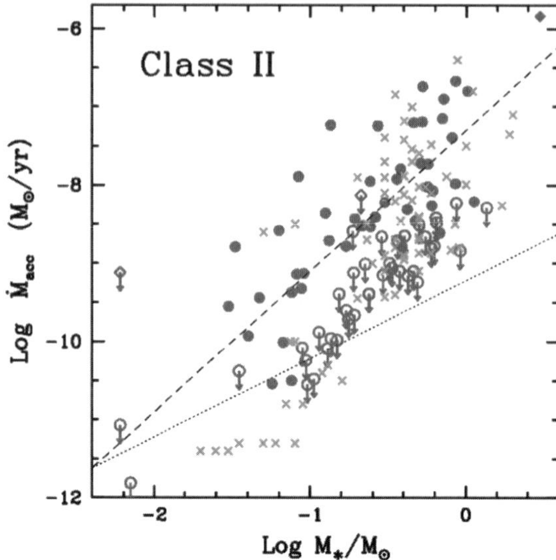

Fig. 1. Mass accretion rate, $\dot{M}_{\mathrm{acc}}$, derived from the luminosity of the hydrogen recombination lines vs. the mass of the star,I $M_\star$, (adapted from [18]). Dots show $\dot{M}_{\mathrm{acc}}$ measurements from Paβ (filled: detections, empty: upper limits); diamonds are measurements from Brγ (filled: detections, empty: upper limits); crosses are objects in Taurus. The dashed line shows the relation $\dot{M}_{\mathrm{acc}} \propto M_\star^{1.8}$; the dotted line plots, for comparison, the relation $\dot{M}_{\mathrm{acc}} \propto M_\star$

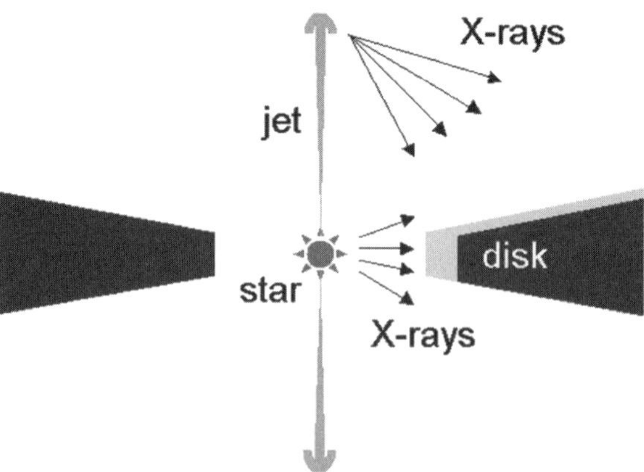

Fig. 2. Schematic view of a YSO with its bipolar jet. The central protostar illuminates the central region of the disk. The head of the jet, located above the disk, illuminates the outer disk regions shielded from the stellar X-rays

discussed in the literature, the ionization of the disk surface may induce disk-accretion instabilities (see [1]) and/or alter the disk chemistry (see [11]).

3 The Discovery of X-rays from Protostellar Jets

The first convincing evidence of X-ray emission originating from protostellar jets was reported by Pravdo and collaborators in 2001 [21] and by Favata and collaborators a few months later [6]. The first team discovered an X-ray-emitting jet in the Orion nebula by analyzing the data collected with the Chandra X-ray observatory in October 2000. Favata et al. [6] discovered an X-ray-emitting jet in the dark cloud L1551 in Taurus by analyzing the data collected with XMM-*Newton* in September 2000. In subsequent years, many other X-ray sources associated with protostellar jets have been discovered, prompting the interest of the scientific community to study this new class of X-ray sources. In the following sections, we briefly review the main results obtained from the data analysis and discuss the observational features characterizing these sources.

3.1 Confirmed X-ray-Emitting Jets

In 2001, Pravdo et al. [21] reported the discovery of X-ray emission from one of the brightest HH objects, namely, HH 2. They observed the region of Orion including HH 1 and HH 2 and several X-ray-emitting YSOs for ~ 21 ksec with the ACIS camera on board the *Chandra* X-ray observatory ([19]; see also [27]) on October 2000. They identified X-ray sources on the *Chandra*/ACIS image in two bands, [0.1–2] keV and [2–8] keV, and discovered a new X-ray source (without a known stellar optical counterpart) near the location of HH 2 in the soft band only. The source is quite faint, and only 11 photons have been detected within a 4-pixel radius corresponding to a diameter of 4 arcsec. Figure 3 shows a close-up view of the *Chandra*/ACIS field-of-view (FoV) including HH 2 as observed with HST (left panel) and *Chandra*/ACIS (right panel).

Pravdo et al. [21] verified that the X-rays are unlikely to originate from a point-like source, considering the so-called encircled energy plots for the HH 2 X-ray source and for two stars located nearby (see Fig. 4). The encircled energy plot shows the fractional intensity from a circle with the center coincident with the center of the source vs. the radius of the circle. The authors found that the plot derived for HH 2 is very different from those derived for the two stars: A Kolmogorov-Smirnov test showed a 99% probability that the spatial distribution of photons from HH 2 is different from that of the stars.

In spite of the low statistics, Pravdo et al. [21] have also shown that the spectrum of the HH 2 source is unique in this field. X-ray events were detected only below 2 keV, and the observed low resolution X-ray spectrum can be fitted with a thermal plasma model with temperature $T\sim 1$ MK. These authors concluded, therefore, that HH 2 contains shock-heated material located at or near its leading edge with a temperature of about 1 MK.

A few months after the discovery of X-ray emission from HH 2, Favata et al. [6] reported the evidence of X-ray emission from another Herbig-Haro

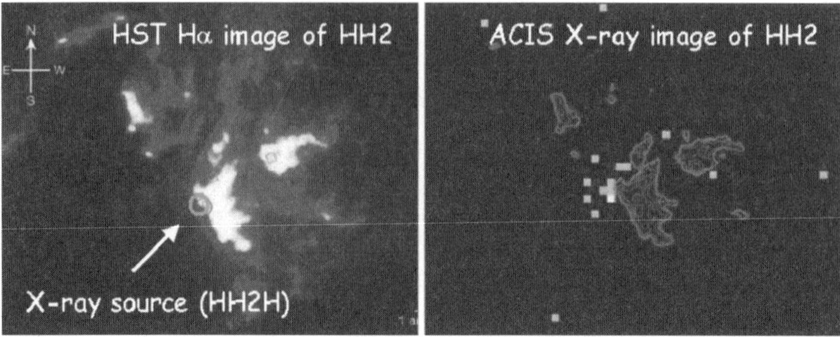

Fig. 3. Close-up view of the *Chandra*/ACIS field-of-view including HH 2 (adapted from [21]). **Left panel**: Hα image taken with the Hubble Space Telescope (HST). **Right panel**: Corresponding *Chandra*/ACIS observation of the same FoV. The superimposed contours are the Hα emission and the filled squares mark the individual X-ray events. The X-ray source is located near the edge of the optical knot

object, namely, HH 154. These authors observed the region, including the dark cloud L1551 in Taurus for ~50 ksec withc EPIC on board XMM-*Newton* [14], in September 2000, and analyzed the energy band ranging between 0.3 and 8 keV. The left panel in Fig. 5 shows the region, including L1551 IRS5 and HH 154 as seen in the X-rays with the XMM-*Newton*/EPIC camera. The right panel shows a close-up view of HH 154 as seen with HST. The leftmost X-ray point-like source (x1) is positionally coincident with the embedded source L1551 IRS5 and its jet.

For the source x1, 42 photons were detected within a circle 45 arcsec in diameter. The full width at half power of the point spread function for XMM-

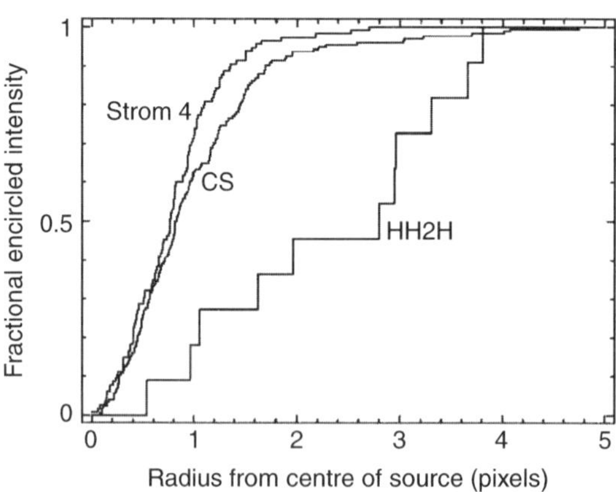

Fig. 4. Encircled energy plots for the HH 2 X-ray source and for two stars located nearby in the [0.1–2] keV energy band (adapted from [21])

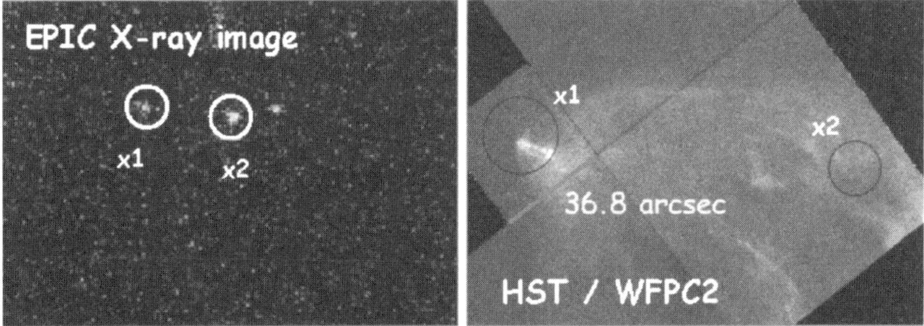

Fig. 5. (Left panel) The region of L1551 IRS5 as seen in X-rays with the XMM-*Newton*/EPIC camera (adapted from [6]); **(Right panel)** A close-up view of L1551 IRS5 and HH 154 as seen in a 1800 s *R-band* CCD image obtained with the HST/WFPC2 camera. The size of the small detector (the WFPC2-PC chip) on the left part of the HST image is $36''8$, while the size of the X-ray image is 9.3×6.3 arcminutes. The position of the X-ray sources x1 and x2 visible in the left panel is indicated on the *R-band* image by the circles

Newton/EPIC is $14''$, significantly larger than the size of the jet (whose visible length is about $10''$; with the XMM-*Newton* data, it is not possible to locate the precise site of the X-ray emission within the jet structure.

The low statistics allowed a limited amount of spectral information to be derived for the source. Favata et al. [6] found that the resulting spectrum is soft and can be reasonably described with a moderately absorbed thermal spectrum. In particular, they found that the emitting plasma has a temperature of ~ 4 MK, with a moderate value of absorption corresponding to an extinction $A_V \sim 7$ mag. The X-ray luminosity of the source was estimated to be $\sim 3 \times 10^{29}$ erg s^{-1}.

The small absorbing column density required to fit the X-ray spectrum (corresponding to an extinction of ~ 7 magnitudes) allows us to exclude the X-ray emission's association with the protostar, which suffers an extinction of about 150 mag. However, with the XMM-*Newton* data, no inference is possible on the precise location of the X-ray emission within the jet structure. Favata et al. [6] suggested that the X-ray emission could originate directly from a shock associated with the so-called knot D located about 10 arcsec from the protostar.

Bally and collaborators [2] observed the region including L1551 IRS5 and HH 154 with *Chandra*/ACIS, exploiting the high spatial resolution of this instrument. Figure 6 shows an optical image of L1551 IRS5 and HH 154 (left panel) and a close-up view with *Chandra*/ACIS of IRS5. These authors found that the X-ray emission from the vicinity of IRS5 arises from a compact source located at the base of the HH 154 jet and not from knot D, as supposed by Favata et al. [6].

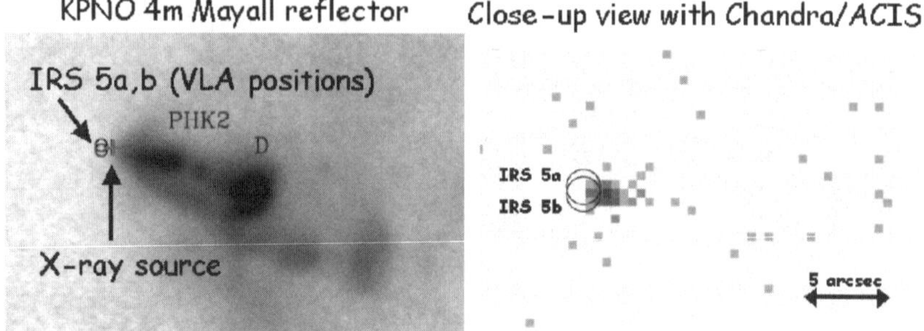

Fig. 6. Left panel: The region of L1551 IRS5 in optical (the sum of Hα and [S II] images, as seen through 80 Å passband filters with the KPNO 4 m Mayall reflector. **Right panel**: Close-up view of the X-ray source near L1551 IRS5, as seen with *Chandra*/ACIS. The circles show the protostellar positions, whereas the cross (left panel) marks the location of the X-ray source (adapted from [2])

Bally et al. [2] extracted 47 counts from a circle with 1.2 arcsec radius and found that (1) the X-ray source is displaced from IRS5 by about 0.5–1 arcseconds to the south–west along the jet axis, and (2) the source is slightly extended along the HH 154 axis. They also found that the resulting spectrum is well described by a moderately absorbed thermal spectrum with a temperature around 6 MK and with an extinction below 8 mag, confirming the results of Favata et al. [6]. Again, these authors found that the X-ray spectrum is inconsistent with the absorption required to obscure IRS5 and, therefore, the X-ray emission originates from the base of the jet.

In the last few years, many other X-ray-emitting jets have been discovered. A list of the confirmed X-ray-emitting jets at the present time is shown in Table 1, which summarizes their main characteristics: X-ray luminosity, temperature of the emitting plasma, absorption column density, distance of the source, and the corresponding references.

3.2 X-rays from the Base of the Jets?

In addition to the X-ray-emitting jets listed in Table 1, there are some other cases in which X-ray emission from the base of the jets is suggested by the evidence of soft X-ray excess in the X-ray spectra of YSOs (e.g. the beehive proplyd [15] and DG Tau [13]). In these cases, the authors found that

1. the X-ray spectra require two thermal components with different absorption column density;
2. the soft component at few million degrees suffers less extinction than the hard one;
3. the hard component at tens of millions of degrees has characteristics in common with the coronal X-ray emission of T Tauri stars.

These results suggest that the soft component is probably due to shocks at the base of the jet that can be detected given the low absorption there, whereas the hard component, strongly absorbed, is due to the stellar corona.

In the case of DG Tau, the above scenario is supported by the analysis of the X-ray images. In fact, Güdel et al. [13] noted that the *Chandra*/ACIS images reveal a weak excess of soft counts ranging between 0.5 and 2 keV at distances of 2–4 arcsec from the protostar along the optical jet (see Fig. 7).

These results, together with the evidence that the X-ray emission associated to HH 154 originates close to the base of the jet (see previous section and [2]), lead one to question whether jets are more active closer to the stars. As discussed in Sect. 2, the consequences would be far-reaching: These large-scale X-ray sources may efficiently ionize larger parts of the circumstellar environment (in particular the disk surface) than the central protostar alone. This, in turn, would affect the physics and chemistry of the accretion disk.

3.3 Variability in the X-ray Band

The analysis of time variability of the X-ray emission from jets is particularly difficult, given the faintness and the distance of these sources (see Table 1). In this respect, probably HH 154 is the most suitable source to study, as it is nearby (at $\sim$140 pc) and is observable through a moderate absorption column density ($A_V \sim$7 mag), whereas the parent star is strongly absorbed.

On the other hand, the analysis of variability in the X-ray band can be a key diagnostic to discriminate among the different physical mechanisms proposed. For instance, it could be possible to detect the proper motion of the X-ray source, or to study its evolving morphology, or to detect variations of temperature of the emitting plasma.

Recently, Favata et al. [7] reported the first study of long-term variability of the X-ray source associated with a jet, by comparing different *Chan-*

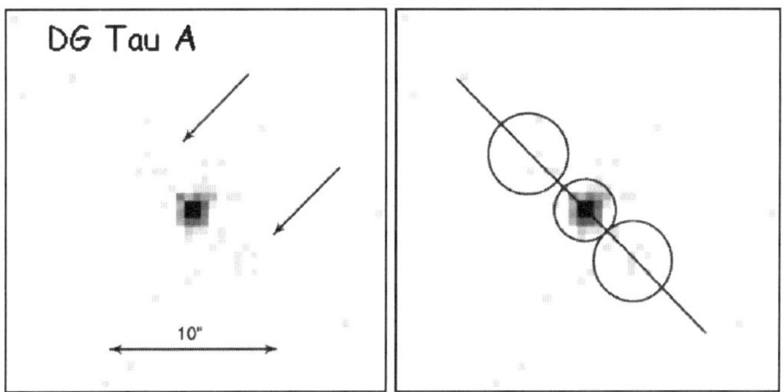

Fig. 7. (*Left panel*) *Chandra*/ACIS image of DG Tau A in the [0.6–1.7] energy band; (*Right panel*) The same, with extraction regions for DG Tau A and extensions overplotted. The image shows faint extensions (*see arrows*) at position angles of $\approx 45°$ and $\approx 225°$ (adapted from [13])

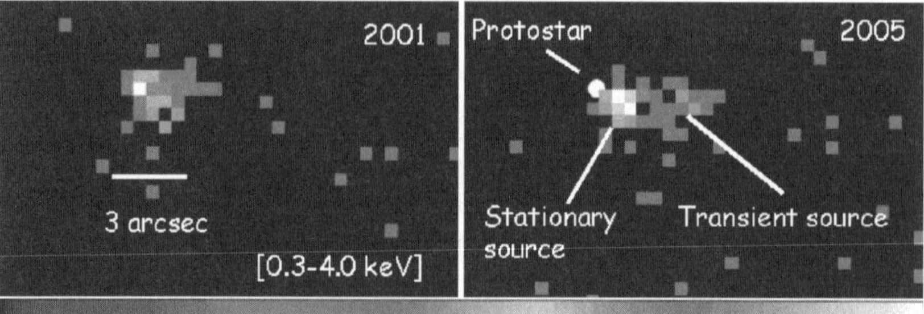

Fig. 8. X-ray source associated with HH 154, in the 2001 (**left**) and 2005 (**right**) *Chandra*/ACIS observations (adapted from [7]). The filled white circle in the 2005 observation marks the position of the protostar L1551 IRS5

dra/ACIS observations of HH 154 collected at different epochs, in July 2001 and in October 2005. Figure 8 shows the X-ray images in the [0.3–4.0] keV band collected with *Chandra*/ACIS in the two epochs.

Already for the 2001 observation, the jet X-ray source might show a hint of being extended (as discussed by Bally et al. [2]). Favata et al. [7] found that in the 2005 observation not only was the extended component confirmed but also the morphology had changed, with the size of the extended component clearly increasing, with respect to the 2001 observation. In particular, the authors identified two different components: a brighter point-like component with no detectable motion and a weaker component, expanding in size by about 300 AU over 4 years, leading to a proper motion of about 500 kms^{-1}. Favata et al. [7] also found limited spectral and X-ray luminosity variability over the period analyzed.

Table 1. Relevant physical quantities observed in confirmed X-ray-emitting HH objects, where L_X is the reported X-ray luminosity, T and N_H are, respectively, the best fit parameters derived from spectral analysis for the temperature and for the interstellar absorption column density, and D is the distance to the object observed. The last column reports the relevant references

Object	L_X [10^{29} erg s^{-1}]	T [MK]	N_H [10^{22} cm^{-2}]	D [pc]	References
HH 2	5.2	2.7	< 0.09	480	[21]
HH 154	3.0	2.0–7.0	1.40	140	[6]
					[2]
HH 80/81	450	1.5	0.44	1700	[22]
HH 168	1.1	5.8	0.40	730	[23]
HH 210	10.	0.8–3.8	0.80	450	[12]

4 The Nature of the X-ray Emission from Jets

From the analysis of the X-ray observations of jets, the following evidence have been collected.

- The X-ray sources associated with jets are attenuated by a considerably lower column density than that responsible for hiding the central protostars.
- The X-ray sources appear displaced from the position of the central protostars of few arcsec and, in some cases, their presence is suggested by a soft excess in the X-ray spectra of the central protostars.
- The X-ray emission is, in general, much softer than that expected from the central protostars.
- In the case of HH 154 (the only source for which a morphological analysis was performed), the X-ray emission appears extended along the jet axis.
- In the case of HH 154 (the only source for which long-term time variability analysis was performed), the morphology of the X-ray source evolves and shows an expanding component.

All these facts lead to the conclusion that a direct association with the central protostars is strongly unlikely and that the X-ray emission has to originate close to the base of the jets. The obvious question is, "What is the mechanism driving the X-ray emission associated with the protostellar jets?"

Several models have been proposed, but a detailed analysis has been performed only for very few. Bally et al. [2] suggested a few scenarios to explain the X-ray emission from HH 154. Among these, they proposed that the origin

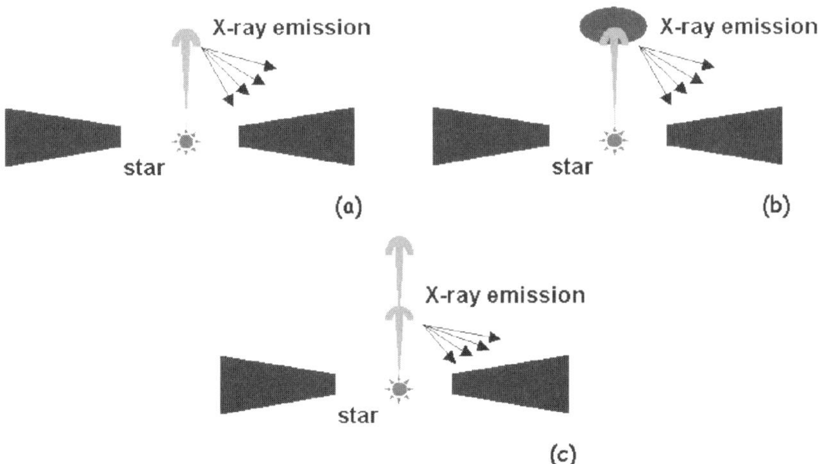

Fig. 9. (a) Interaction of a continuous jet with the homogeneous circumstellar medium; (b) Interaction of a continuous jet with inhomogeneities of the circumstellar medium; (c) Auto-interactions of a pulsed jet

of the X-ray emission could be similar to that observed in Seyfert 2 galaxies. In this case, the X-rays produced by the central protostar are reflected into our line-of-sight by a scattering layer located about 100–200 AU from the parent star. Another scenario proposed is that near the location of jet collimation the dense medium or the magnetic field could act like a nozzle leading to quasi-stationary shocks emitting X-rays.

Raga et al. [24] suggested that the X-ray emission could originate from the interaction of the protostellar jet with the surrounding medium. In the simplest case, proposed by these authors, the interaction of a continuous jet with the homogeneous circumstellar medium could lead, under particular conditions, to X-ray emission at the head of the jet (Fig. 9a). In this case, the X-ray source should be characterized by a proper motion that could be measured with the instruments in operation.

The X-ray emission may also arise from the interaction of the jet with inhomogeneities of the circumstellar medium (Fig. 9b). In this case, the X-ray emission should arise during the jet/cloud interaction and, therefore, it should be a transient phenomenon observable for a relatively short time interval.

Finally, X-rays may also originate in the case of a jet with variable outflow velocity (a pulsed jet, see Fig. 9c). In this case, internal shocks emitting in the X-ray band may form due to auto-interactions of jet material ejected at different epochs. Also in this case, the X-ray emission would be a transient phenomenon.

Advanced numerical modeling is a powerful tool to study the nature of X-ray emission from protostellar jets. The calculations include the most relevant physical mechanisms (e.g. thermal conduction, radiative losses, etc.), and the model results can be compared with observations. In the following, we discuss an example based on this approach.

4.1 Hydrodynamic Modeling

Bonito et al. [3] investigated the simple case of a continuous jet traveling through an homogeneous medium with the following purpose: (i) to infer the configuration(s) that can give rise to X-ray emission, (ii) to determine the range of parameters consistent with observations, and (iii) to get an insight into the jet's physical conditions (see also [4] for an extensive exploration of the parameters influencing the jet/ambient interaction). To this aim, the authors developed a two-dimensional hydrodynamic model (adopting a cylindrical coordinate system) that follows the jet propagation. This includes the effects of optically thin radiative losses [16, 25] and thermal conduction [26] with saturation effects [5]. The model solves the time-dependent fluid equations for the conservation of mass, momentum, and energy. The authors used the FLASH code [10], an adaptive mesh refinement multiphysics code for astrophysical plasmas. From the numerical simulations, synthetic spectra were produced, as would be collected with *Chandra*/ACIS or XMM-*Newton*/EPIC. A comparison with the observations is thus possible.

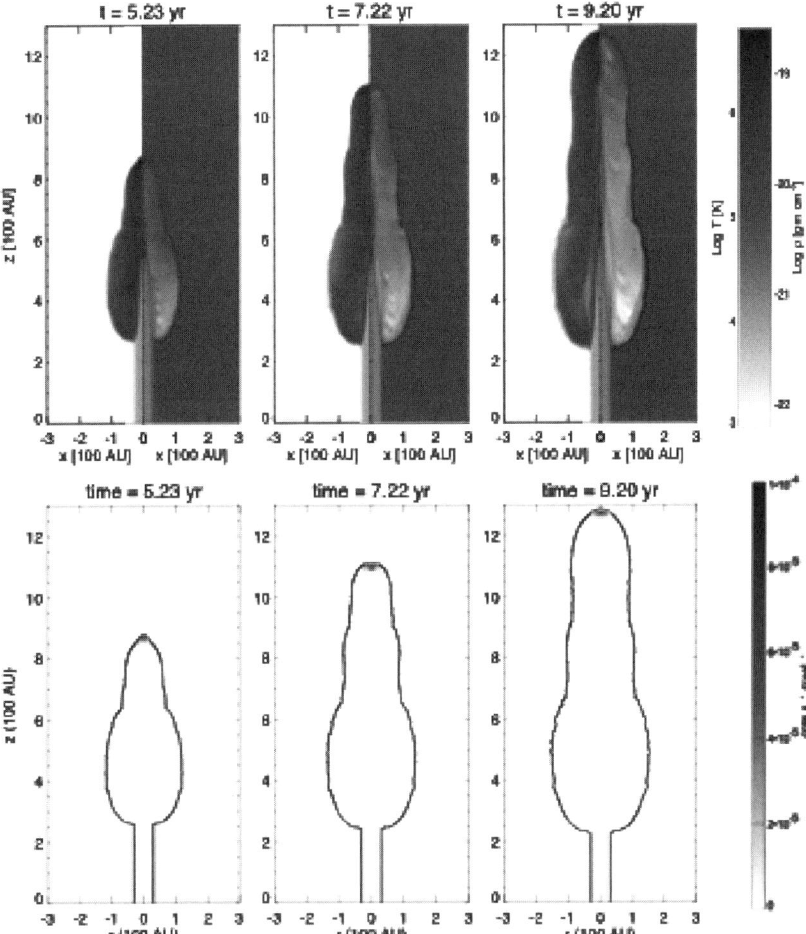

Fig. 10. (Upper panels) 2-D cuts through the rz plane of the jet temperature (on the left-hand side) and density (on the right-hand side), in a logarithmic scale, at three different evolutionary stages; (**Lower panels**) synthetic X-ray emission integrated along the line-of-sight and on macro-pixels with a size of $\sim$10 AU (six times better that the *Chandra*/ACIS spatial resolution), as would be observed with *Chandra*/ACIS. The contour plot marks the region occupied by the jet and by the cocoon. The X-ray source is localized in a blob behind the bow shock, moving with measurable proper motion; its linear size is $\sim$30 AU (adapted from [3])

This analysis showed that the continuous jet model explains the observed X-ray emission from protostellar jets in a natural way. In particular, the results derived from the data analysis of HH 154 (i.e. X-ray luminosity of the source, temperature and emission measure of the emitting plasma, absorbing column density, etc.) can be reproduced, without any ad hoc assumption, by a protostellar jet that is less dense than the ambient medium [3, 4]. The jet is characterized by a density $n_j = 500$ cm^{-3} and a temperature $T_j = 10^4$ K and travels through an ambient medium 10 times denser, according to Fridlund & Liseau [8]; the jet has high Mach number, M = 300, corresponding to an initial velocity $v_j \sim 1.4 \times 10^3$ kms^{-1}.

The upper panels in Fig. 10 (adapted from [3]) show the evolution of temperature (on the left) and density (on the right) derived from the model, reproducing the observations of HH 154. The lower panels show the predicted evolution of the X-ray emission, integrated along the line-of-sight with a bin size $\sim$10 AU, as would be observed with *Chandra*/ACIS[3]. The *Chandra* resolution corresponds to $\sim$60 AU at the distance of HH 154 ($\sim$140 pc).

Bonito et al. [3] found that a cocoon with an almost uniform (because of the thermal conduction) temperature up to $T\approx0.7$ MK surrounds the jet. The cocoon temperature decreases in time, and a cooler and denser shell forms due to the radiative cooling. They also found that a dense and hot blob with temperature of over 4 MK and density $n_b \sim 10^4$ cm^{-3} is localized just behind the shock front.

As concerns the predicted evolution of the X-ray emission (lower panels in Fig. 10), Bonito et al. [3] found that the X-ray emission originates from the hot and dense blob localized just behind the shock front. Even with *Chandra*, this blob cannot be spatially resolved and will be detected as a point-like source. On the other hand, the model clearly predicts a proper motion of the X-ray

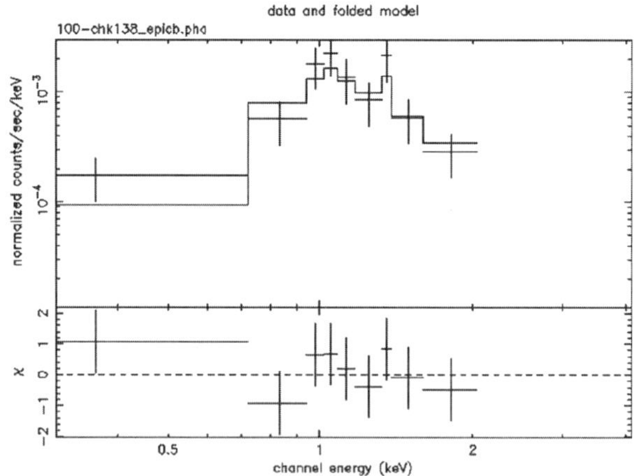

Fig. 11. Upper panel: synthesized spectrum as would be collected with XMM-*Newton*/EPIC, 25 years from the beginning of the jet-ambient interaction (adapted from [3]). The MEKAL best fitting spectrum is superimposed. The **bottom panel** shows the contribution of each bin to the total χ

---

[3] The synthesis of the emission takes into account the *Chandra*/ACIS instrumental response and the interstellar medium absorption column density with values according to those measured by Favata et al. [6] and Bally et al. [2] for HH 154.

Table 2. Best-fit parameters to the XMM-*Newton*/EPIC simulated X-ray spectrum derived from the hydrodynamic model (shown in Fig. 11; [3]) and to the XMM-*Newton*/EPIC data analyzed by Favata et al. [6]

	Hydrodynamic model	XMM-*Newton*/EPIC observations
N_H (cm^{-2})	1.5 ± 0.3	1.4 ± 0.4
T (MK)	3.4 ± 1.2	4.0 ± 2.5
F_X (erg cm^2 s^{-1})	1.4×10^{-13}	1.3×10^{-13}

point-like source, with an average velocity $v_{sh} \approx 500$ km s^{-1} corresponding to ~ 0.7 arcsec yr^{-1}, detectable with *Chandra*.

Bonito et al. [3] synthesized also the spectra as would be observed with XMM-*Newton*/EPIC, in order to compare their analysis with the results obtained by Favata et al. [6]. Figure 11 shows an example of EPIC focal plane spectrum synthesized from the model, 25 years from the beginning of the jet/ambient interaction. These authors found that all the spectra are well fitted with the emission from an optically thin plasma at a single temperature. The parameters of the fit derived from their model (Table 2) are consistent with those obtained from the observations [6]. They also found that the X-ray luminosity shows small variations around 3×10^{29} erg s^{-1}, according to the observations ([2], [6]).

Although Bonito et al. [3] found that the simple continuous jet model can reproduce the spectral characteristics observed in most X-ray-emitting jets (see also [4]), the evolving morphology observed in HH 154 [7] is not compatible with any of the simple models proposed: (i) a continuous jet propagating through an homogeneous medium [24], (ii) scattered stellar X-ray light [2], and (iii) static shocks resulting from the stellar wind hitting the cavity or the disk [2]. Alternative models are therefore required. For instance one could consider episodic ejection of plasma bullets from the jet's source, interaction of the jet with inhomogeneities of the circumstellar medium, or a de Laval nozzle at the base of the jet.

5 Conclusions and Future Prospects

X-ray-emitting protostellar jets have been discovered only recently and have been detected only with limited statistics. This is mainly due to intrinsic limitations that make the detection of X-rays from protostellar jets difficult. An important limitation is that the known sources are relatively distant: The closest one (HH 154) is at ~ 140 pc from us, and the others are located at more than 400 pc (see Table 1). Another limitation is that these sources have intrinsic luminosity in the X-ray band, ranging between 10^{29} erg s^{-1} and 10^{30} erg s^{-1} so that at the distances to which they are located, the X-ray observatories in operation can collect just a few tens of photons with exposure times of the order of 50 ksec. Briefly, X-ray-emitting jets are faint sources and, as a consequence, only limited spectral and morphological analysis can be performed.

On the other hand, studying the jet emission properties in the X-ray band can be very important. X-ray emission reveals the hottest parts of the jet, allowing a new development in observations of outflows from YSOs. The X-ray data open up the possibility of analyzing energetic HH objects and make it possibile for us to of detect fast, strong shocks associated with outflows. Moreover, the analysis of X-ray data can provide an unique probe of the high energy phenomena taking place close to the jet acceleration site (most of the X-ray sources associated to jets are located at the base of the jet) and can provide a direct determination of the scale of density changes in the immediate circumstellar environment.

To date, the X-ray observations have shown that HH 154 is an unique laboratory to study the nature of the jets' X-ray emission, as the parent star is hidden by a very large absorbing column density and the jet emerges with limited absorption. As a consequence, the X-ray emission in this source can be studied very close to the jet acceleration site, without being blinded by the stellar light. In addition, HH 154 is the nearest known X-ray-emitting jet and, this makes the morphological studies possible to a high level of spatial detail. For instance, as shown by Favata et al. [7], it is possible to detect the proper motion of the X-ray source on time-scales of the order of few years.

In the light of these considerations, deep *Chandra*/ACIS observations (with exposure times of the order of 500 ksec) may be a very useful tool to study, for instance, the evolving morphology of the two X-ray components recently identified in HH 154 (the bright stationary component and the faint evolving one; see [7]). Among other things, these observations could allow us to measure the plasma temperatures and motions of both components and to test the proposed mechanism(s) of X-ray emission.

In the future, we plan to develop a detailed magneto-hydrodynamic model describing a jet with a de Laval nozzle at its base and to compare the model results with the X-ray observations. To explore whether a de Laval nozzle can lead to X-ray emission, we have already built a simple hydrodynamic model providing clear predictions for what would be observed with *Chandra*/ACIS. The model simulates a tube with a nozzle and is based on the acceleration of an initially subsonic plasma flow to supersonic speeds through a narrowing tube, which later opens allowing the flow to escape. In the case of an over-expanded supersonic flow at the exit of the nozzle, a strong stationary shock forms downstream, and weaker and weaker shocks (related to reflections from the tube walls) form further downstream in the escaping plasma, as shown in the upper panel of Fig. 12. The bottom graph in the figure shows the central temperature profile, where it is possible to identify the shocks' position. The first shock at the acceleration site is cleanly visible near the jet's origin at about 500 AU (100 AU = 0.7 arcsec at the distance of HH 154), followed by

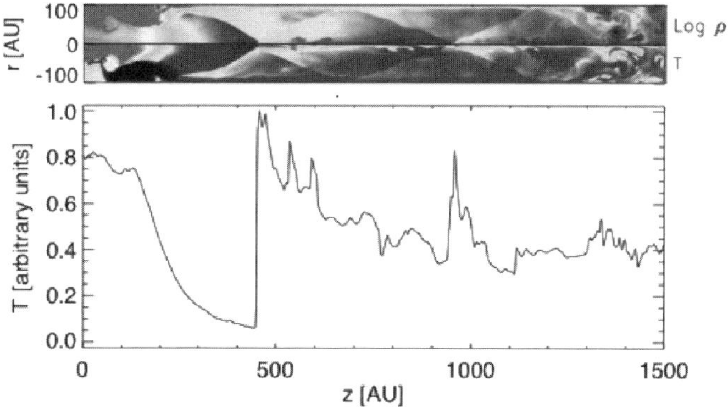

Fig. 12. (**Upper panel**) 2-D cuts through the rz plane of the jet density (**top**) and temperature (**bottom**) 5 years after the initial acceleration; (**Lower panel**) temperature profile along the jet axis (in arbitrary units)

a weaker one at about 900 AU. The absolute scale for the temperature values depends on the jet's initial parameters, in particular speed and density; the emission measure and temperature observed for HH 154 can be reproduced with values for the jet's speed and density compatible with the ones observationally determined (a few hundreds km/s for the speed, 10^2–10^4 cm^{-3} for the density).

Acknowledgements

We are grateful to R. Bonito, M. Fridlund, G. Micela and G. Peres for useful discussions on the properties of X-ray emission from protostellar jets. We thank the JETSET network for inviting us to present this lecture at the Elba JETSET school. This work was supported by Ministero dell'Istruzione, dell'Università e della Ricerca and by Istituto Nazionale di Astrofisica.

References

1. Balbus, S.A., Hawley, J.F.: Astrophys. J. **376**, 214 (1991)
2. Bally, J., Feigelson, F., Reipurth B.: Astrophys. J. **584**, 843 (2003)
3. Bonito, R., Orlando, S., Peres, G., Favata, F., Rosner R.: Astron. Astrophys. **424**, L1 (2004)
4. Bonito, R., Orlando, S., Peres, G., Favata, F., Rosner R.: Astron. Astrophys. preprint http://dx.doi.org/10.1051/0004-6361:20065236, (2006)
5. Cowie, L., McKee C.: Astrophys. J. **211**, 135 (1977)
6. Favata, F., Fridlund, C.V.M., Micela, G., Sciortino, S., Kaas, A.A.: Astron. Astrophys. **386**, 204 (2002)

7. Favata, F., Bonito, R., Micela, G., Fridlund, C.V.M., Orlando, S., Sciortino, S., Peres G.: Astron. Astrophys. **450**, L17 (2006)
8. Fridlund, C.V.M., Liseau, R.: Astrophys. J. **499**, L75 (1998)
9. Fridlund, C.V.M., Liseau, R., Djupvik, A.A., Huldtgren, M., White, G.J., Favata, F. Giardino, G.: Astron. Astrophys. **436**, 983 (2005)
10. Fryxell, B., Olson, K., Ricker, P., et al.: Astrophys. J. **131**, 273 (2000)
11. Glassgold, A.E., Najita, J., Igea, J.: Astrophys. J. **615**, 972 (2004)
12. Grosso, N., Feigelson, E.D., Getman, K.V., Kastner, J.H., Bally, J., McCaughrean, M.J.: Astron. Astrophys. **448**, L29 (2006)
13. Güdel, M., Skinner, S.L., Briggs, K.R., Audard, M., Arzner, K., Telleschi, A.: Astrophys. J. **626**, L53 (2006)
14. Jansen, F., Lumb, D., Altieri, B., et al.: Astron. Astrophys. **365**, L1 (2001)
15. Kastner, J.H., Franz, G., Grosso, N., Bally, J., McCaughrean, M.J., Getman, K., Feigelson, E.D., Schulz, N.S.: Astrophys. J. Suppl. **160**, 411 (2006)
16. Mewe, R., Gronenschild, E., van der Oord, G.: Astron. Astrophys. 62, 197 (1985)
17. Muzerolle, J., Hillenbrand, L., Calvet, N., Briceño, C., Hartmann, L.: Astrophys. J. **592**, 266 (2003)
18. Natta, A., Testi, L., Randich, S.: Astron. Astrophys. **452**, 245 (2006)
19. O'dell, S.L., Weisskopf, M.C.: In Proc. SPIE X-Ray Optics, Instruments, and Missions, vol. 3444, ed by Hoover, R.B., Walker, A.B. pp. 2–18 (1998)
20. Pravdo, S.H., Marshall, F.E.: Astrophys. J. **248**, 591 (1981)
21. Pravdo, S.H., Feigelson, E., Garmire, G., Maeda, Y., Tsuboi, Y., Bally, J.: Nature **413**, 708 (2001)
22. Pravdo, S.H., Tsuboi, Y., Maeda, Y.: Astrophys. J. **605**, 259 (2004)
23. Pravdo, S.H., Tsuboi, Y.: Astrophys. J. **626**, 272 (2005)
24. Raga, A., Noriega-Crespo, A., Velazquez, P.: Astrophys. J. **576**, L149 (2002)
25. Raymond, J., Smith, B.: Astrophys. J. **35**, 419 (1977)
26. Spitzer, L., Physics of Fully Ionized Gases. Interscience, New York (1962)
27. Weisskopf, M.C., Brinkman, B., Canizares, C., et al.: PASP **114**, 1 (2002)
28. Zel'dovich, Y.B., Raizer, Y.P.: Physics of Shock Waves and High-Temperature Hydrodynamic Phenomena. Academic Press, New York (1966)

Interferometry: Technique and Applications

Principles of Interferometry

N. Jackson

University of Manchester, Jodrell Bank Observatory
neal.j.jackson@manchester.ac.uk

Abstract. Interferometry is an important technique for achieving high angular resolution in astronomical observations. It has been a standard technique at centimetre wavelengths for decades, and is now beginning to make a major impact at shorter wavelengths with advances in technology. In this article I describe the basic theory of interference, and show how it may be applied to practical interferometer systems. I then describe methods for image production from interferometers and for dealing with wavefront corruption by the atmosphere. Finally, I give an overview of some of the major interferometric facilities at long wavelengths operating today and of projected future telescopes.

In this review, I outline the principles of interferometry and describe how these are put into practice in a range of modern interferometric telescopes. The basic philosophy of the review is to emphasize a pictorial approach to the subject, rather than delve into detailed mathematical derivations. Much more rigorous treatment of interferometry can be found in [18]. In addition, the US

N. Jackson: *Principles of Interferometry*, Lect. Notes Phys. **742**, 193–218 (2008)
DOI 10.1007/978-3-540-68032-1_9

National Radio Astronomy Observatory (NRAO) hosts a lecture series every three years, in interferometry, published in [16].

Throughout, I concentrate on applications of interferometry to interferometers at long (wave-regime) wavelengths, with occasional excursions into methods used in the new generation of optical interferometers. The final section is an overview of current interferometers working in the meter-to-millimeter band. The principles behind interferometers at all wavebands, however, are very similar, including the whole of Sect. 1 and most of Sect. 2. A recent full and complementary review of optical interferometry is given by [10], and other reviews in these proceedings cover the VLT interferometer in detail.

1 Basics: Young's Slits and Fourier Transforms

1.1 Young's Slits

Interferometry begins with the Young's slits fringe pattern (Fig. 1). With a single point source emitting coherent radiation, interference fringes are observed, with constructive and destructive interference observed as the relative delay of the two interfering rays changes; the separation of the fringes is λ/d, the wavelength of the light divided by the slit separation.

If the source is made wider (Fig. 1b), we can think of it as a sequence of point sources each of which emits radiation, which is uncorrelated with the emission from the others. It follows that the total interference intensity pattern is the sum of the individual patterns. Since an angular displacement in the source produces an equal angular displacement in the fringe pattern, as the source size approaches λ/d the fringe patterns will add to give a constant illumination (Fig. 1c). In this case, the fringe visibility (defined as the difference between maximum and minimum intensity, normalized by the sum of maximum and minimum intensity) drops to zero. Conversely, when the angular size of the source is $\ll \lambda/d$, the fringe visibility is 1; this corresponds to a situation in which the source size is smaller than the angular resolution of the interferometer, and only an upper limit of order λ/d can be obtained on it.[1]

Now suppose that the slit spacing d is decreased. For the same size of source, this produces less "washing-out" of the fringes, because the same displacement of the source now produces much less displacement of the fringe patterns as a fraction of the fringe separation λ/d (Fig. 1d). The smaller the slit separation, the larger the source size that can be probed using interferometry.

The situation is summarized in Fig. 2. If we plot, for a given source distribution, the way in which visibility varies with slit separation, it can be

---

[1] In practice, the fact that the visibility function begins to decrease as soon as the source extends significantly often allows some information to be derived down to at least $0.5\lambda/d$ and in some cases further.

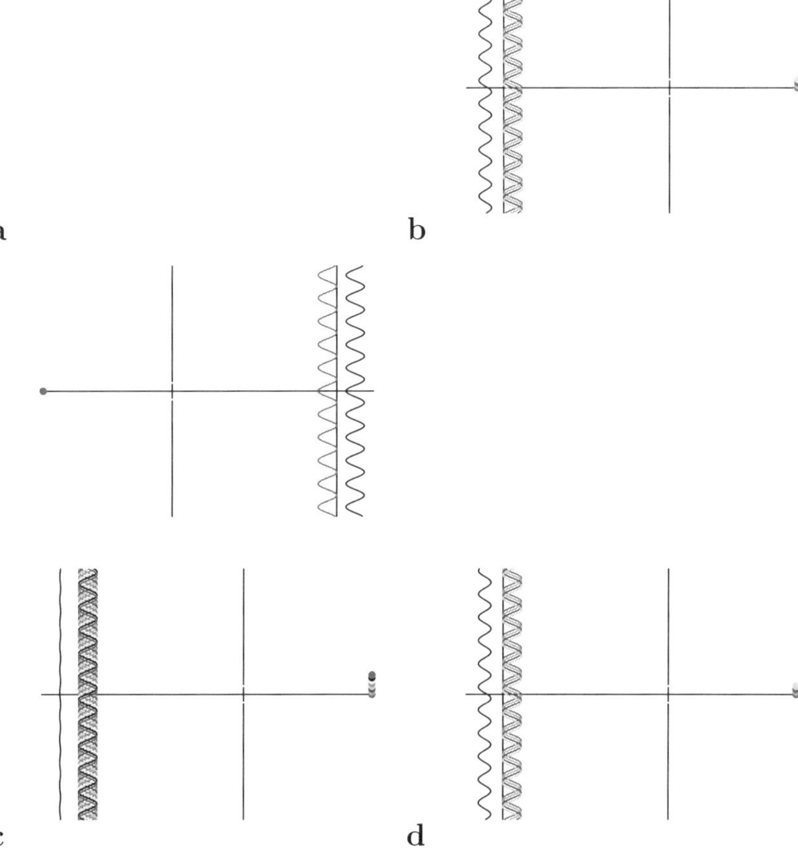

Fig. 1. Young's slits in various situations. In each panel, the source is shown on the left, and on the right of the slit are shown the fringe patterns separately for each part of the source and then the added fringe pattern; (**a**). The basic two-slit pattern, showing fringes an angular distance λ/d apart; (**b**). The effect of increasing the source size. An angular shift of the source position by θ shifts the fringe patterns by θ the other way. Since the patterns come from mutually incoherent sources, the intensity patterns add to give a pattern of reduced visibility; (**c**). When the size of the source reaches λ/d, the fringes add to give zero visibility; (d) If the slit spacing is then reduced, the fringe spacing increases, and the same size of source is still able to give visible fringes: the source would need to be increased in size to λ/d_{new} in order to wash out the fringes

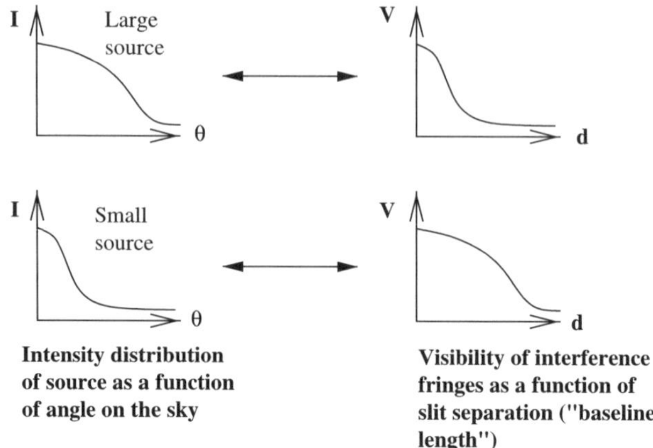

Fig. 2. Relation between source brightness as a function of angular distance and visibility of interference fringes as a function of slit separation (baseline length)

seen that for small sources the visibility remains high out to large slit separation (in the limit of point sources, to infinite slit separation), whereas large sources produce visibility patterns which fall off quickly as the slit separation increases.

The relation between $I(\theta)$ and $V(d)$ represented here is one which maps a large Gaussian into a small Gaussian, and vice versa, and it is fairly obvious that it is a Fourier transform;[2] this relationship is the basis of the whole discussion that follows.

This relationship was applied early in the history of radio interferometers to find the sizes of quasars, which were known to be exceedingly small. The method adopted was to use one fixed telescope and one movable telescope and measure the visibility function, using electronic combination of the signals over baselines up to $\sim$150 km [1]. As the telescopes were moved further apart, the visibility finally fell below unity at large separations, allowing the angular size to be calculated as λ/d.

1.2 Application to Real Interferometers

The Young's slit experiment discussed so far involves sampling two parts of a plane wave generated a large distance away, delaying one wave with respect to the other, and generating the interference pattern as a function of delay. There are many situations in which exactly the same thing is being done. In Fig. 3, for example, a plane wave from a source at infinity is sampled by two telescopes separated by a vector **B**. The path delay between the two waves is

---

[2] This is known as the Van Cittert-Zernicke theorem. Readers requiring a more rigorous derivation are referred to [3] a classic textbook on optics, or the introduction to radio astronomy by [5].

given by $\mathbf{B} \cdot \mathbf{s}$, where $\mathbf{s}$ is the unit vector in the direction of the source, and the phase delay is therefore given by $k\,\mathbf{B} \cdot \mathbf{s}$, where $k \equiv 2\pi/\lambda$.

Consider a point source. If the electric field received by the first telescope is E, that received by the second is just $Ee^{ik\mathbf{B} \cdot \mathbf{s}}$, because of the phase delay. We can combine these signals, in a way analogous to the screen in Young's slits, by multiplying them together electronically[3] or, in the case of optical systems, by using a Michelson or Fizeau interferometer system to combine the beams. If we then add the fringe patterns over different parts of the source, we obtain the response of the interferometer R as

$$R = \int I(\sigma) e^{ik\mathbf{B} \cdot (\mathbf{s}+\sigma)} \mathrm{d}\sigma,$$

where $\mathbf{s}+\sigma$ is the vector in the direction of a particular small part of the source with an intensity $I(\mathrm{d}\sigma)$. Noting that σ is parallel to the projected baseline vector $\mathbf{b}$ (Fig. 3) and so $\mathbf{B} \cdot \sigma = \mathbf{b} \cdot \sigma$, we then have

$$R = e^{ik\mathbf{B} \cdot \mathbf{s}} \int I(\sigma) e^{ik\mathbf{b} \cdot \sigma} \mathrm{d}\sigma,$$

where the $e^{ik\mathbf{B} \cdot \mathbf{s}}$ term is solely dependent on the array geometry and has, therefore, been removed from the integral.

What we, therefore, have is a series of fringes, whose amplitude is given by the Fourier transform of the source intensity distribution. In practice, steps are usually taken to get rid of the fringes using a phase rotation whose rate is

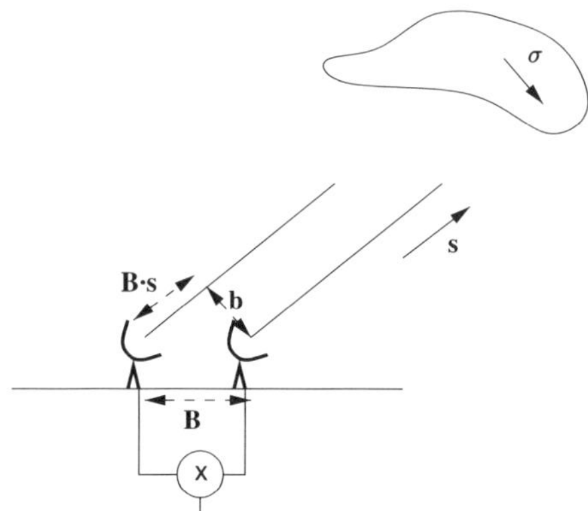

Fig. 3. Basic diagram of an interferometer with baseline vector $\mathbf{B}$ observing a source in a direction with unit vector $\mathbf{s}$

---

[3] This is not exactly the same as the Young's slits screen, which adds the electric fields and then forms the intensity using $I = (E_1 + E_2)^*(E_1 + E_2)$, but the result is almost the same, apart from a constant offset term in the addition case. Not having this term is useful, because we do not have to worry about the offset term being constant with time.

known (as both **B** and **s** are known). This is done in optical interferometers by use of accurate delay lines to compensate for the path difference, and in radio interferometers by the insertion of electronic delays. We are then left with the Fourier transform response only, which conveys information about the source. The response is a complex quantity which contains an amplitude and a phase; both are interesting.

Due to the fact that the signal from an interferometer results from the correlation of signals from two telescopes, interferometers have the advantage of much lower sensitivity to interference because most interference does not correlate. Thus, the only interference which causes a serious problem is that which saturates or disables the receiver. Such interference can be dealt with by dividing the observing band into spectral channels and removing any affected channels.

1.3 The u, v Plane

A further step is to decompose both σ and **b** into Cartesian coordinates. The decomposition of σ is easy, as it is just a vector in the sky plane: $\sigma = \sigma_x \mathbf{i} + \sigma_y \mathbf{j}$, where **i** and **j** are unit vectors in the east–west and north–south directions, respectively. This then suggests a decomposition of **b** into $u\mathbf{i} + v\mathbf{j}$, so that $\mathbf{b} \cdot \sigma = ux + vy$. The response after fringe stopping then becomes

$$R(u, v) = \iint I(x, y) e^{2\pi i (ux + vy)} \mathrm{d}x \mathrm{d}y,$$

a much more explicit 2-D Fourier transform. Note that u and v are defined in units of wavelength, hence the k in the previous expressions has become 2π.

The physical interpretation of the decomposition of **b** is fairly straight-forward. Imagine sitting on the source (Fig. 4); then the projected baseline vector appears as a line drawn on the earth. This can be decomposed into a component parallel to the equator at its nearest point to the source, and a component parallel to the line between this point and the north pole. These components are u and v, and they change as the earth rotates. Specifically, they trace out an ellipse in u, v space during one earth rotation (Fig. 4).

This change in u and v is useful, as we see if we consider how the response function $R(u, v)$ tells us what is on the sky. Figure 5 shows the basic Fourier transform relation in a diagram. A double source of separation 1 radian produces stripes in the u, v plane of separation 1 wavelength. Since the Fourier transform gives an inverse relation between distances in the two spaces, a double source of separation a arcseconds gives a series of stripes of separation $206265/a$ wavelengths. Superposed on this is the track of the u, v ellipse, and over a day there are, therefore, variations in the interferometer response as the interferometer follows the elliptical track over these stripes. Studying this variation in amplitude (and phase) response over the period, we could work backwards to deduce the separation and orientation of the stripes and, by taking the Fourier transform, recover the source structure.

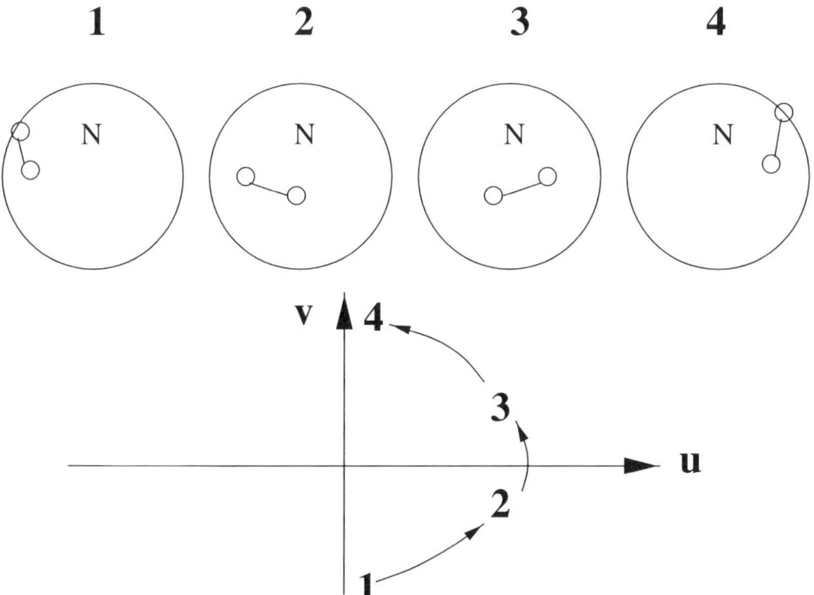

Fig. 4. Schematic diagram showing the baseline between two telescopes as the earth rotates. The E–W and N–S components of this vector give u and v. For an east–west baseline, $v{=}0$ at source transit

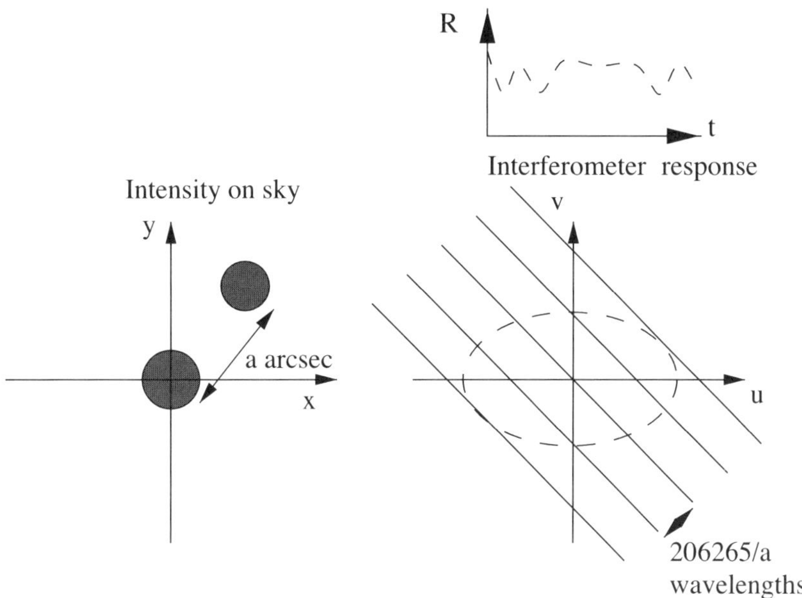

Fig. 5. Diagram showing the interferometer response as a function of u and v for a double source on the sky

The u, v track has a semimajor axis in the u direction of $\frac{L}{\lambda}\cos\delta$, and a semiminor axis in the v direction of $\frac{L}{\lambda}\cos\delta\sin D$.[4] In these expressions, L is the baseline length, D is the declination of the source, and δ the declination of the baseline; the latter quantity is the declination of the point on the celestial sphere to which the baseline vector points.

The resolution of which a baseline is capable is given by the inverse of the maximum extent of the u, v ellipse, namely λ/L. The point-spread function of an image has dimensions, which are the inverse of the spread in the u, v plane of the images being used, which means that for sources close to the equator (where $D \rightarrow 0$ and the ellipses collapse to straight lines) the point-spread function of an interferometer is typically less ideal, although images of the sky can be made with care. Figure 6 shows the u, v tracks for the MERLIN interferometer array, which contains baselines from 6 km to 250 km at a range of orientations. Note the gradual change from circular u, v tracks to nearly linear tracks as the source declination decreases.

If many baselines are present, many simultaneous measurements can be made in the u, v plane. The more completely the Fourier plane is filled, the easier it is to obtain a faithful reproduction of the sky intensity distribution in an interferometric image.

1.4 A Cautionary Tale

Interferometric (Fourier) imaging has important differences from direct imaging. The most important difference can be deduced by going back to the Young's slits setup: Long baselines record small-scale structures in the source very well, but are *insensitive* to large-scale structures, because once the source becomes larger than λ/L the fringes wash out and do not return as the source size increases. An example of this is shown in Figs. 7 and 8.

Figure 7 (top panels) shows a simulated source of $12''$ extent, mapped using an array whose u, v coverage gives a maximum baseline length corresponding to a resolution of $3''$. The source structure is recovered reasonably well, as the range of baselines present cover spatial scales from the resolution up to approximately 10 times larger scales. Now suppose that we get greedy and decide that we would like to observe at ten times higher resolution. This is no problem – we move the telescopes to spacings ten times greater and repeat the observation. This indeed gives a map with 300-mas resolution, on which there is no sign of the source (Fig. 7, bottom panels).

Let us then try to smooth the image and recover the structure. If we try this, what we actually recover is shown in Fig. 8. The awful realization dawns

---

[4] Spherical trigonometry can be used to show that the ellipse is parametrized by the equations: $u = \frac{L}{\lambda}\cos\delta\sin(H-h)$, $v = \frac{L}{\lambda}(\sin\delta\cos D - \cos\delta\sin D\cos(H-h))$ (e.g. [12]). Here, H is the hour angle of the source, and h the hour angle of the point on the sky to which the baseline points. For a general baseline, the center of the ellipse is offset by $\frac{L}{\lambda}\sin\delta\cos D$ from the origin in the v-direction.

at this point that by using a long set of baselines, we have not recorded the structure on large spatial scales at all and have lost it irretrievably. The moral is that interferometer arrays should be chosen carefully to match resolution to the spatial scales required by any particular astrophysical problem.[5]

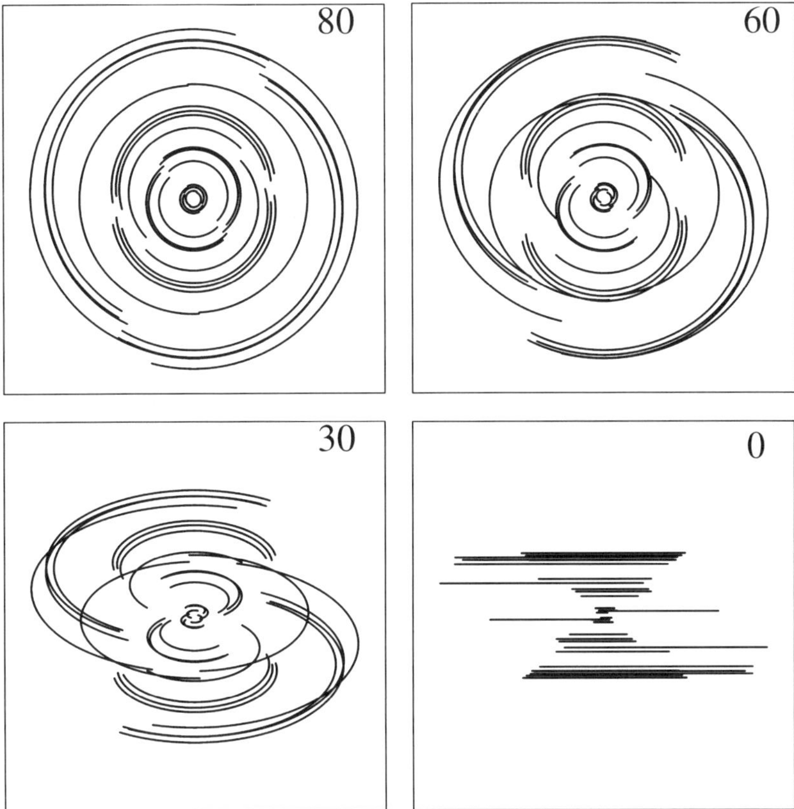

Fig. 6. u, v tracks for the MERLIN interferometer for sources at four different declinations: $80°$, $60°$, $30°$, and $0°$

---

[5] In fact, using resolution higher than required often causes even worse problems. This is because, for any given array, higher resolution demands observing at higher frequencies, which in turn imposes penalties in source brightness for typical steep-spectrum radio sources and in generally worse system performance at high frequencies. Some interferometers, in particular the VLA, allow different resolution images at the same frequency by regularly moving the telescopes between different configurations, from compact to more extended.

1.5   Field of View of Interferometric Images

Primary Beam

Once again, we can go back to Young's double slits to deduce another fundamental limitation of the interferometric image. If the slits are widened, the aperture distribution no longer consists of two delta functions, but of two delta functions convolved with a single wide slit. It follows from the convolution theorem that the interference pattern, being the Fourier transform of the aperture distribution, consists of the original two-slit fringe pattern multiplied by the Fourier transform of a wide slit, namely a sinc function. The sinc func-

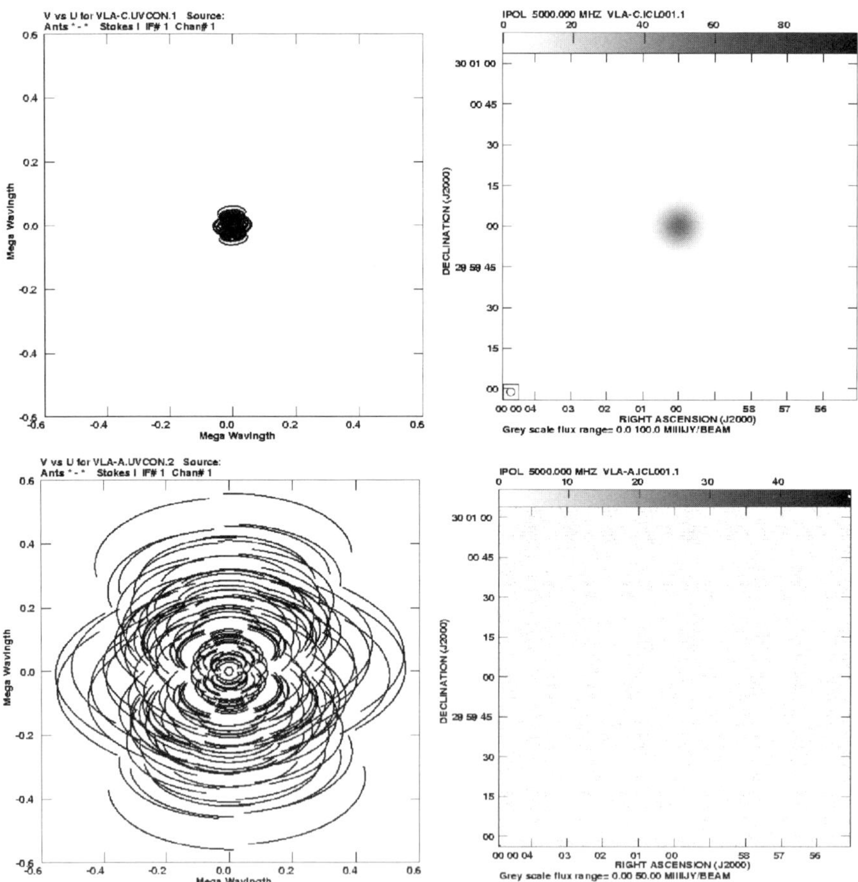

Fig. 7. Simulations of observations of a large Gaussian source with a set of short baselines, giving low resolution matched to the size of the source (top left, u, v coverage, top right, resulting image) and a set of long baselines giving higher resolution (bottom left, u, v coverage, bottom right, resulting image)

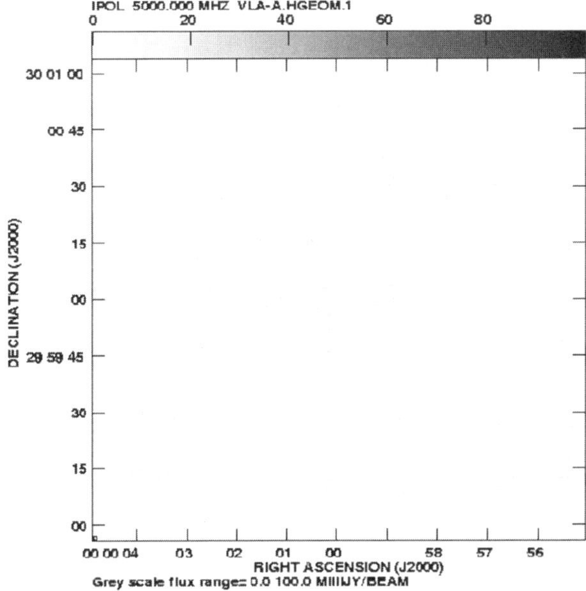

Fig. 8. Smoothed image of a long-baseline observation of a diffuse source. Nothing is visible

tion has a width inversely proportional to the width (w) of the slits, and the fringes disappear at delays greater than the width of the sinc function, λ/w.

Now in an interferometer, going further away from the center of the field of view just corresponds to a different delay from that which obtains at the center. The width of the slits translates directly to the size of each interferometer element, and the field of view in radians is the wavelength of the light being studied divided by the diameter of the elements.

Wavelength Ranges

Again going back to Young's slits, it is easy to see that other effects may intervene before the primary beam limit is reached. The most serious of these is that the radiation is not monochromatic. We can consider each single frequency separately and add the resulting fringe patterns which have different separation (λ/d) between maxima. The result is that in the center of the fringe pattern, full-visibility fringes are seen, since here the delay is zero. At larger values of delay, further up the screen, the interference fringes from different colors add in such a way as to reduce the visibility to zero even for a point source. Once again, the effects at large delay translate directly to the interferometer, and a range of wavelengths in the interferometer causes loss of response at the edge of a field of view where the delay between the interfering waves is different.[6] If the bandwidth is $\Delta\lambda$, then the field of view is given by

$$\Delta\theta = (\lambda/\Delta\lambda)(\lambda/L),$$

or the beam size divided by the fractional bandwidth. In order to achieve reasonable signal-to-noise, many interferometers use large fractional bandwidth, implying a very restricted field of view. The solution is simply to divide the signal into many frequency channels and correlate each separately. The cost is greater complexity of the correlator and larger datasets, but in most modern interferometer systems computing and hardware are advanced enough that it is usually possible to image the full primary beam.

Other Effects

Two other effects should be mentioned briefly; see [16] for further details. The first is that a limitation on the field of view is imposed by the integration time per data point, because the values of u and v change during a finite integration time. This gives a roughly tangential smearing in the u, v plane, which becomes worse further out; Fourier transforms this into a tangential smearing in the sky plane which becomes worse further out. The result is that amplitude is lost at the edge of the field. The second effect is that the sky is not flat and that instead of using a 2-D Fourier transform we should have used a 3-D transform, with an extra phase term of the form $\sqrt{(1 - x^2 - y^2)}$ added to the transform. Unlike the bandwidth and integration time effects, the effect of non-flatness is curable after the event by additional processing.

2 Producing the Image

2.1 Deconvolution

So far we have circumvented the major problem, which is that the interferometer response function has not been measured over the whole u, v plane. To do this at a single frequency would require enough telescopes to provide baselines at all possible separations and orientations, an expensive operation with substantial planning implications. Lack of this information means that the number of different images consistent with the data is infinite, since we could in principle fill in the unmeasured parts of the u, v plane in an infinite number of ways.

---

[6] For a large Gaussian-shaped bandwidth, the fringe response as a function of angular distance from the center is a small Gaussian, and for a small Gaussian-shaped bandwidth, the fringe response varies as a broad Gaussian with angular distance. This completes another "proof" of a Fourier transform relation between these two quantities.

The basic problem is that we want the image $I(x, y)$ resulting from the full u, v response function $I(u, v)$,

$$I(x, y) = \iint I(u, v)e^{2\pi i(ux+vy)}\,dudv$$

but instead have the "dirty image"

$$I_{\mathrm{D}}(x, y) = \iint I(u, v)S(u, v)e^{2\pi i(ux+vy)}\,dudv,$$

which results from the intervention of the sampling function $S(u, v)$, which is 1 in parts of the u, v plane where we have sampled and zero where we havennot.

We recognize the right-hand side of the last equation as a Fourier transform, where the argument is the product of two functions I and S. We can, therefore, use the convolution theorem to write

$$I_{\mathrm{D}}(x, y) = I(x, y) * B(x, y),$$

where

$$B(x, y) = \iint S(u, v)e^{2\pi i(ux+vy)}\,dudv,$$

the "dirty beam" is the Fourier transform of the sampling function. Since we know where the telescopes are and can do spherical trigonometry and Fourier transforms, the sampling function and, hence, dirty beam are accurately known. Recovering the image $I(x, y)$ is therefore a classical deconvolution problem, for which we need to supply additional information in order to do the deconvolution.

Clean

The first way to do this is by using the algorithm known as CLEAN [8], which amounts to a brute force deconvolution. The basic algorithm begins by detecting the brightest point in the dirty map, shifting the dirty beam to this point, and scaling and subtracting off the dirty beam.[7] At each subtraction, the flux and position subtracted are noted, until the map from which the dirty beams have been subtracted (known as the residual map) consists only of noise. At this point, the subtracted fluxes are convolved with a restoring beam selected by the user and added back into the field of noise to give a final "CLEAN map" from which the sidelobes of the dirty beam have been removed. The usual procedure is to make the CLEAN beam of the same dimensions as the central spike in the dirty beam. This is a logical procedure, since the dirty beam is the Fourier transform of the sampling function, and the further

---

[7] In practice, a fraction – typically 5–10% – of the dirty beam is subtracted to improve stability. This fraction is known as the "loop gain".

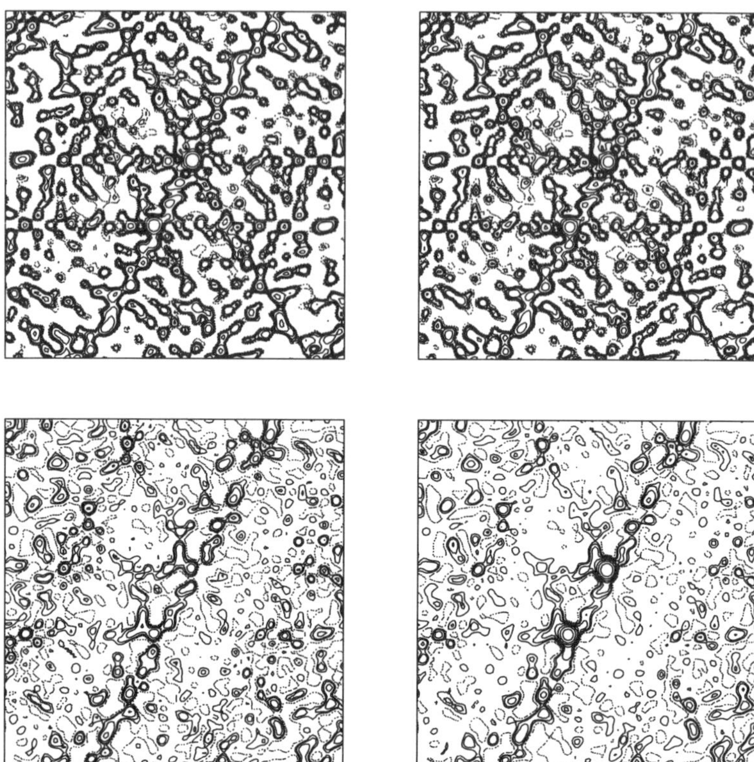

Fig. 9. CLEANing procedure applied to a radio source consisting of two point source components observed with the VLA. All maps are contoured at the same level: (**Top left**) the dirty map; (**Top right**) the residual map after 10 iterations of CLEAN, in which a small amount of flux has been removed at each iteration; (**Bottom left**) the residual map after 100 iterations of CLEAN: note the removal of most of the dirty beam structure; (Bottom right) the CLEAN map after some further CLEANing, formed by the addition of the point source components back into the final residual map. Further CLEANing does not give significant improvement; although the basic source structure is visible, there are some clear artefacts remaining. Their causes and cure are addressed in Sect. 3

out in the u, v plane the sampling function has non-zero values, the higher the resolution we are justified in using. It is possible to use a CLEAN beam smaller than the formal resolution (a process known as super-resolution) at increasing risk of introducing incorrect structure in the map. Figure 9 shows an example of the CLEAN procedure in action.

An important decision in the CLEANing process is the weighting to be applied to the data, as the recorded data are not uniformly distributed across the u, v plane and in practice is usually concentrated towards the center. One option is "natural weighting," in which all data points are treated equally. Statistically, this provides the best signal-to-noise in the final image, but because of the central concentration of the u, v data the sampling function is more centrally concentrated, and its Fourier transform, the dirty beam in the sky plane, is therefore more extended. The result is worse resolution in the final map. An alternative option is "uniform weighting," in which equal weights are applied to each u, v grid, giving increased resolution at the expense of weighting down data at small u and v and degrading the signal-to-noise.

The additional information that has been supplied to the deconvolution problem by CLEAN is the assumption that the sky consists of a finite number of point sources or alternatively that most of the sky is empty. Not surprisingly, therefore, CLEAN works very well for simple sources, but can occasionally fail on very large amorphous sources of low surface brightness.

Maximum Entropy

A second deconvolution method is altogether different in philosophy, and is known as the Maximum Entropy Method, or MEM (e.g. [4]). The starting point is to consider possible images of the sky and prefer those which are more likely. The most preferred image is a completely uniform distribution, which gives the maximum entropy (or minimum information); however, such an image is normally inconsistent with the data. If we consider images which are progressively less likely to be produced by chance, sooner or later we encounter an image which still occurs relatively often but is nevertheless consistent with the data. The process, therefore, corresponds to a joint minimization, including the goodness of fit to the data and the maximum effective smoothness (usually parametrized in forms such as $\Sigma p_i \ln p_i$, where the p_i's are the individual pixel values).

2.2 Sensitivity

The sensitivity (r.m.s. noise) of a wave-regime (meter-centimeter) interferometer is given by

$$S = \frac{\sqrt{2}k_\mathrm{B}T_\mathrm{sys}}{A\eta\sqrt{n_b\Delta\nu t_\mathrm{int}}},$$

where T_sys is the system temperature, A is the area of each antenna, η is the aperture efficiency, n_b is the number of baselines, $\Delta\nu$ is the observing bandwidth, and t_int is the integration time. The units are $\mathrm{Wm^{-2}\,Hz^{-1}}$, but because the Boltzmann constant k_B is uncomfortably small the usual unit is the Jansky, where $1\ \mathrm{Jy} \equiv 10^{-26}\ \mathrm{Wm^{-2}\,Hz^{-1}}$. For extended sources the sensitivity is Janskys per beam area. The sensitivity of many modern interferometers after a few minutes of integration is around $100\,\mu\mathrm{Jy/beam}$.

A couple of terms in the equation deserve comment. We can define "temperatures" in this context in terms of the temperature of a black body which would provide the equivalent received power of radiation at the observing frequency.[8] The noise contribution to a radio interferometer is provided mainly by the receivers (contributing typically 30–50 K), spillover from thermal emission from the ground, and ultimately by the 3 K contribution from the cosmic microwave background. The aperture efficiency η can be varied according to how the aperture is illuminated (i.e. the relative weight given to radiation reaching the feed from different parts of the antenna).

In an optical interferometer, the formulas are somewhat different, because we are collecting photons. There are a number of detailed differences. The first is that it is impossible to clone photons in the same way that electrical signals can be reproduced indefinitely, so every time a beam is split, signal-to-noise is lost. For example, if the array consists of 11 elements, the beam from each element must be split 10 times, losing signal-to-noise, if we wish to interfere all the beams to produce fringes on all baselines. Second, the practical limit is always imposed by the fact that we need a reasonable number of photons in one isoplanactic patch (the area over which the atmospheric corruption is approximately the same) in one atmospheric coherence time (the timescale of variation of atmospheric corruption). Third, optical interferometers typically contain a large number of reflecting elements, with a some light loss at each reflection.

Although it is probably fair to say that the problems have been more difficult than anticipated, major progress is now being made. The use of adaptive optics on individual telescopes means that the wavefront can be corrected over the whole aperture, increasing the coherence patch to the area of the telescope diameter and hence increasing the potential signal. In an optimum site such as that of the Very Large Telescope Interferometer (VLTI) on Paranal mountain in Chile, images of 14th magnitude objects can be made. A list of current optical interferometer systems is given by [10].

3 Dealing with the Atmosphere

Electromagnetic radiation travels to us for billions of years through a nearly perfect vacuum as a nearly perfect plane wave. Unfortunately, the Earth's atmosphere intervenes in the last microsecond to convert a smooth wavefront into a wavefront with phase corrugations, which vary over small spatial scales and on potentially small timescales. At some wavelengths, both the amplitude of the wavefront and the phase are affected.

There are a number of features in the earth's atmosphere that corrupt the wavefronts. At low radio frequencies, the problem is the ionosphere, consisting of a collection of charged particles capable of shifting phases below the

---

[8] In the wave regime, we are in the Rayleigh-Jeans part of the Planck spectrum, and the specific intensity at a given frequency (in units of $\mathrm{Wm}^{-2}\,\mathrm{Hz}^{-1}\,\mathrm{sr}^{-1}$) can be written as $2k_\mathrm{B}T/\lambda^2$.

plasma frequency (typically a few hundred MHz) and which responds to solar activity, producing most disturbance at times of solar maximum. At higher radio frequencies, the problems are mainly due to water vapor, which produces phase rotations on "coherence timescales" of minutes above 10 GHz. Successively shorter coherence times are seen as the frequency increases, until not only the phase but also the amplitude is affected. Observations at higher radio frequencies, such as the ≥ 30 GHz observations typically used to observe the peak of the CMB radiation, are usually done from high mountains above most of the water vapor; some experiments in this region are planned at the South Pole where the water vapor is frozen out.

In the infrared and optical region of the electromagnetic spectrum, the coherence times are typically much shorter. The transverse length scale of phase fluctuations is given by the Fried parameter r_0, and these atmospheric fluctuations are blown across any given line of sight by tropospheric winds. The resulting phase and amplitude fluctuations have characteristic timescales of tens of milliseconds, requiring corrections to be applied on short timescales, which are now being achieved.

3.1 Closure Quantities

Ignoring atmospheric phase fluctuations is not an option, as the response function of the interferometer is directly affected by them and their effect is to wipe out the fringes. The phase on any individual baseline is not a good observable, since atmospheric errors e_1 and e_2 on two telescopes produce a resultant of the form $e_1 - e_2$ in the response function when the signals are correlated.[9]

We can observe, however, that if we have a triangle of three telescopes, and if we measure the interferometer phase response on each baseline, we obtain three phases responses containing the phase error terms $e_1 - e_2$, $e_2 - e_3$ and $e_3 - e_1$. These add to zero, leaving only information on the astronomical structure. Although we have slightly fewer constraints, by modeling these "closure phases" we can obtain constraints on the phases of the response function in the u, v plane and hence deduce the source structure.

A similar quantity can be derived for amplitudes. Since amplitudes are multiplicative, four telescopes are needed in order to use the baseline amplitudes to form the quantity $A_{12}A_{34}/A_{13}A_{24}$. The error on A_{12} is the product of the amplitude error terms, $a_1 a_2$, and once again the errors cancel out.

Closure mapping has been used for many years. It has been successfully used in optical interferometry, even from sites with considerable phase fluctuations such as the situation of the COAST optical interferometer. Despite the less-than-ideal atmospheric conditions, maps have been produced of bright stars such as Capella and Betelgeuse [20]. This gives a clue to the main limitation, however; use of this method requires that the sources be bright. This

---

[9] Correlation involves an operation of the form $< E_1 E_2^* > = A_1 \mathrm{e}^{i\phi_1} A_2 \mathrm{e}^{-i\phi_2}$, so amplitudes multiply and phases add or subtract.

is because what is required is sufficient signal on the source in an atmospheric coherence time to separate atmospheric and source phases.

3.2 Self-calibration

The assumption of closure mapping is that errors in amplitude and phase are separable by telescope and that no additional errors are introduced into the response function of each baseline separately. This is not precisely true, and in practice the biggest problem is usually mismatched bandpasses in the correlator, which give baseline-dependent errors. Great care is usually taken to minimize these, resulting in such errors being a few tenths of a percent.

The assumption of only telescope-based error is used in the procedure known as self-calibration [6], which uses the data, together with a guessed model, to determine the phase and amplitude corrections on each telescope. Suppose we have visibilities V_{ij} on baselines between telescopes i and j, and we call the telescope complex gains g_i and g_j. Suppose also that we have a model whose Fourier transform predicts visibilities V_{ij}^M. Then we write the equation

$$V_{ij} = g_i g_j V_{ij}^M$$

for all i,j, and use a least-squares solution to determine the g_is. The process is repeated by replacing the original data with $V_{ij}/g_i g_j$, mapping and CLEANing the new data to produce a new model, Fourier transforming to give a new set of model visibilities $V_{ij}^{M'}$, and repeating the process until it converges and the g_i corrections are close to 1.

At first sight, this looks an uncomfortably Self-Referential procedure. A model which may or may not look like the sky has been used to correct the visibility data, and we have then used a model derived from the corrected data to determine further corrections. One answer is that the procedure works. Simulated data can be created, phase and amplitude errors added, and the original sky map is recovered by self-calibration. The underlying reason for its success is that the problem is overconstrained, because in any one integration the number of unknowns is proportional to the number of telescopes, n, and the number of constraints is proportional to the larger number of baselines, $n(n-1)/2$. Figure 10 shows an example of self-calibration in action.

There are a number of caveats in practice. The most important is that the source needs to contain a point source bright enough to be visible on all baselines at $> 3\sigma$ in one coherence time – for many interferometer arrays, this means of the order of 10–20 mJy. The reason is that the least-squares fit in which the g_is are computed degenerates into a noisy mess if the V_{ij}s are noisy. Because the corrections change over a coherence time, it is not possible to integrate for a long time in order to build up signal-to-noise to do the correction.

A second caution concerns the order in which the phase and amplitude corrections are built up. Since atmospheric effects on phase are nearly always

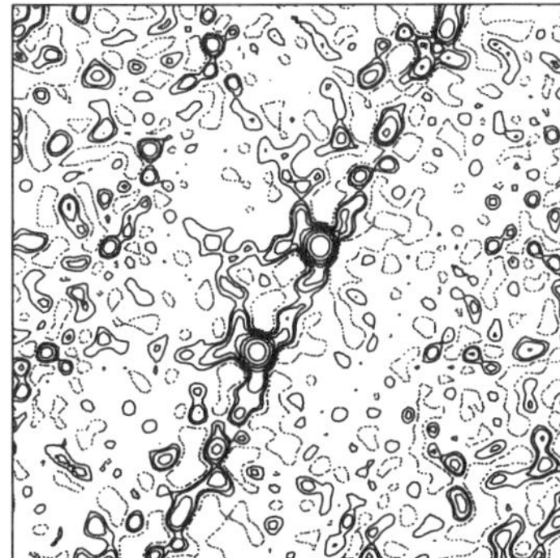 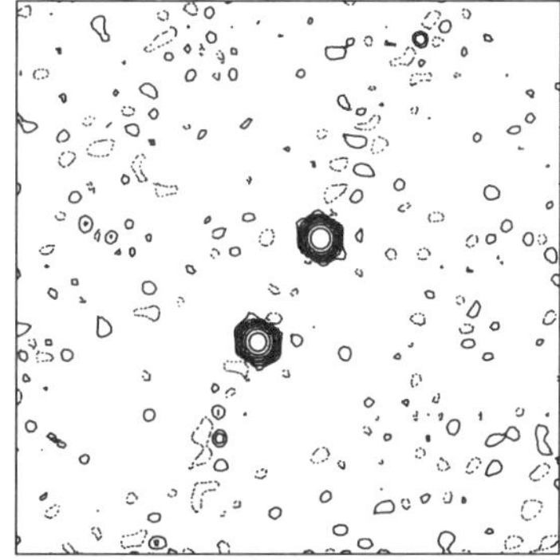

Fig. 10. The same radio source as in Fig. 9. The left panel shows the deconvolved map after CLEAN only. On the right is the same map after one iteration of phase self-calibration and further CLEANing. The maps are contoured at the same level, but the impact of self-calibration in removing artefacts due to phase corruption is obvious

more major than amplitude effects, it is usually better to begin by computing only the phase part of the complex gains, using a time interval shorter than the phase coherence time, and only then to correct the amplitude part. This halves the number of free parameters in the early part of the process and thereby makes it much more stable. Once the phases are determined, the amplitudes can be corrected over a longer timescale as (at least at a few GHz) the amplitude corruption does not change as quickly. Indeed it is often a good idea, particularly in sparsely filled arrays or for relatively weak sources, not to use too short a correction time for the amplitudes. For most VLBI experiments 15–30 minutes is probably safe.

3.3 Phase Calibration

A more direct way to correct the atmospheric phase corruption is to calibrate it directly, rather than sort it out after the event. In this approach, a calibrator a source of known structure is observed periodically. Because the source structure is known, it can be Fourier transformed and removed from the interferometer response function. Any residual phase structure must be atmospheric and can be interpolated and removed from observations of the target.

This approach is attractive because, provided the calibrator source is strong enough to allow good signal-to-noise per baseline per atmospheric coherence time, there is in principle no limit on the brightness of the target source. Phase calibration, otherwise known as "phase referencing," is therefore widely used. There are two caveats: first, the target source must be in the same isoplanactic patch (that is, the target and phase calibrator must be close enough that the atmospheric phase corruption is similar for both), and the switching must be done with in a period not greater than the atmospheric coherence time.

4 Interferometers in Practice

4.1 Very Brief History

Much of the development of connected-element interferometry was done by groups in the UK, USA, the Netherlands and Australia, and a Nobel Prize was awarded to Martin Ryle in Cambridge for development of the technique. This group used arrays of dipoles, and later connected dish antennas together forming the One-Mile Telescope, to discover and investigate radio sources [13]. The earliest large radio source catalogue, 3C, contained the first two known quasars, 3C48 and 3C273. The successor to the One-Mile Telescope, the 5-km Telescope (now known as the Ryle Telescope, [14]) was built soon afterwards.

At the same time, other interferometer arrays were being built, including the Westerbork Synthesis Radio Telescope (WSRT) in the Netherlands [2], the Bologna Cross, arrays at Molonglo in Australia, and Ooty in India. The MER-LIN six-telescope interferometer array, based at Jodrell Bank, pioneered the extension of connected-element interferometers to longer baselines of around 200 km giving higher resolution [7]. Also, in the late 1970s, the Very Large Array (VLA, [17]) was built. This instrument has 27 25-m diameter telescopes, giving high sensitivity, and has a maximum baseline of 36 km. It is arranged in a Y-shape, and the antennas can be moved to four different configurations with baselines shorter by successive factors of 3.

4.2 More Current and Future Interferometer Systems

A brief overview of some long-wavelength interferometer systems follows (see also Fig. 11). Again, the new optical systems, such as the VLTI, COAST, and Keck interferometers are covered by the article by [10].

VLBI

An important subset of interferometer arrays are those operating on very long baselines, known as VLBI (very long baseline interferometer) arrays. The original such collaboration was the European VLBI Network (EVN) whose founding member telescopes included the 76-m Jodrell Bank telescope in the UK, the WSRT in the Netherlands,[10] the 100-m Effelsberg antenna near Bonn,

---

[10] The WSRT is itself a connected-element interferometer, but it is possible to insert phase delays into the arm of each element in such a way as to use the interferometer as a single telescope with effective area of the sum of the individual telescope areas (a process known as "phasing up").

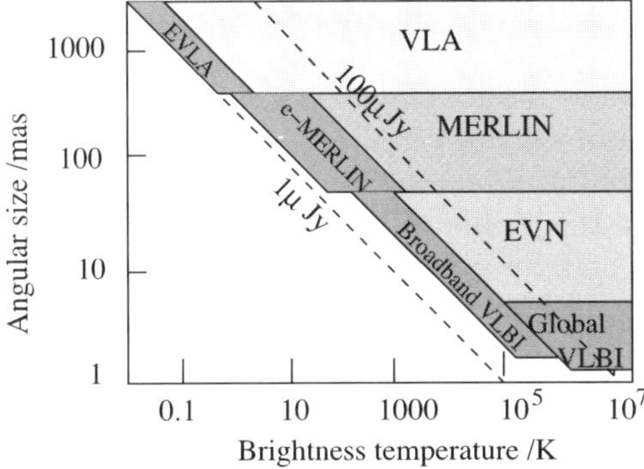

Fig. 11. Range of angular source size (set by the resolution) and brightness temperature of sources, with the northern interferometer array suited to the observation at 5 GHz. The lower limit to each range is set by resolution, and the upper limit by the insensitivity of interferometers to sources larger than that visible to their shortest baselines. The upgrades described in the text are included in the figure. The strength of the sources is plotted as "brightness temperature" which is related to flux density S in Jy by the equation $S = 2k_{\mathrm{B}}T\Omega/\lambda^2$, where Ω is the beam solid angle. Reproduced from the MERLIN website www.jb.man.ac.uk/∼merlin

Germany, the Onsala 25-m telescope in Sweden, and the 32-m Bologna telescope in Italy. Further telescopes now part of the EVN include the Torun telescope in Poland, and further antennas in Italy, Spain, and China. A 10-telescope array, the Very Long Baseline Array (VLBA) was subsequently built in the USA with somewhat higher resolution but lower sensitivity. It is possible to combine the EVN and VLBA into a global VLBI array with high sensitivity (about $10\,\mu$Jy/beam after 12 hours) and a resolution of 1 mas about 50 times higher than that of the Hubble Space Telescope. It is also possible to increase the baseline still further by launching a radio antenna into space. This was done experimentally by use of the TDRSS satellite [9] and later by the dedicated VLBI satellite VSOP/Halca, launched by the Japanese space agency in 1997 and which observed until 2003. A larger mission (VSOP-2) is currently funded, which will feature a larger telescope to be launched in 2012.

VLBI observations are usually not combined into visibilities at the time of observation. Instead, they are often recorded on tape, together with accurate time stamps from a maser clock, and shipped to a central processor for correlating, usually the JIVE facility in the Netherlands or the NRAO correlator in Socorro. Recently, with the availability of increased Internet bandwidths, it has been possible to send the signals to the central correlator online, a process known as e-VLBI. In principle, the only limit to this technique is the speed of the correlator and the availability of the huge Internet bandwidths required.

GMRT

The Giant Meter-wave Radio Telescope [15] is located in India, near the city of Pune. It consists of thirty 45-m antennas, and the resulting large collecting area gives very high sensitivity between 50 MHz and 1420 MHz; the frequency range is limited at the high end by the fact that the antennas are of mesh rather than solid metal. It is particularly suitable for relatively high-resolution, sensitive imaging at low frequencies including the redshifted neutral hydrogen line.

ATCA

The Australia Telescope Compact Array is the most significant long-wavelength interferometer system in the southern hemisphere. It consists of six 22-m antennas over baselines of up to 6 km, with good high-frequency performance.

Fibres and Sensitivity: EVLA and e-MERLIN

Important upgrades to Earth-based interferometers are currently under way. The major programmes involve an increase in sensitivity by using higher bandwidth. Currently signals are transmitted using transmission lines or microwave links with a limited bandwidth, typically a few tens of MHz. Both MERLIN and the VLA are being upgraded within the next few years (becoming, respectively, e-MERLIN and the EVLA) by the addition of optical fiber links between telescopes. These links can carry signals of $\sim$2 GHz of bandwidth, resulting in a factor of 5–10 in improvement in signal-to-noise, and in both cases for a small fraction of the cost (in current dollars) of the original arrays. The increased bandwidth has another benefit when observing a broadband source, namely an improvement in u, v plane coverage. Since the u, v plane is measured in wavelengths, a wide band means that for a given baseline, a range of positions in the u, v plane are measured simultaneously. This means that even an array with a small number of elements, such as MERLIN with six telescopes, can cover essentially the entire u, v plane and thus deliver very high image fidelity (Fig. 12). With complete aperture coverage, radio interferometers will be able to produce the same level of detail as in current direct optical images.

LOFAR and the MWA

At the low-frequency end of the radio spectrum, it is possible to increase sensitivity using the fact that huge collecting areas can be achieved relatively cheaply. This is being exploited by the Low Frequency Array (LOFAR, [11]) being built in the Netherlands and by the Mileura Widefield Array (MWA) currently undergoing demonstrator tests in Western Australia. LOFAR will

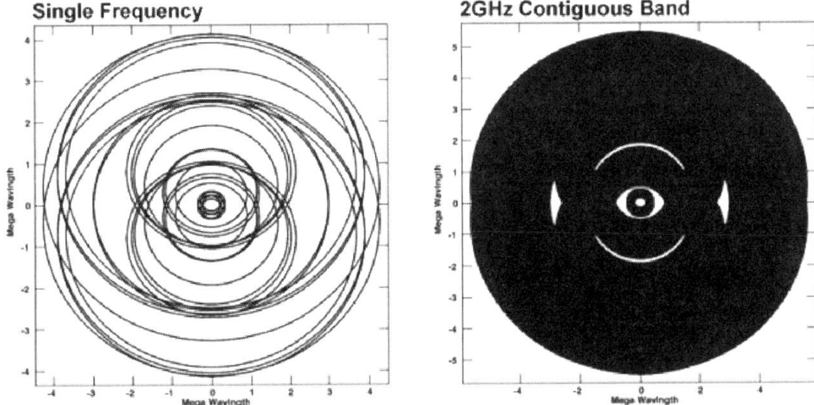

Fig. 12. The difference between u, v coverage of MERLIN, a six-telescope interferometer array, using a single frequency (**left**) and a 2-GHz bandwidth at 5 GHz (**right**). The essentially complete coverage of the u, v plane allows extremely high fidelity imaging of much more complex structure than hitherto possible. The processing power needed is considerably greater, because of possible spectral index gradients across the source, but this too is essentially a solved problem. The image is reproduced from the e-MERLIN science case

consist initially of about 50 elements, each consisting of a field of 50-m diameter filled with dipole antennas. This will give high sensitivity up to 240 MHz and a resolution of a few arcseconds, with a possibility of higher resolution if long baselines are added. At such low frequencies, the field of view is very wide and survey speeds are correspondingly high. It is also possible to manipulate the phases applied to the antennas to form multiple beams on the sky, effectively allowing the telescope to look in a number of directions at once. The major science goals include the detection of the epoch of reionization in redshifted neutral hydrogen, the production of sensitive wide-field surveys, and the monitoring of transient sources which becomes possible with rapid sky coverage.[11] The difficulties include the large amount of processing power required, wide-field problems with removing sidelobes from bright sources at large angles, and more seriously the calibration problems associated with dealing with rapidly varying phase corruption from the ionosphere and (particularly in the LOFAR case) the necessity for very efficient excision of radio-frequency interference.

---

[11] Another widefield telescope at somewhat higher frequencies is the Allen Telescope Array, currently being built at a site in New Mexico, USA. When complete it will consist of three hundred 6-m dishes and have a large survey speed by virtue of the sensitivity from the large number of elements combined with the large primary beam of the small individual elements.

ALMA

At the other end of the frequency range, the Atacama Large Millimetre Array (ALMA) is a new interferometer being built on the Chajnantor plateau in Chile, a dry and high (5000-m) site close to the Bolivian border. The choice of site is due to the effects previously mentioned of the water vapor in the lower atmosphere on the amplitude of radio signals. ALMA will operate between 30 and 950 GHz, in the windows permitted by the small quantity of atmospheric water vapor which remains above it. Its strength lies in the wide variety of molecular astrophysics and chemistry which can be probed at these frequencies, due to the huge number of molecular lines in the millimeter and submillimeter bands; it will be able to probe areas of star formation very sensitively, as well as detect molecular gas on cosmological scales from distant galaxies.

CMB Observations

A specialized niche in interferometry is occupied by experiments which are detecting structure in the Cosmic Microwave Background. These experiments tend to use short baselines, because the power in the CMB fluctuations is observed at significant strength on scales from arcminutes up to degrees and frequencies from a few tens of GHz upwards due to the fact that the CMB radiation has a thermal spectrum with a temperature of 3K and a consequent spectral peak at 300 GHz. Most, like ALMA, operate from high, dry sites such as the Chajnantor plateau itself, the island of Tenerife, or the South Pole.

Square Kilometer Array

After the bandwidth of interferometers has been increased to a maximum ($\Delta\nu/\nu \sim 1$), there is only one option for increasing sensitivity, namely, increasing the collecting area. In the wave regime, the noise level becomes better as A^{-1} rather than the $A^{-1/2}$ obtained by increasing the collecting area of a photon collector so increased telescope acreage is rewarded particularly spectacularly in the low-frequency part of the electromagnetic spectrum.

The idea of a very large interferometric telescope arose 15 years ago (e.g. [19]) and is now under detailed design study, with a view to full-scale construction in either western Australia or South Africa in the middle part of the next decade. There are a number of technological challenges that are being solved by various parts of the community using existing interferometers, including large-area coverage, multiple beams and large-scale data processing and correlation (e.g. LOFAR), transmission by long distances along optical fiber (e.g. e-MERLIN, EVLA), the possibility of high-resolution imaging using very long baselines and real-time correlation (e.g. eVLBI) and the problems of high frequencies (e.g. ALMA).

Although the SKA is some years in the future, the scientific potential is huge. For example, it will be able to see the faint emission of neutral hydrogen at cosmologically significant distances, give precision tests of cosmology and the star formation history of the universe by galaxy counts, resolve stellar disks and protoplanetary systems, search for extraterrestrial intelligence, and provide very sensitive tests of general relativity using pulsar studies.

Acknowledgement

I thank the organizers of the workshop for their excellent organization and hospitality. It was supported by the Jetset EU research training network programme MRTN-CT-2004-005592. Work at Jodrell Bank on interferometry and gravitational lenses is supported by the EU network MRTN-CT-2004-505183 "ANGLES". I thank the Kavli Institute for Theoretical Physics for hospitality during the writing of this review, and Sir Francis Graham-Smith for a careful reading of the manuscript.

References

1. Adgie, R.L., Gent, H., Slee, O.B., Frost, A.B., Palmer, H.P., Rowson, B.: Nature **208**, 275 (1965)
2. Allen, R.J., Hamaker, J.P., Wellington, K.J.: A&A **31**, 1 (1974)
3. Born, M., Wolf, E.: Principles of Optics. Cambridge University Press, ISBN: 0-521-63921-2
4. Bryan, R.K., Skilling, J.: MNRAS **191**, 69 (1980)
5. Burke, B., Graham-Smith, F.: Cambridge University Press, ISBN 0-521-00517-5 (2002)
6. Cornwell, T., Wilkinson P.N.: MNRAS **196**, 1067 (1981)
7. Davies, J.G., Anderson, B.A., Morison, I.: Nature **288**, 64 (1980)
8. Hogbom, J.A.: Astron. Astrophys. Suppl. **15**, 417 (1974)
9. Linfield, R.P., et al.: Astrophys. J. **336**, 1105 (1989)
10. Monnier, J.: Rep. Prog. Phys. **66**, 789 (2003)
11. Röttgering, H.: IAU Coll. 199, Shanghai, China, ed. Williams P. et al., Cambridge University Press, p. 281 (2005)
12. Rowson, B.: MNRAS **125**, 177 (1963)
13. Ryle, M.: Nature **194**, 517 (1962)
14. Ryle, M.: Nature **239**, 435 (1972)
15. Swarup, G., Ananthakrishnan, S., Kapahi, V.K., Rao, A.P, Subrahmanya, C.R., Kulkarni, V.K.: Curr. Sci. **60**, 2 (1991)
16. Taylor, G.B., Carilli, C., Perley, R.A.: Astronomical Society of the Pacific Conf. Ser., vol. 180 (1998)
17. Thompson, A.R., Clark, B.G., Wade, C.M., Napier, J.R.: Astrophys. J. Supp. **44**, 151 (1980)
18. Thompson, A.R., Moran, J.M., Swenson, G.W.: ISBN 0-471-25492-4, Wiley eds. (2001)

19. Wilkinson, P.N.: Radio interferometry: Theory, techniques, and applications. In: Proceedings of the 131st IAU Colloquium, Socorro, Astronomical Society of the Pacific, p. 428. (1991)
20. Young, J.S., et al.: Proc. National Astronomy Meeting, Milton Keynes ed., (2004), available from www.mrao.cam.ac.uk/telescopes/coast

(Sub)mm Interferometry Applications in Star Formation Research

H. Beuther[1]

Max-Planck Institute for Astronomy, Königstuhl 17, 69117 Heidelberg, Germany
beuther@mpia.de

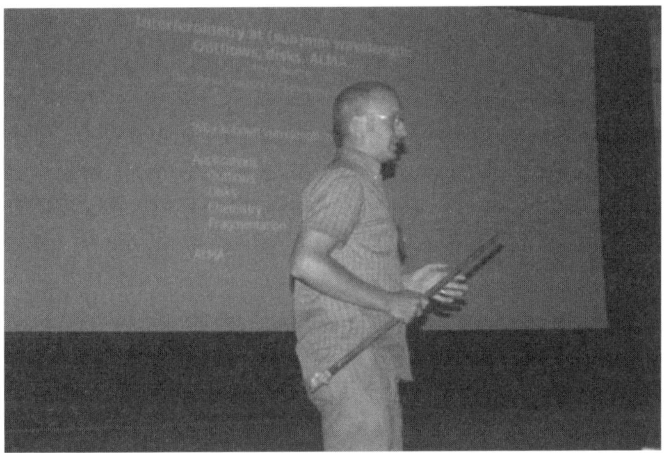

Abstract. Interferometry at (sub)mm wavelengths is one of the most important tools to study the physical and chemical properties of the youngest and most embedded stages of star formation. High spatial resolution is needed to resolve small scale structure e.g. details of accretion disks or molecular outflows and the sub(mm) wavelength bands are important because the young and deeply embedded stages of star formation are characterized by cold gas and dust temperatures of the order of 10K. In this article I will discuss some of the most important applications of sub(mm) interferometry and in addition I will give an overview of the sub(mm) array ALMA.

H. Beuther: *(Sub)mm Interferometry Applications in Star Formation Research*, Lect. Notes Phys. **742**, 219–239 (2008)
DOI 10.1007/978-3-540-68032-1_10   © Springer-Verlag Berlin Heidelberg 2008

1 Introduction

Interferometry at (sub)mm wavelengths is one of the most important tools for studying the physical and chemical properties of the youngest and most embedded stages of star formation. High spatial resolution is needed to resolve small-scale structures, e.g., details of accretion disks, molecular outflow, or the planet-formation processes in low-mass star formation. Furthermore, in high-mass star formation research high spatial resolution is similarly important because the regions are on average more distant, and massive star formation proceeds always in a clustered mode. The (sub)mm wavelengths bands are important because young and deeply embedded stages of star formation are characterized by cold gas and dust temperatures of the order $10\,\mathrm{K}$. At such low temperatures, the peak of the Planck-blackbody curve is at $\lambda_{\mathrm{max}}[\mathrm{mm}] = 2.9/\mathrm{T} = 290\,\mu\mathrm{m}$ (Wien's law), and one can easily observe the whole Rayleigh-Jeans tail at (sub)mm wavelengths. Therefore, the combined requirement of high spatial resolution and (sub)mm wavelengths focus the research interest to (sub)mm interferometry.

While early exploratory interferometry at even longer wavelengths had already started shortly after World War II, real imaging interferometry rather began at cm wavelengths in the 1970s, with instruments like the the Very Large Array (VLA) in the USA and the Westerbork Synthesis Array in the Netherlands (in the 1960s). During the 1980s, technical progress allowed it to go to even shorter wavelengths, and mm facilities like the Plateau de Bure Interferometer (PdBI) in Europe, the Berkeley-Illinois-Maryland-Array (BIMA) and the Owens Valley Radio Observatory (OVRO) in the USA, and the Nobeyama Millimeter Array (NMA) in Japan were built to explore the mm wavelength range at high angular resolution. Only a few years ago, in 2003, the Submillimeter Array (SMA) managed to reach the submm window with interferometric techniques, and the Atacama Large Millimeter Array (ALMA) will cover all earth-accessible (sub)mm wavelengths windows from the next decade onwards.

Each (sub)mm and cm wavelengths window has its own advantages and disadvantages, and one has to select the right regime according to the scientific questions. For example, at cm wavelengths, the free-free emission is strong, one finds important maser transitions (e.g., H_2O, CH_3OH, OH, etc.), a few interesting molecules have easy accessible spectral lines there (e.g., OH, NH_3), and one can observe the 21-cm line of neutral hydrogen there. Going to the mm regime, the dust continuum emission rises strongly, and one can observe the dense dust and gas cores. Furthermore, the mm regime harbors the low-energy lines of one of the most important astrophysical molecules Carbon Monoxide (CO $J = 1-0,\ 2-1$). In addition to CO, many more complex molecules have important transitions in the mm regime, and one can trace various physical and chemical properties with the large range of molecular line transitions. Stepping to the submm window, the dust and line emission rises even more strongly ($S_{\mathrm{cont}} \propto \nu^{\alpha}$ with $2 < \alpha < 4$, $S_{\mathrm{line}} \propto \nu^5$). Regarding the temperature and density regimes many molecules are sensitive to, the mm regime rather favors spectral lines emitted from the colder gas ($10 < T < 50\,\mathrm{K}$), whereas one finds in the submm regime mainly lines from warmer and denser gas ($T > 50\,\mathrm{K}$). However, this differentiation is only on average valid, since one finds extremely highly excited lines already at cm wavelengths (e.g., high (J, K) NH_3 inversion lines), and other low-level lines can be observed in the submm regime (e.g., some CH_3OH lines).

The scientific questions possible to target with (sub)mm interferometry in star-formation research cover nearly all physical and chemical processes one may think of. Here, I will present some of the most important applications in the framework of this European network: obviously molecular outflows, then the closely related accretion disks, furthermore, the question of fragmentation and early sub-structure formation, as well as astrochemical applications. Last but not least, I will give an outlook for the future (sub)mm array ALMA.

2 Molecular Outflow

2.1 Low-mass Outflows

Shocks associated with molecular outflows and jets had first been detected in the 1950s [32, 34]. Later in the 1970s, the first non-Gaussian line-wing emission from molecular CO was detected [44]. While molecular outflows were originally not predicted by theory, after their observational detection it was obvious that they are a crucial ingredient to removing angular momentum during star formation. Today, it is well established that star-forming regions of all masses and luminosities drive molecular outflows; however, it is still debatable whether the outflow driving mechanisms are always the same (e.g., [2]). In low-mass star formation, the current paradigm includes the formation of a centrifugally supported accretion disk and associated outflows/jets driven by magneto-centrifugal acceleration (e.g., [54, 55]).

The textbook example of a low-mass outflow is the one observed toward the class 0 source HH 211. The region was observed in CO(1–0) with the PdBI covering both outflow lobes in a large mosaic [30]. They found a collimated jet-like component at velocities ¿ 10 km/s with respect to the v_{lsr}, and a less collimated, spatially broader component at lower velocities. This morphology was interpreted in the framework of jet entrainment where a collimated high-velocity jet entrains additional gas from the surrounding envelope. More recent follow-up observations in several transitions of SiO and higher excited CO lines revealed the jet-like nature of this outflow in even greater detail (Fig. 1, [36, 52]). Corresponding SiO knots are closer to the driving source in more highly excited lines, indicating higher temperatures at the knot shock fronts. Furthermore, SiO(8–7)/(5–4) line ratios decrease with distance from the driving source, which is indicative of a density gradient in the surround gas core. Last but not least, it was found that the SiO emission exhibits always higher velocities than CO at the same projected position of the molecular outflow (Fig. 1, [52]). This is interpreted again in the jet entrainment picture, where the SiO emission is closer to the axis of the primary beam and hence at higher velocities, whereas the CO emission traces entrained gas further outside. However, one has to keep in mind that even such jet-like molecular components are not the primary jet ejected by the protostar-disk system but that even the observed jet-like molecular gas is already entrained gas. Observed optical jets show far higher velocities of the order a few 100 km/s [49]. Different jet-entrainment processes have been proposed, e.g., entrainment via bow-shocks at the head of the jet, turbulent entrainment via Kelvin-Helmholtz instabilities or wide-angle winds, for more details see a recent discussion of the various entrainment possibilities [2].

While the jet entrainment scenario is relatively well established today for low-mass star formation, other interesting topics are under discussion. For example, optical slit spectroscopy observations toward DG Tau found rotation of the optical jet around its outflow axis [3], and complementary mm inter-

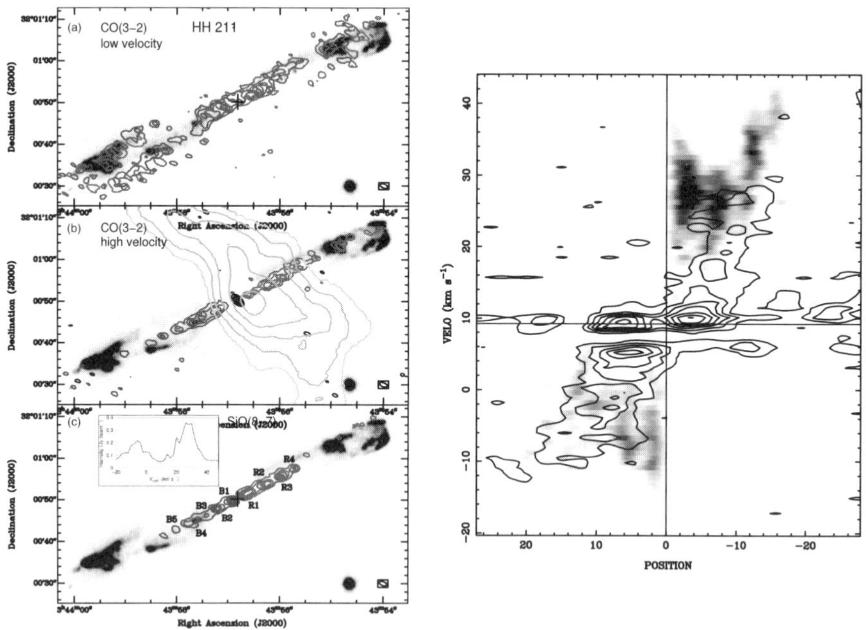

Fig. 1. The **left** panel presents the recent SMA observations of the molecular jet observed at submm wavelengths in the SiO(8–7) and CO(3–2) lines [52]. Blue and red contours show the line emission as labeled in each panel and the gray-scale outlines the shocked H_2 emission. The gray-contours in the middle panel outline NH_3 emission, and the SiO(8–7) spectrum in the bottom panel is averaged over the central $10''$ of the jet. The **right** panel shows the corresponding position–velocity diagram with SiO(8–7) in gray-scale and CO(3–2) in contours

ferometer investigations revealed the corresponding accretion disk around the protostar with the same rotation orientation as the jet [70]. This is strong support for the tight disk-jet connection and that the jets have to be launched from the protostar-disk interface. The main two competing theories for the jet launching are magneto-centrifugal disk winds emanated from an extended inner disk region (e.g., [54]) and the X-wind theory predicting that the jet-launching region is close to the inner disk truncation radius, at the so-called X-point [64]. Reference [25] presented new jet rotation observations that did not actually trace the disk-launching region but allowed an extrapolation of the distance from the protostar along the disk from which the jet is most likely to be launched. Although these observations are not absolutely conclusive, they are consistent with current disk-wind models favoring a more extended disk-region as jet-launching site.

2.2 High-mass Outflows

The picture of molecular outflows from high-mass star-forming regions has changed considerably over the last few years, mainly, based on interferometric observations at mm wavelengths. Since the mid-90s, increasing evidence arose that molecular outflows are ubiquitous phenomena in high-mass star formation as well (e.g., [11, 62]). However, early single-dish mapping studies of high-mass outflows claimed that the collimation degree of massive outflows is lower than known for Low-mass outflows (e.g., [57, 61]). This was interpreted as support for alternative formation scenarios going to higher mass stars (e.g., [18, 67]). In contrast to this claim, single-dish observations with better angular resolution revealed that the observed collimation degrees of massive outflows are consistent with those from their low-mass counterparts as soon as the larger distances and lower spatial resolution of such regions are properly taken into account [11]. As the next step, interferometric observations of some high-mass star-forming regions clearly resolved the previously chaolically appearing single-dish observations into multiple outflow systems from various members of the evolving high-mass protocluster (e.g., [10, 12, 27, 28, 29, 68]). Figure 2 shows the multiple collimated outflows in the massive star-forming region IRAS 05358+3543 as an example. These observations support a picture of massive star formation, where the qualitative physical processes are similar to those in low-mass star formation, i.e., accretion processes mediated by accretion disks and associated outflows/jets. The main differences are (1) the clustered mode of high-mass star formation, which produces observational problems and the need for high-spatial resolution and (2) that the most relevant physical processes are quantitatively much stronger, i.e., higher accretion rates, higher outflow rates and masses, larger luminosities, more UV radiation, etc. On a cautionary note, it should be stressed that the observed molecular outflows rarely exceed regions where the central objects are more massive than $30\,M_\odot$. Therefore, going to higher mass regions, we may encounter more severe changes than so far observed.

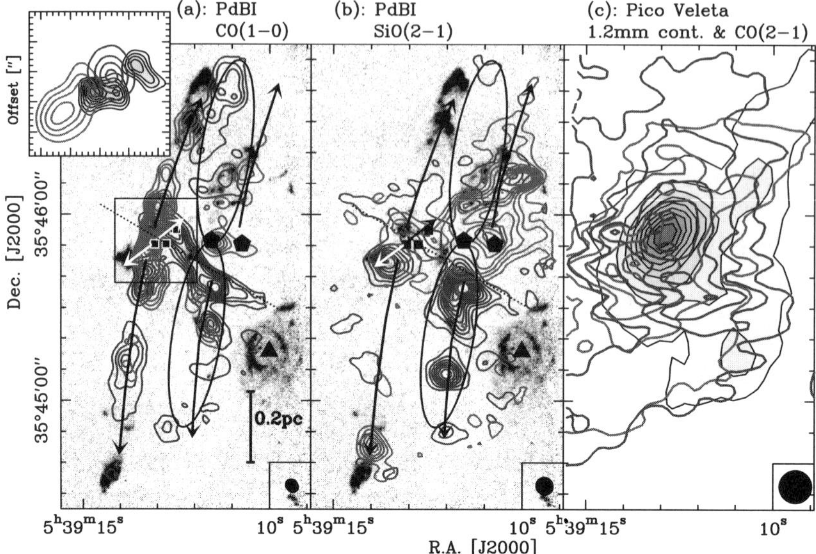

Fig. 2. The left and middle panel present PdBI observations toward the young massive star-forming region IRAS 05358+3543 [10]. The gray-scale shows the shocked H_2 emission and the blue/red contours the CO(1–0) and SiO(2–1) emission, respectively. The arrows and large ellipses guide the eye to the various outflow directions, and the pentagons show additional $H^{13}CO^+(1-0)$ peak positions. The top-left inlay zooms into the marker box with high-velocity CO emission in contours and 3 mm continuum in gray-scale. The right panel presents the corresponding 1.2 mm continuum (gray-scale) and blue/red CO(2–1) emission (contours) observed previously with the IRAM 30m single-dish telescope

3 Accretion Disks

Angular momentum conservation and rotation of molecular cores always predicted that during star formation gas and dust have to assemble in disk-like structures around the central protostar which is forming. However, evidence was lacking for a long time. First observational indications for the existence of accretion disks was based on single-dish multi-wavelengths observations of TTauri stars. For example, analyzing the spectral energy distributions of a sample of low-mass protostars from optical to mm wavelength, the mm dust continuum emission implied such high column densities that in spherical symmetry one would not have detected the sources in the optical or infrared [5]. However, since these sources were optical visible TTauri stars, spherical symmetry was excluded and disk-symmetry appeared the most likely way to solve this problem. Then in the mid-90th, Hubble Space Telescope observations revealed accretion disks in Orion as absorption shadows against the strong background radiation (e.g., [48]).

Since accretion disks are dense, flattened dust and gas condensations, they have on average relatively low temperatures of the order of a few 10 K. Therefore, if one wants to learn more about their structure, dynamics and kinematics, the mm and submm bands are the wavelength regimes of choice. Furthermore, with disk sizes of the order 100 AU at typical Taurus distances of 150 pc, a spatial resolution of $\leq 1''$ is necessary to resolve any substructure. Although disks in massive star formation are likely larger of the order 1000 AU, their average distance is ≥ 2 kpc, and again one needs sub-arcsecond spatial resolution to study such objects. Hence (sub)mm interferometry is the tool of choice.

Observational studies have focused over recent years on several issues in accretion disk studies; among them are (1) dust evolution, the formation of larger grains and the way to planet formation, (2) kinematic studies of disks and their stability,and (3) chemical properties and chemical complexity in accretion disks. While chemistry will be discussed in §5, important results for topics (1) and (2) will be summarized here.

It is known for centuries that the planets of our solar system are located approximately within the ecliptic plane originating from the initial accretion disk, and that their velocity structure can be well explained by the Keplerian laws. Because the formation of planets is so tightly linked with accretion disks, dust evolution and the formation of larger objects is essential for our understanding of the earth history. One way to study dust properties is to observe accretion disks at different wavelength and then investigate their spectral energy distribution. In the Rayleigh-Jeans limit, which is a good approximation at mm wavelength, the mm continuum flux S scales with $\nu^{2+\beta}$, where ν is the frequency and β the dust opacity index ($\tau \propto \nu^{\beta}$). For normal interstellar grains β is ~2 [35]. With increasing grain sizes this index β decreases continuously. Single-dish studies of a sample of accretion disk candidates found lower values of β and already claimed grain growth in these sources [5]. However, with single-dish studies alone, the results are ambiguous because a decreasing spectral index can also be mimicked by an increasing optical depth within the disk. Only spatially resolved interferometric disk studies later could determine the disk density structure, infer their optical depth, and hence better determine that grain growth actually takes place in accretion disks [50, 4]. Incorporating more sophisticated dust and radiative transfer models, one can infer density distributions, temperature distributions, and dust composition in even better detail (e.g., [20, 50, 59, 78]).

An interesting new way to study dust and disk properties is interferometry at mid-infrared wavelength. The Very Large Telescope Interferometer (VLTI) offers the MIDI instrument [45], which allows two-baseline interferometry at wavelengths between 8 and 12 µm covering a strong silicate band. In one such study, it was found that the inner disk regions has a higher degree of crystallization than usually observed in the interstellar medium [73].

As the evolutionary next step, the dust around debris disks can be studied at (sub)mm wavelengths as well. For example, Wilner et al. [75] observed the

1.3 mm dust continuum emission toward the Vega debris disk, and they found two dust condensations between 60 and 100 AU from the exiting star. Such a dust distribution can be explained by the dynamical influence of an unseen planet of a few Jupiter masses in a highly eccentric orbit that traps dust in principal mean motion resonances [75].

To investigate disk kinematic properties, one needs spectral line observations of the disks. The best accessible target for such kind of studies are TTauri stars in the relatively evolved class II stage of protostellar evolution [1]. At this stage, most of the original protostellar envelope has already been dispersed, and the accretion disks remains rather isolated and undisturbed for observational studies (In younger regions, one suffers from confusion between disk and envelope contributions to the observed spectral lines.). A good example of CO emission from a protostellar disk is the study of DM Tau with the PdBI (Fig. 3, [31]). They resolve the velocity structure of the disk finding a velocity distribution along the disk axis, consistent with Keplerian rotation (i.e., the centrifugal force equals the gravitational force $F_{\mathrm{cen}} = \frac{mv^2}{r} = F_{\mathrm{g}} = \frac{Gm_*m}{r^2} \rightarrow v = \sqrt{\frac{Gm_*}{r}}$). This also implies that the observed linewidth increases getting closer to the central star (see Fig. 3). In a more statistical sense, larger samples of accretion disk sources were observed at high spatial resolution, and typical Keplerian velocity structure was found in many objects (e.g., [65]).

However, different disk structures have also been observed. For example, the disk around the prototypical Herbig Ae star AB Aurigae shows a central depression in the cold dust and gas emission and a non-Keplerian velocity profile ($v = r^{-0.4\pm0.01}$) [46, 53]. Possible explanations for such deviations from more typical disk structures could be either the formation of a low-mass companion or planet in the inner disk or a much earlier evolutionary phase, where the Keplerian motion is not yet established. Hence, non-Keplerian velocity structures are expected for very young as well as more evolved accretion disks.

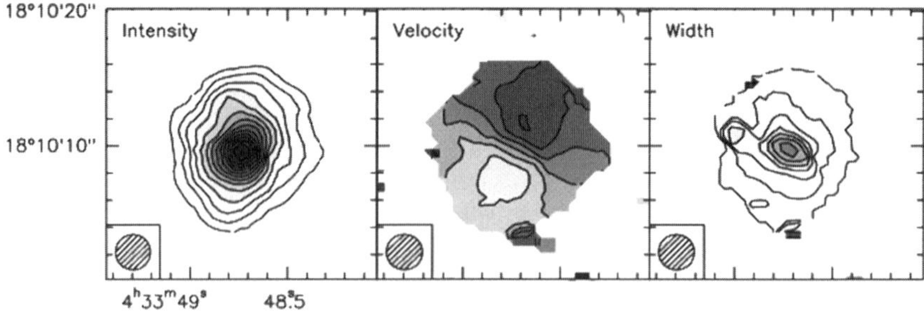

Fig. 3. CO(1–0) emission from the Keplerian protostellar disk DM Tauri [31]. The left panel shows the integrated intensity, the middle panel the velocity field, and the right panel the line width

While the picture for low-mass accretion disks has tremendously evolved over the last decade, progress in massive disk studies has been significantly slower. This is again partly due to the on average larger distances of massive star-forming regions, and due to the clustered mode of massive star formation that complicates the spatial isolation of massive accretion disks. However, an additional problem arises from the fact that massive star formation proceeds much faster than low-mass star formation (of the order 10^5 yr compared to a few times 10^6 years) and that evolved massive stars will relatively quickly dissipate the original accretion disks. Therefore, it is unlikely to find a massive accretion disk in a similarly evolved and exposed state like the low-mass TTauri stars, but we have to search for these objects in the much younger and deeply embedded evolutionary stages. This implies that molecular line as well as dust continuum emission are usually not solely attributable to an accretion disk but that we always have additional contributions from the larger scale core and envelope emission. In many cases, this envelope emission can dominate any line and continuum study. One technical advantage of interferometers is that they filter out the emission on large spatial scales because they are not sensitive to structures θ corresponding to baseline lengths D below the shortest separation of two baselines ($\theta \sim \frac{\lambda}{D}$; for more details see, e.g., [71, 69]). Therefore, we can filter out large parts of the envelope emission; nevertheless, not all emission is usually filtered out, and one still suffers from confusion between genuine disk and surrounding core emission. To overcome thes difficulties, one is interested in observing molecular lines that are usually weak or not found in the envelope but strong in the inner disk regions. Typical hot Core molecules like CH_3CN, $HCOOCH_3$, $C^{34}S$, $HN^{13}C$, or torsionally excited CH_3OH transitions are good candidates for such lines. However, observations over the last few years have shown that in many sources exclusively one or the other tracer only allows rotational studies, whereas other candidate spectral lines are either too weak, optically thick or possibly have not been formed in the chemical network yet. For more details on these problems, see a summary in a recent review [8]. These difficulties complicate statistical studies of massive accretion disks, and our current knowledge is based on only a few selected examples.

One of the best-known massive disk candidates is within the high-mass star-forming region IRAS 20126+4104 [21, 22, 24, 79] (also see Fig. 4). Observations in various spectral lines clearly identify a velocity gradient perpendicular to the molecular outflow/jet, and the velocity structure is consistent with Keplerian rotation [24]. Based on the Keplerian motion, the mass of the central object is estimated to be "only" $7\,M_\odot$, hence a large fraction of the observed luminosity of the order $10^4\,L_\odot$ has to be due to accretion luminosity [42]. [Based on the existing data, it is suggested that this source is still actively accreting and may well form a much more massive star at the END of its evolution [24]]. For most other massive star-forming regions of higher luminosity, no clear signatures of Keplerian rotation have been found so far. The current interpretation of larger scale rotation signatures is that such larger

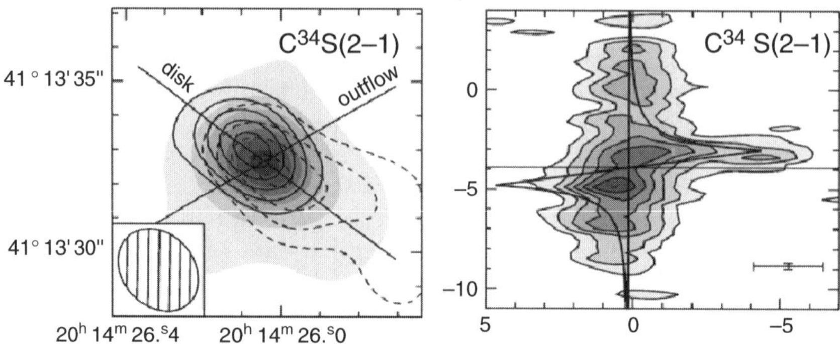

Fig. 4. Keplerian disk emission toward the HMPO IRAS 20126+4104 [24]. The left panel shows in gray-scale the 3 mm continuum emission and in full and dashed contours the blue- and red-shifted $C^{34}S(2-1)$ emission. The outflow and disk axis are indicated. The right panel presents a position velocity diagram (X-axis offset, Y-axis velocity). The full line shows the expected emission of a Keplerian disk

scale entities with sizes of a few 1000 AU rather resemble rotating, probably infalling toroid that may harbor more genuine accretion disk at their so far unresolved centers [23]. The likely most extreme case of rotating and infalling signatures has been observed toward the well-known very luminous ($\sim 10^6\,L_\odot$) ultracompact HII region G10.6. Rotational infalling signatures were observed on larger scales in the molecular gas (NH$_3$ lines [66]), and on much smaller scales similar signatures were identified in the ionized gas (H$_\alpha$ recombination line, [39]). These observations likely trace a larger scale in-spiraling structure that continues from the molecular to the ionized gas. In classical HII region theory that would not have been possible because the formed HII region should expand by the pressure and not allow any further infall and accretion. The observations of G10.6 allow a different scenario, where gravity still dominates the structure of the UCHII region (not yet the pressure), and hence accretion through the evolving HII regions may be possible for some time [40, 41].

4 Fragmentation

Since its first determination in 1955 [58], the rather universal validity of the Initial Mass Function (IMF) has been confirmed in many ways. For a recent compilation see [26]. One of the essential questions in general star formation research is how and at what evolutionary stage the shape of the IMF gets established. Turbulent fragmentation theories predict that the turbulence right at the onset of fragmentation processes already produces core mass functions with the same power law like the IMF and that by applying star formation efficiencies these core mass functions directly convert into the later observed IMF [47, 51]. Contrary to that, other groups argue that the the early star-forming gas clumps fragment down to many cores of more or less the Jeans mass ($\sim 0.5\,M_\odot$) and that the IMF will be formed from that stage via competitive accretion from previously unbound gas [17, 19].

In the low-mass regime, dust continuum studies with bolometer arrays installed on single-dish telescopes revealed the core mass distributions in nearby regions like ρ Ophiuchus and Orion. The power-law distributions $\mathrm{d}N/\mathrm{d}M \propto M^{-\alpha}$ of the pre-stellar dust condensations resemble the Salpeter IMF with typical values of α between 2 and 2.5. While this finding supports the turbulent fragmentation scenario, it only covers a small range of the IMF, and it is necessary to expand such studies to higher-mass regions. Several groups investigated with the same single-dish instruments samples of massive star-forming regions, and the derived cumulative mass distributions of their samples resemble the Salpeter IMF as well [6, 56, 63, 74]. However, since these massive star-forming regions are at about an order of magnitude larger distances, these single-dish studies do not resolve individual protostars but they average over the whole forming protocluster. Hence, these cumulative mass distributions rather resemble protocluster mass functions, and it is far from clear whether they can set any constraints on the formation of the IMF. Therefore, again high spatial resolution is required to resolve individual protostars within the forming protoclusters. So far, only one study resolved and imaged enough sub-sources within a massive star-forming region that a derivation of a core mass function appeared meaningful. The two massive gas condensation in IRAS 19410+3543 were resolved in 1.3 mm continuum observations with the PdBI into 24 sub-sources (Fig. 5, [9]), and the resulting core mass function again has a power-law distribution with $\alpha \sim 2.5$. Although the statistics are still poor, and a few caveats have to be taken into account in this analysis ("Are all dust continuum peaks of protostellar nature"? "Is the assumption of the same temperature for all sub-sources justified?)", it is exciting that all studies so far find core mass distributions resembling the IMF, indicating that turbulent fragmentation processes are really important. However, especially in the high-mass regime, we cannot draw any definitive conclusions yet. Larger source samples observed at high angular resolution are required to base the results on solid grounds. Furthermore, it will be necessary to investigate the temperature structure of the sub-sources in detail, and we need additional spectral line data to determine complementary virial masses and thus establish that the observed sub-sources are really bound and not transient structures.

Another observational curiosity of several high-mass star-forming regions that appear very similar from previous single-dish observations (They are at similar distances, should be at approximately the same evolutionary stage, and have comparable luminosities and other massive star-forming tracers like masers and outflows.) is that they reveal very different sub-structures when observed at high angular resolution. Figure 6 shows a few interferometric example studies of regions, where one would have expected comparable fragmentation results. However, some of the regions show only a single massive

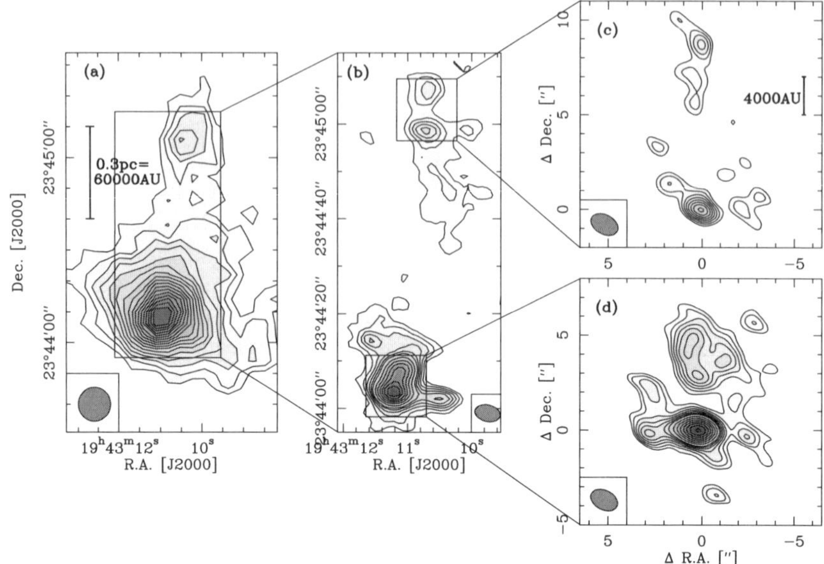

Fig. 5. Single-dish and interferometer mm dust continuum observations toward the young high-mass star-forming region IRAS 19410+2336 [9]. The left panel shows the 1.2 mm map obtained with the IRAM 30 m telescope. The middle and right panel present 3 and 1.3 mm continuum images from the PdBI with increasing spatial resolution. The synthesized beams are shown at the bottom-left of each panel. The resolution difference between the left and right panels are an order of magnitude

central source, even at the highest angular resolution (Fig. 6 top row), whereas other sources exhibit many sub-sources as expected from a star-forming cluster (Fig. 6 bottom row). While the latter is less of a surprise since we know that massive stars almost always form in clusters, finding only a single peak in other regions is more difficult to explain. While all these studies are sensitivity limited and usually are not capable of tracing any condensation below approximately $1\,M_\odot$, one would nevertheless expect to find at least some intermediate-mass objects in the vicinity of the central object. One important additional fact is that toward most regions deeply embedded near-infrared clusters have been found, and hence they are no isolated objects. The existence of an embedded near-infrared protocluster in the vicinity of a massive forming star may be interpreted in the direction that the low-mass stars form first and the high-mass objects later (see also [43]). However, currently we do not understand why various apparently similar massive star-forming regions show such a diverse fragmentation behavior. Is it only an observational bias and the selected sources may in fact be not as similar as we believe, or could there be different paths massive star-forming regions can fragment in the first place?

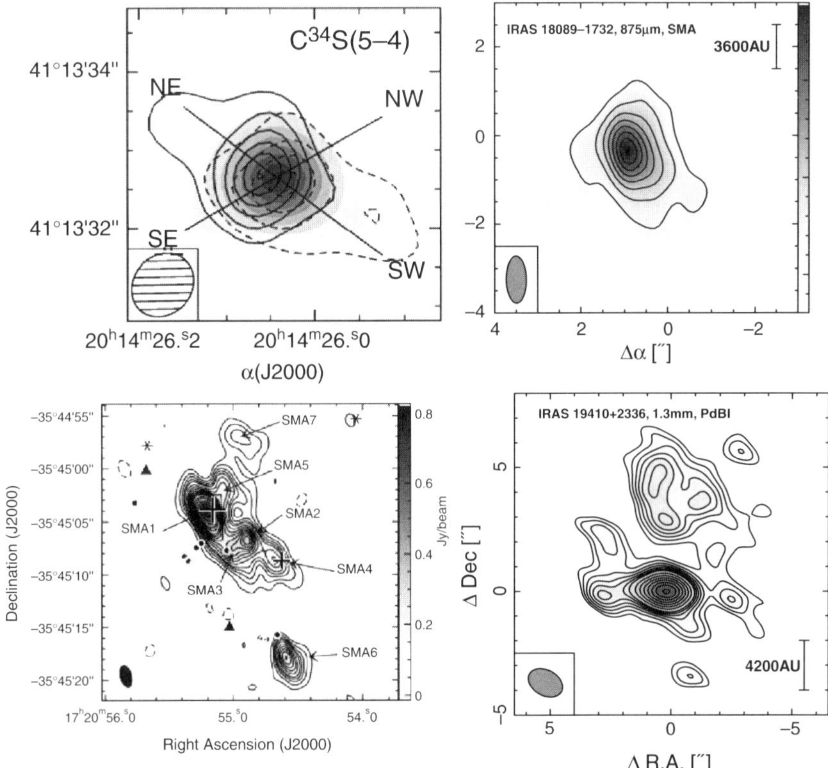

Fig. 6. Interferometric high-spatial-resolution (sub)mm dust continuum images of sources at comparable evolutionary stage and luminosity but with very different fragmentation characteristics. The top-left panel shows IRAS 20126+4104 (grey: 1.3 mm continuum, contours C^{34}S(5-2), 1.7 kpc, [24]), the top-right is IRAS 18089-1732 (875 μm, 3.6 kpc, [15]), the bottom-left is NGC6334I(N) (1.3 mm, 1.7 kpc, [38], the bottom-left is IRAS 19410+2336 (1.3 mm, 2.1 kpc, [9]). The synthesized beams are shown at the bottom-left of each image

5 Chemistry

Astrochemistry has become an increasingly interesting is a steadily rising topic over recent years. Large-scale chemical mapping surveys of molecular clouds (e.g., [7, 72] and line surveys toward high-mass hot molecular cores [33, 60], both conducted with single-dish instruments, have already been conducted for more than two decades. However, studying the chemical small-scale diversity of hot molecular cores has been a difficult task because the spectral bandpasses available for most existing interferometers were mostly too small to cover enough molecular lines from various species in a reasonable amount of time (a noteworthy exception is [16]). This has changed significantly since the advent of the SMA with its two times 2 GHz bandwidth in the upper and lower sideband [37], and similar capabilities to be available very soon at the PdBI and CARMA. As a prime showcase of the spatial chemical complexity of hot molecular cores, a recent investigation of Orion-KL will be presented here.

The Orion-KL region at a distance of 450 pc is the closest and prototypical hot molecular core. With the advent of the SMA, it obviously became one of the early targets of this instrument, and the region was mapped at arcsec spatial resolution in the $865\,\mu m$ submm wavelength band [14]. The $4\,GHz$ bandpass covered more than 150 spectral lines from at least 13 species, 6 isotopologues, and 5 vibrational excited states within a single observation run. From a chemical point of view, the observations included nitrogen- as well as oxygen-bearing species, shock-tracer like SiO and SO, more than 40 CH_3OH lines, and many complex molecules like CH_3CH_2CN or $HCOOCH_3$. While this chemical diversity is interesting in itself, it also allows to study various physical properties in detail, e.g., the SiO data can be used for a kinematic analysis of the molecular outflow(s) in the region, and the large number of CH_3OH lines (ground state as well as torsionally excited $v_t = 1, 2$ lines) are an ideal tool to investigate the temperature structure of the whole region. Furthermore, the data also allowed to derived the first sub-arcsecond resolution dust continuum map of this region [13], clearly differentiating the central power house source I from the hot molecular core, detecting source n for the first time in cold dust emission, and identifying a new protostellar source SMA1, which may be the driver of one of the outflows in this complex star-forming region.

Figure 7 shows a small number of line images from the whole dataset to outline the chemical complexity of the Orion-KL hot molecular core. While SiO traces the collimated outflow structure centered on source I, all other species show a more complex morphology. Nitrogen-bearing species like CH_3CN or CH_3CH_2CN exhibit the kind of horse-show morphology which was know for the hot core already from early NH_3 observations (e.g., [76]). In contrast, the oxygen-bearing molecules, foremost CH_3OH, are relatively speaking weaker there, but they show a strong additional peak toward the south-west, the so-called compact ridge. This southern region is believed to be the interface of one of the molecular outflows (as traced in SiO) with the ambient cloud, and hence the CH_3OH is strongly excited in this shocked inter-face region. Other molecules, e.g., $C^{34}S$, show a mixture of these morphologies and sometimes even an additional peak position toward the north-west associated with the infrared-source IRC6. The main lesson one may take home from this spatial molecular differences is that massive star-forming regions are far more complex than one generally assumes and that one always has to take into account various physical and chemical processes like heating, outflows, shocks or chemical evolution if one wants to model such regions accurately. While single-dish surveys give a good average molecular inventory of such regions, they do not allow to set more detailed constraints. Interferometric imaging surveys are needed to investigate their structures in depth.

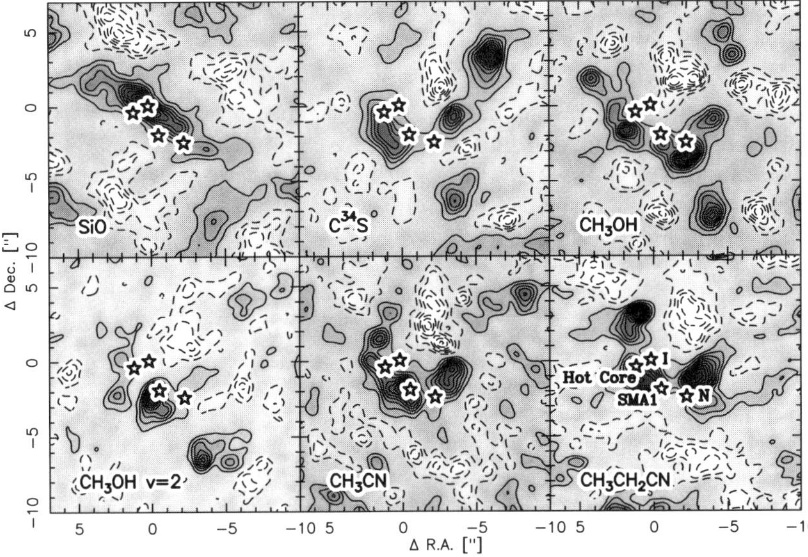

Fig. 7. SMA submm spectral line observations toward Orion-KL [14]. The figure shows representative images from various molecular species labeled in each panel. Full contours show positive emission, dashed contours negative features due to missing short spacings. The stars mark the locations of source I, the hot Core peak position, the newly identified source SMA1, and source n (see bottom-right panel)

6 Outlook and Future

Interferometry at mm wavelength has been conducted for more than one and a half decades by now, but only recently with the advent of the SMA, the submm spectral window started to be accessible for high-spatial-resolution observation. Both wavelength regimes can be considered close siblings where a range of scientific questions can be targeted by both but where also some other questions are unique for each band. For example, hot molecular cores are by definition strong in the submm regime whereas colder and younger regions are more easily studied in the mm band. Therefore, none of the bands is obsolete, but they complement each other well. The main facilities in the northern hemisphere are and will be the SMA, the PdBI, and CARMA, and they will be the sites of many exciting science over the coming decade.

Nevertheless, the upcoming Atacama Large Millimeter Array (ALMA), which will be built within the next few years in Chile by a worldwide collaboration of North-America, Europe, and Japan, will supersede the already existing instruments in many ways by orders of magnitude. At an altitude of ∼5000 m, ALMA will consist of fifty 12 m dishes (25 provided by Europe and 25 by the USA), combined with a few 7 m antennas provided by the Japanese partners to supplement the short spacing information (the Atacama Compact Array, ACA). The baselines range will be between 15 m and 15 km, and the receivers are expected to cover all earth-accessible spectral windows between 30 and 950 GHz. The spectral first-light bands are expected to be around 100, 230, 345, and 690 GHz, and the bandpass will be broad with 8 GHz and dual polarization capability. Combining these specs, the anticipated 350 GHz continuum sensitivity is 1.4 mJy in 1 second integration time. For more detailed descriptions of the specs and the scientific capabilities of ALMA, one may visit http://www.eso.org/projects/alma/.

The requirements on the ALMA capabilities have been mainly driven by three scientific topics:

- Detect CO or CII in a normal galaxy at $z=3$ in less than 24 hours ($\rightarrow$ sensitivity).
- Image protostars and proto-planetary disks around sun-like stars at 150 pc (Taurus) to study kinematics, chemistry, magnetic fields and tidal gaps created by forming planets ($\rightarrow$ spatial resolution down to 10 mas).
- Good imaging quality down to $0.1''$ resolution ($\rightarrow$ spatial resolution and uv-coverage/number of antennas).

In the framework of the topics discussed in this chapter, a few likely highlights should be mentioned. With the highest spatial resolution, we will be able to probe the central launching regions of the outflow and jets and hence observationally much better constrain the physical processes governing these energetic processes of energy and momentum transfer. As outlined above, disk/planet research has been one of the main drivers for ALMA, and we will witness the planet-formation processes within the young accretion disks (e.g., [77]). For massive accretion disk studies, the advent of ALMA will be important in many ways: The high spatial resolution will help to better disentangle the clustered sub-structure of the regions, the even broader bandpasses will allow to sample many spectral lines simultaneously and thus give a broad range of analysis tools within single observations, and the high sensitivity will give access to the weaker, optically thin and less abundant lines, which are likely the best candidate lines to differentiate the genuine disk emission from the surrounding envelopes. Regarding future fragmentation studies, ALMA will allow to study larger source-samples in a consistent manner, which is needed to draw statistically significant constraints. Furthermore, the instantaneous extremely good uv-coverage provides an order of magnitude better image fidelity, which is very important to derive credible core-mass distributions in young massive star-forming regions. Last but not least, spectral imaging surveys will gain significantly more momentum: Again based on the broad bandpasses and high sensitivity, such surveys can be done in a far more consistent manner for large source samples covering various evolutionary stages and extending from the high-mass to the low-mass regime. This will help differentiate in a much better way the chemical evolution depending on mass as well as on time. Furthermore, combining the good uv-coverage of the main ALMA array with the short spacing information from the ACA, we will not have the problem of negative features caused by the missing short spacings as shown in Fig. 7. These missing flux problem so far severely hampered correct column density determinations. Overcoming these difficulties is crucial for accurate abundance determinations which is one of the important parameters in any evolutionary chemical network. In addition, it is expected that spectral imaging surveys will greatly help identify new molecules and thus potentially pave the way to find the first bio-molecules in space.

Although this chapter could only roughly outline the potential of (sub)mm interferometry, it has to be stressed that this field is extremely vibrant and, many exciting questions are waiting to be investigated. The progress over the last two decades has been enormous, and currently we have a number of excellent facilities to study many important astrophysical/astrochemical questions. In addition to this, the future is bright with ALMA on the horizon. This new facility will add another quantum leap to (sub)mm interferometry and probably to nearly all fields of astrophysical research.

H.B. acknowledges financial support by the Emmy-Noether-Program of the Deutsche Forschungsgemeinschaft (DFG, grant BE2578).

References

1. Andre, P., Ward-Thompson, D., Barsony, M.: From Prestellar Cores to Protostars: The Initial Conditions of Star Formation. Protostars and Planets IV. p. 59 (2000)
2. Arce, H., Shepherd, D., Gueth, F., Lee, C.F., Bachiller, R., Rosen, A., Beuther, H.: Molecular Outflows from Low- to High-Mass Star Formation. In: PPV, astro-ph/0603071 (2006)
3. Bacciotti, F., Ray, T.P., Mundt, R., Eislöffel, J., Solf, J.: Hubble space telescope/STIS spectroscopy of the optical outflow from DG Tauri: Indications for rotation in the initial jet channel. Astrophys. J. **576**, 222–231 (2002)
4. Beckwith, S.V.W., Henning, T., Nakagawa, Y.: Dust Properties and Assembly of Large Particles in Protoplanetary Disks. Protostars and Planets IV. p. 533 (2000)
5. Beckwith, S.V.W., Sargent, A.I., Chini, R.S., Guesten, R.: A survey for circumstellar disks around young stellar objects. Astron. J. **99**, 924–945 (1990)
6. Beltrán, M.T., Brand, J., Cesaroni, R., Fontani, F., Pezzuto, S., Testi, L., Molinari, S.: Search for massive protostar candidates in the southern hemisphere. II. Dust continuum emission. Astron. Astrophys. **447**, 221–233 (2006)
7. Bergin, E.A., Langer, W.D.: Chemical evolution in preprotostellar and protostellar cores. Astrophys. J. **486**, 316 (1997)
8. Beuther, H.: Physics and chemistry of hot molecular cores. In: IAU proceedings 237, astro-ph/0610411 (2006)
9. Beuther, H., Schilke, P.: Fragmentation in massive star formation. Science **303**, 1167–1169 (2004)
10. Beuther, H., Schilke, P., Gueth, F., McCaughrean, M., Andersen, M., Sridharan, T.K., Menten, K.M.: IRAS 05358+3543: Multiple outflows at the earliest stages of massive star formation. Astron. Astrophys. **387**, 931–943 (2002)

11. Beuther, H., Schilke, P., Sridharan, T.K., Menten, K.M., Walmsley, C.M., Wyrowski, F.: Massive molecular outflows. Astron. Astrophys. **383**, 892–904 (2002)
12. Beuther, H., Schilke, P., Stanke, T.: Multiple outflows in IRAS 19410+2336. Astron. Astrophys. **408**, 601–610 (2003)
13. Beuther, H., Zhang, Q., Greenhill, L.J., Reid, M.J., Wilner, D., Keto, E., Marrone, D., Ho, P.T.P., Moran, J.M., Rao, R., Shinnaga, H., Liu, S.Y.: Subarcsecond submillimeter continuum observations of Orion KL. Astrophys. J. **616**, L31–L34 (2004)
14. Beuther, H., Zhang, Q., Greenhill, L.J., Reid, M.J., Wilner, D., Keto, E., Shinnaga, H., Ho, P.T.P., Moran, J.M., Liu, S.Y., Chang, C.M.: Line imaging of Orion KL at 865 μm with the submillimeter array. Astrophys. J. **632**, 355–370 (2005)
15. Beuther, H., Zhang, Q., Sridharan, T.K., Chen, Y.: Testing the massive disk scenario for IRAS 18089-1732. Astrophys. J. **628**, 800–810 (2005)
16. Blake, G.A., Mundy, L.G., Carlstrom, J.E., Padin, S., Scott, S.L., Scoville, N.Z., Woody, D.P.: A lambda = 1.3 millimeter aperture synthesis molecular line survey of Orion Kleinmann-Low. Astrophys. J. **472**, L49 (1996)
17. Bonnell, I., Larson, R., Zinncker, H.: The Origin of the Initial Mass Function. In: PPV, astro-ph/0603447 (2006)
18. Bonnell, I.A., Bate, M.R., Zinnecker, H.: On the formation of massive stars. MNRAS **298**, 93–102 (1998)
19. Bonnell, I.A., Vine, S.G., Bate, M.R.: Massive star formation: nurture, not nature. MNRAS **349**, 735–741 (2004)
20. Calvet, N., D'Alessio, P., Hartmann, L., Wilner, D., Walsh, A., Sitko, M.: Evidence for a developing gap in a 10 myr old protoplanetary disk. Astrophys. J. **568**, 1008–1016 (2002)
21. Cesaroni, R., Felli, M., Jenness, T., Neri, R., Olmi, L., Robberto, M., Testi, L., Walmsley, C.M.: Unveiling the disk-jet system in the massive (proto)star IRAS 20126+4104. Astron. Astrophys. **345**, 949–964 (1999)
22. Cesaroni, R., Felli, M., Testi, L., Walmsley, C.M., Olmi, L.: The diskoutflow system around the high-mass (proto)star IRAS 20126+4104. Astron. Astrophys. **325**, 725–744 (1997)
23. Cesaroni, R., Galli, D., Lodato, G., Walmsley, C., Zhang, Q.: Disks around Young O-B (Proto)Stars: Observations and theory. In: PPV, astro-ph/0603093 (2006)
24. Cesaroni, R., Neri, R., Olmi, L., Testi, L., Walmsley, C.M., Hofner, P.: A study of the Keplerian accretion disk and precessing outflow in the massive protostar IRAS 20126+4104. Astron. Astrophys. **434**, 1039–1054 (2005)
25. Coffey, D., Bacciotti, F., Woitas, J., Ray, T.P., Eislöffel, J.: Rotation of jets from young stars: New clues from the Hubble space telescope imaging spectograph. Astrophys. J. **604**, 758 (2004)
26. Corbelli, E., Palla, F., Zinnecker, H. (eds.): The Initial Mass Function 50 Years Later (2005) Astrophysics & Space Science **327**, Springer
27. Davis, C.J., Varricatt, W.P., Todd, S.P., Ramsay Howat, S.K.: Collimated molecular jets from high-mass young stars: IRAS 18151-1208. Astron. Astrophys. **425**, 981–995 (2004)
28. Garay, G., Brooks, K.J., Mardones, D., Norris, R.P.: A triple radio continuum source associated with IRAS 16547-4247: A collimated stellar wind emanating from a massive protostar. Astrophys. J. **587**, 739–747 (2003)

29. Gibb, A.G., Hoare, M.G., Little, L.T., Wright, M.C.H.: A detailed study of G35.2-0.7N: Collimated outflows in a cluster of high-mass young stellar objects. MNRAS **339**, 1011–1024 (2003)
30. Gueth, F., Guilloteau, S.: The jet-driven molecular outflow of HH 211. Astron. Astrophys. **343**, 571–584 (1999)
31. Guilloteau, S., Dutrey, A.: Physical parameters of the Keplerian protoplanetary disk of DM Tauri. Astron. Astrophys. **339**, 467–476 (1998)
32. Haro, G.: Herbig's nebulous objects near NGC 1999. Astrophys. J. **115**, p. 572 (1952)
33. Hatchell, J., Thompson, M.A., Millar, T.J., MacDonald, G.H.: A survey of molecular line emission towards ultracompact HII regions. Astron. Astrophys. Suppl. **133**, 29–49 (1998)
34. Herbig, G.H.: The spectra of two nebulous objects near NGC 1999. Astrophys. J. **113**, pp. 697–699 (1951)
35. Hildebrand, R.H.: The determination of cloud masses and dust characteristics from submillimetre thermal emission. QJRAS **24**, 267 (1983)
36. Hirano, N., Liu, S.Y., Shang, H., Ho, P.T.P., Huang, H.C., Kuan, Y.J., McCaughrean, M.J., Zhang, Q.: SiO J = 5–4 in the HH 211 protostellar jet imaged with the submillimeter array. Astrophys. J. **636**, L141–L144 (2006)
37. Ho, P.T.P., Moran, J.M., Lo, K.Y.: The submillimeter array. Astrophys. J. **616**, L1–L6 (2004)
38. Hunter, T.R., Brogan, C.L., Megeath, S.T., Menten, K.M., Beuther, H., Thorwirth, S.: Millimeter multiplicity in NGC 6334 I and I(N). Astrophys. J. **649**, 888–893 (2006)
39. Keto, E.: An ionized accretion flow in the ultracompact H II region G10.6-0.4. Astrophys. J. **568**, 754–760 (2002)
40. Keto, E.: On the evolution of ultracompact H II regions. Astrophys. J. **580**, 980–986 (2002)
41. Keto, E.: The formation of massive stars by accretion through trapped hypercompact H II regions. Astrophys. J. **599**, 1196–1206 (2003)
42. Krumholz, M., Klein, R., McKee, C.: Radiation-hydrodynamic simulations of collapse and frgamentation in massive protostellar cores. ArXiv Astrophys. e-prints: astro-ph/0609798 (2006)
43. Kumar, M.S.N., Keto, E., Clerkin, E.: The youngest stellar clusters. Clusters associated with massive protostellar candidates. Astron. Astrophys. **449**, 1033–1041 (2006)
44. Kwan, J., Scoville, N.: The nature of the broad molecular line emission at the Kleinmann-Low nebula. Astrophys. J. **210**, L39–L43 (1976)
45. Leinert, C.: Scientific observations with MIDI on the VLTI: present and future. In: Traub W.A. (ed.) New Frontiers in Stellar Interferometry, Proceedings of SPIE Volume 5491, p. 19. The International Society for Optical Engineering, Belling-ham, WA (2004)
46. Lin, S.Y., Ohashi, N., Lim, J., Ho, P.T.P., Fukagawa, M., Tamura, M.: Possible molecular spiral arms in the protoplanetary disk of AB aurigae. Astrophys. J. **645**, 1297–1304 (2006)
47. Mac Low, M., Klessen, R.S.: Control of star formation by supersonic turbulence. Rev. Mod. Phys. **76**, 125–194 (2004)
48. McCaughrean, M.J., O'dell, C.R.: Direct imaging of circumstellar disks in the Orion Nebula. Astron. J. **111**, pp. 1977–1986 (1996)

49. Mundt, R., Buehrke, T., Solf, J., Ray, T.P., Raga, A.C.: Optical jets and outflows in the HL Tauri region. Astron. Astrophys. **232**, 37–61 (1990)
50. Natta, A., Testi, L.: Grain growth in circumstellar disks. In: Johnstone, D., Adams, F.C., Lin, D.N.C., Neufeeld, D.A. Ostriker, E.C. (eds.) ASP Conf. Ser. 323: Star Formation in the Interstellar medium: In honor of David Hollenbach, p. 279 (2004)
51. Padoan, P., Nordlund, A.: The stellar initial mass function from turbulent fragmentation. Astrophys. J. **576**, 870–879 (2002)
52. Palau, A., Ho, P.T.P., Zhang, Q., Estalella, R., Hirano, N., Shang, H., Lee, C.F., Bourke, T.L., Beuther, H., Kuan, Y.J.: Submillimeter emission from the hot molecular jet HH 211. Astrophys. J. **636**, L137–L140 (2006)
53. Piétu, V., Guilloteau, S., Dutrey, A.: Sub-arcsec imaging of the AB Aur molecular disk and envelope at millimeter wavelengths: A non Keplerian disk. Astron. Astrophys. **443**, 945–954 (2005)
54. Pudritz, R., Ouyed, R., Fendt, C., Brandenburg, A.: Disk Winds, Jets and Outflows: Theoretical and Computational Foundations. In: PPV, astroph/ 0603592 (2006)
55. Ray, T., Dougados, R., Bacciotti, F., Eisloeffel, J., Chrysostomou, A.: Toward Resolving the Outflow Engine: An Observational Perspective. In: PPV, astroph/0605597 (2006)
56. Reid, M.A., Wilson, C.D.: High-mass star formation. I. The mass distribution of submillimeter clumps in NGC 7538. Astrophys. J. **625**, 891–905 (2005)
57. Richer, J.S., Shepherd, D.S., Cabrit, S., Bachiller, R., Churchwell, E.: Molecular Outflows from Young Stellar Objects. Protostars and Planets IV. p. 867 (2000)
58. Salpeter, E.E.: The luminosity function and stellar evolution. Astrophys. J. **121**, 161 (1955)
59. Schegerer, A., Wolf, S., Voshchinnikov, N.V., Przygodda, F., Kessler- Silacci, J.E.: Analysis of the dust evolution in the circumstellar disks of T Tauri stars. Astron. Astrophys. **456**, 535–548 (2006)
60. Schilke, P., Groesbeck, T.D., Blake, G.A., Phillips, T.G.: A line survey of orion KL from 325 to 360 GHz. Astrophys. J. Suppl. **108**, 301 (1997)
61. Shepherd, D.S., Churchwell, E.: Bipolar molecular outflows in massive star formation regions. Astrophys. J. **472**, 225 (1996)
62. Shepherd, D.S., Churchwell, E.: High-velocity molecular gas from high-mass star formation regions. Astrophys. J. **457**, 267 (1996)
63. Shirley, Y.L., Evans, N.J., Young, K.E., Knez, C., Jaffe, D.T.: A CS J=5–4 mapping survey toward high-mass star-forming cores associated with water masers. Astrophys. J. Suppl. **149**, 375–403 (2003)
64. Shu, F.H., Najita, J.R., Shang, H., Li, Z.Y.: X-winds Theory and Observations. Protostars and Planets IV. p. 789 (2000)
65. Simon, M., Dutrey, A., Guilloteau, S.: Dynamical masses of T Tauri stars and calibration of pre-main-sequence evolution. Astrophys. J. **545**, 1034–1043 (2000)
66. Sollins, P.K., Zhang, Q., Keto, E., Ho, P.T.P.: Spherical infall in G10.6-0.4: Accretion through an ultracompact H II region. Astrophys. J. **624**, L49–L52 (2005)
67. Stahler, S.W., Palla, F., Ho, P.T.P.: The formation of massive stars. Protostars and Planets IV. p. 327 (2000)
68. Su, Y., Zhang, Q., Lim, J.: Bipolar molecular outflows from high-mass protostars. Astrophys. J. **604**, 258–271 (2004)

69. Taylor, G.B., Carilli, C.L., Perley, R.A. (eds.): Synthesis Imaging in Radio Astronomy II (1999)
70. Testi, L., Bacciotti, F., Sargent, A.I., Ray, T.P., Eislöffel, J.: The kinematic relationship between disk and jet in the DG Tauri system. Astron. Astrophys. **394**, L31–L34 (2002)
71. Thompson, A.R., Moran, J.M., Swenson, Jr., G.W.: Interferometry and Synthesis in Radio Astronomy. 2nd ed. Wiley, New York (2001)
72. Ungerechts, H., Bergin, E.A., Goldsmith, P.F., Irvine, W.M., Schloerb, F.P., Snell, R.L.: Chemical and physical gradients along the OMC-1 Ridge. Astrophys. J. **482**, pp. 245–266 (1997)
73. van Boekel, R., Min, M., Leinert, C., Waters, L.B.F.M., Richichi, A., Chesneau, O., Dominik, C., Jaffe, W., Dutrey, A., Graser, U., Henning, T., de Jong, J., Köhler, R., de Koter, A., Lopez, B., Malbet, F., Morel, S., Paresce, F., Perrin, G., Preibisch, T., Przygodda, F., Schöller, M., Wittkowski, M.: The building blocks of planets within the 'terrestrial' region of protoplanetary disks. Nature **432**, 479–482 (2004)
74. Williams, S.J., Fuller, G.A., Sridharan, T.K.: The circumstellar environments of high-mass protostellar objects. I. Submillimetre continuum emission. Astron. Astrophys. **417**, 115–133 (2004)
75. Wilner, D.J., Holman, M.J., Kuchner, M.J., Ho, P.T.P.: Structure in the dusty debris around vega. Astrophys. J. **569**, L115–L119 (2002)
76. Wilson, T.L., Gaume, R.A., Gensheimer, P., Johnston, K.J.: Kinematics, kinetic temperatures, and column densities of NH3 in the Orion hot core. Astrophys. J. **538**, 665–674 (2000)
77. Wolf, S., D'Angelo, G.: On the observability of giant protoplanets in circumstellar disks. Astrophys. J. **619**, 1114–1122 (2005)
78. Wolf, S., Padgett, D.L., Stapelfeldt, K.R.: The circumstellar disk of the butterfly star in taurus. Astrophys. J. **588**, 373–386 (2003)
79. Zhang, Q., Hunter, T.R., Sridharan, T.K.: A rotating disk around a high-mass young star. Astrophys. J. **505**, L151–L154 (1998)

Observing Young Stellar Objects with Very Large Telescope Interferometer

F. Malbet

Laboratoire d'Astrophysique de Grenoble, Unité mixte de recherche 5571
CNRS/Université J. Fourier, BP 53, F-38041 Grenoble cedex 9, France
`Fabien.Malbet@obs.ujf-grenoble.fr`

Abstract. One of the techniques used to probe the formation of stars and planets, is optical long baseline interferometry. This technique offers milli-arcsecond spatial resolution. The Very Large Telescope Interferometer (VLTI) has been built to offer this new observational facility to any astronomer. In this review, I detail the specificity of high angular resolution observations in the field of stellar and planetary formation. I present a short introduction to interferometry, followed by an extensive presentation of the VLTI. Finally I give some advice to those who would like to use the instruments of the VLTI.

F. Malbet: *Observing Young Stellar Objects with Very Large Telescope Interferometer*, Lect. Notes Phys. **742**, 241–255 (2008)
DOI 10.1007/978-3-540-68032-1_11 © Springer-Verlag Berlin Heidelberg 2008

It has always been a human dream to possess extraordinary powers. In two fairy tales, *Sechse kommen durch die ganze Welt*[1] and *Die sechs Diener*[2], the brothers J. and W. Grimm wrote stories of a group of men who have special powers and who go through life in the world. One of them has a sharp vision which allows him to see a *ring sticking on a pointed stone in the depth of the Red Sea*. In astronomy, and especially in the field of stars and planet formation, astronomers have the same dreams: to probe the very close environment of young stars where they believe planets form and jets are launched. The Hubble Space Telescope and the 10-m class telescopes equipped with adaptive optics allow the astronomical community to take an important step forward. However, reaching a spatial resolution corresponding to less than an Earth orbit at the closest star forming region requires even sharper instruments. Optical long-baseline interferometry allows astronomers to probe these systems at such high angular resolution.

1 Optical Interferometric Observations of Young Stellar Objects at AU Scales

In the scenario of the formation of stars, one important stage is when the star is formed but is still accreting matter through an equatorial disk. To probe the physical phenomena that occur at the AU scale (where temperature ranges between 300 K and 3000 K), one needs to observe in the near and mid-infrared wavelengths but with milli-arcsecond scale. This is the stage when planets are believed to form and is the type of subject that requires infrared interferometry.

Long baseline optical interferometry involves the mixing of light from an astronomical source. The light is collected by several independent telescopes, located several tens or even hundreds of meters apart. The light beams once collected, are overlapped, and form interference patterns. The interference patterns form when the optical path difference between the different arms of the interferometer—taking into account paths from the source down to the detector—is smaller than the coherence length of the incident wave (typically of the order of several microns). The pattern consists of fringes, with a succession between faint intensities (destructive interferences) and bright intensities (constructive interferences). By measuring the contrast of these fringes, i.e., the normalized flux difference between the darkest and brightest regions, one can recover some information about the morphology of the observed astronomical source. Figure 1 illustrates this phenomenon.

Interferometric observations of young stellar objects (YSOs) were performed and are still performed at six different facilities on seven different instruments (see Table 1). We can classify these observations into three different categories:

---

[1] How Six Men Got on in the World.
[2] The Six Servants.

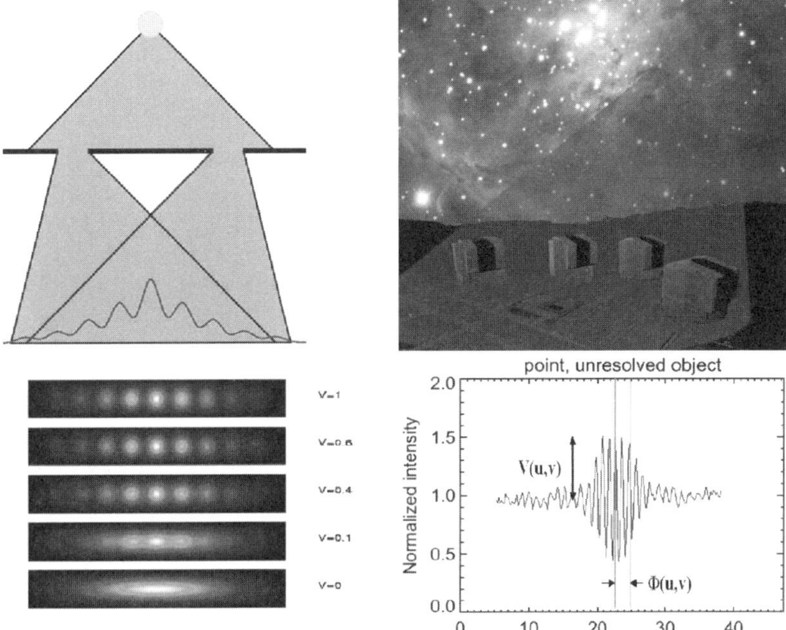

Fig. 1. Principle of interferometry. *(Upper panels)* the Young's slit experiment (**left**) compared to optical interferometry (**right**): In both cases the light travels from a source to plane where the incoming wavefront is split. The telescope's apertures play the same role as the slits. The difference lies in the propagation of light after the sampling plane. In the case of optical interferometry, the instrument controls the propagation of light down to the detectors. At the detector plane, the light beams coming from the two apertures are overlapped; *(Lower panels)*: interference fringes whose contrast changes with the morphology of the source. Left panel shows fringes whose contrast varies from 0 to 1. Right panel displays actual stellar fringes but scanned along the optical path. The measure of the complex visibilities corresponds to the amplitude of the fringes for the visibility amplitude and the position of the fringes in wavelength units for the visibility phase

- **Small-aperture interferometers**: PTI, IOTA, and ISI were the first facilities to be operational for YSO observations in the late 1990s (see Fig. 2). They have mainly provided the capability of measuring visibility amplitudes and lately closure phases. CHARA is the latest, built with apertures of 1 m diameter. The instruments are mainly accessible through team collaboration.

- **Large-aperture interferometers**: KI, VLTI and are facilities with apertures larger than 8 m; soon LBT will offer the same features. The instruments are wide open to the astronomical community through general calls for proposals. Lately, these facilities have significantly increased the number of young objects observed.

Table 1. Interferometers involved in YSO science

Facility	Instrument	Wavelength (microns)	Numbers of apertures	Aperture diameter (m)	Baseline (m)
PTI	V^2	H, K	3	0.4	80–110
IOTA	V^2, CP	H, K	3	0.4	5–38
ISI	heterodyne	11	2 (3)	1.65	4–70
KI	V^2, nulling	K	2	10	80
VLTI/AMBER	V^2, CP (imaging)	1–2.5 /spectral	3 (8)	8.2/1.8	40–130 /8–200
VLTI/MIDI	V^2 (/CP)	8–13 /spectral	2 (4)	8.2/1.8	40–130 /8–200
CHARA	V^2, CP (imaging)	1–2.5 /spectral	2/4 (6)	1	50–350
LBT	imaging, nulling	1–10	2	8.4	6–23

V^2: visibility measurement; CP: closure phase.
Acronyms. PTI: *Palomar Testbed Interferometer*; IOTA: *Infrared and Optical Telescope Array*; ISI: *Infrared Spatial Interferometer*; KI: *Keck Interferometer*; VLTI: *Very Large Telescope Interferometer*; CHARA: *Center for High Angular Resolution Array*; LBT: *Large Binocular Telescope* (not yet operational).

- **Instruments with spectral resolution:** CHARA, MIDI, and AMBER provide spectral resolution from a few hundred up to 10,000, whereas other instruments mainly provide broadband observations. The spectral resolution allows the separation of the various phenomena occurring in the environment of young stars.

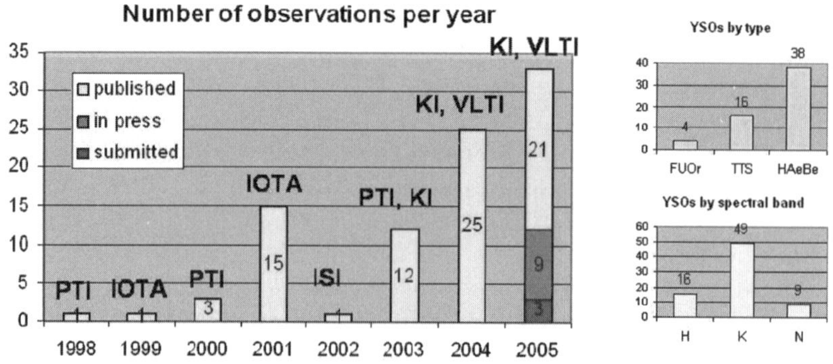

Fig. 2. Young stellar objects observed by interferometry in the period 1998–2005. from [4]

Figure 2 shows that with improved facilities there is an increase in the number of published results. We are in an epoch where interferometry is opening itself to fainter targets allowing YSOs, including the brightest Herbig Ae/Be stars, the fainter T Tauri stars, and the few FU Orionis stars to be observed at milli-arcsecond scales. While these observations are made mainly in the near-infrared wavelength domain, some are made in the mid-infrared. The readers should refer to the review [4] for examples of what can be performed with modern optical interferometric techniques in the field of YSOs.

2 The Very Large Telescope Interferometer: A Unique Site in the World

The *Very Large Telescope Interferometer* (VLTI), is one of the largest facilities available to the astronomical community. Indeed, it is the only one freely available to the European community.

2.1 Brief History of the VLTI

In the early 1980s, interferometry was already an integral part of the VLT project. The VLT aimed to give the largest telescope in the world to the member states of the *European Southern Observatory* (ESO). The science specifications led to the decision to build a telescope of 16-m diameter resulting in a 200-m^2 collecting area. However, it soon appeared that a monolithic 16-m telescope would not be possible and that building a set of four 8-m telescopes would better fulfill all the initial scientific requirements. The initial configuration for these four telescopes was a linear and regular array, but quickly the interferometrist involved in the project showed that the Fourier plane coverage would be better filled with a configuration where the 4 telescopes are at the summit of an irregular trapezium. Everything that was required to make the four telescopes an interferometer, such as delay lines, relay optics, and tunnel, was part of the initial plan.

In the early 1990s, the engineering of the general layout of the site began and the basics of the VLTI infrastructure was built (e.g., the tunnel on the platform). However, in 1993 the council stalled the VLTI project because of financial problems. Luckily, the implementation of the infrastructure continued so that later development of interferometry was not jeopardized.

In 1996, the *Max-Planck Gesellschaft* (MPG) in Germany, the *Centre National de la Recherche Scientifique* (CNRS), and ESO signed a tri-partite agreement in order to resume development of the VLTI by injecting additional resources. A descoping of the VLTI plan was also decided in order to start, as soon as possible, the European interferometry. In 1997, the two instruments MIDI and AMBER were proposed by teams external to ESO and approved by the community. In the meantime, a commissioning instrument called VINCI was built in order to measure the first fringes and to commission the VLTI infrastructure. Additional partners joined the tripartite agreement, funding additional delay lines and a fringe tracker. The implementation of the optical elements required, in addition to the telescopes, started in 2000. In March 2001, ESO obtained the first fringes with VINCI. The first light of MIDI was in December 2002 and the first light of AMBER in March 2004.

2.2 The VLTI Infrastructure

Figure 3 displays a sketch of the VLTI layout. The source of scientific interest lies in the upper right corner and emits its radiation through the propagation of a spherical electromagnetic wave. [At the surface of the Earth, where the telescopes are positioned, the incoming wavefront is flat even if it appears slanted due to the position of the source in the sky].

First, the VLTI samples the wavefront, since only a fraction of it is kept. This fraction corresponds to several (two in the case of Fig. 3) circular disks having the exact telescope diameter and whose separation is called the projected baseline. The baseline of the interferometer is the distance between two telescopes, but, as can be seen in Fig. 3, the separation between the two sam-

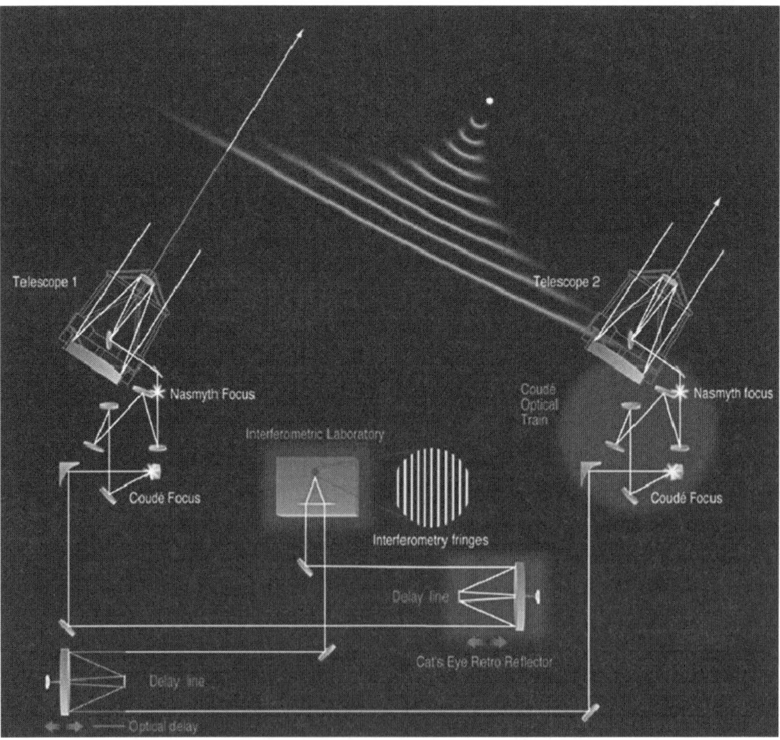

Fig. 3. Sketch of the VLTI optical lay-out (see text for details)

pled parts of the wavefront has to take into account the position of the source in the sky. If at zenith, the projected baseline will exactly match the actual telescope baseline. The projected baseline is important not only in length but also in orientation. As the Earth rotates, or, as the star is moving across the night sky, the projected baseline changes both in length and in angle. The best way to understand the projected baseline is to take the point of view of the celestial source, looking at the interferometer site. When the trajectory of the projected baseline is plotted on a projected sky map, it follows the arc of an ellipse [2]. The position of the projected baseline at a given time scaled to the wavelength of observation is called the (u, v) position. [This is why, it is said that the telescopes of an interferometer sample the (u, v) plane]. This plane is important because the theory of interferometry [2] shows that the coherence measured by the interferometer is the absolute value of the Fourier transform of the intensity distribution of the source onto the sky at these (u, v) points. The more points you get, the better is the (u, v) plane coverage and therefore the better is the quality of the observations.

As can be seen in Fig. 3, the incoming wavefront does not reach the two telescopes at the same time. There is a delay for the beam propagating at the left part of the figure. In order to obtain interferences of the electromagnetic fields sampled by the two different telescopes, the distance traveled must be almost exactly the same (at the scale of a wavelength). This is always the case when the source passes the meridian, but unfortunately it does not stay long! If the two telescopes were on the same pointing mount, like on a large antenna, there would be no problems, but in the case of the VLTI the delay must be compensated on one arm before the meridian or in the other arm after the meridian. This is why there are optical elements called delay lines, which are drawn at the bottom of the sketch. These delay lines act like optical trombones. The light is sent along a direction, toward an optical delay line carriage which reflects the light back exactly in the opposite direction. Therefore, the optical path of the light can be adjusted by moving the carriage. Since the optical path length is the double of the delay line distance, to compensate sidereal optical path delay due to the elevation of the source in the sky, one has to have optical delay lines lengths that reach half the maximum separation of the telescopes. In the case of the VLTI, the maximum optical delay that can be compensated is 120 m.

The beams arriving at the telescopes have to be routed optically toward the delay lines and then toward the instruments that lie in the interferometric lab placed at the center of the VLTI platform. In the instrument, the beams sampled by the different telescopes are mixed together to form interference fringes, possible only if some coherence exists between the beams [2].

In the case of the VLTI, the astronomer has access to a whole facility which comprises

- four 8.2-m telescopes, called *Unit Telescopes* (UTs), all equipped with low-order adaptive optics and which sample the (u, v) plane with 6 baselines ranging between 47 and 130 m;
- four 1.8 m telescopes, called *Auxiliary Telescopes* (ATs), which are relocatable on 30 different stations on the VLTI platform and sample the (u, v) plane with baselines ranging between 8 and 200 m;
- six delay lines;
- a fringe tracker called FINITO which can servo the interferometer on 3 pairs of fringes;
- two instruments: MIDI and AMBER (see next section);
- optical facilities to manage the control of the interferometer;
- a control software [10] which supervises all operations.

In the near future, in addition there will be the PRIMA facility [8] which will allow advanced observations.

2.3 The Instruments: MIDI and AMBER

As mentioned in the previous section, the VLTI hosts two instruments, MIDI and AMBER, whose characteristics are reported in Table 2 (see also Figures 4 and 5).

MIDI is the instrument which operates at mid infrared wavelengths between 8 and 13 μm. The instrument combines two beams in a co-axial scheme [3] using a beam splitter. The fringes are coded temporally using small internal delay lines scanning the optical path delays. They are spectrally dispersed either by a prism or a grism. Since it is operating at wavelengths where the main source of noise is the background thermal emission, MIDI includes a chopping operating mode, and the fringes are obtained as the subtraction of the two outputs from the beam splitter. More details on how to use MIDI is given in [11], and details about the data reduction are given in [1].

AMBER operates at near infrared wavelengths between 1 and 2.5 μm. This instrument can combine three beams in a multi-axial scheme [7] using the overlap of all beams in a non-redundant way. The fringes are, therefore, coded spatially and a thorough internal calibration is required. The fringes are spectrally dispersed by a prism and two gratings leading to 3 spectral resolutions. The main limitation for the moment is vibrations occurring at the telescope level that blur the fringes and limit the capacity of AMBER. The observables of AMBER is like that of MIDI—visibility amplitudes at different wavelengths, but three at once and, in addition, the closure phase information. More details on how to use AMBER can be found in [11], and details about the data reduction are given in [9].

2.4 Status of the VLTI

As the end of 2006, the status of the VLTI for the potential VLTI observers is

Table 2. MIDI and AMBER characteristics

	MIDI	AMBER
Consortium	D/F/NL	F/D/I
PI	C. Leinert (Heidelberg)	R. Petrov (Nice)
UT first fringes	Dec. 2002	Mar. 2004
Beams combined	2	3
Wavelength of operation	8–13 µm	1–2.5 µm
Spectral resolution	30 (prism), 230 (grism)	35 (LR); 1500 (MR); 12000 (HR)
Limiting magnitude UT	N = 4 (current) N = 9 (FT)	K = 7 (current) K = 11 (FT) K = 20 (PRIMA)
Limiting magnitude AT	N = 0.25 (current); N = 5–8 (FT)	K = 6 (current) K = 8 (FT)
Visibility accuracy	< 20% (current); 1–5% (goal)	3% (current); < 1% (goal)
Airy disk FOV	0.26" (UT); 1.14" (AT)	60 mas (UT); 250 mas (AT)
Diffraction limit, 200 m	10 mas	1 mas (J), 2 mas (K)

– all UTs are operational with full adaptive optics system, all six baselines and all four phase closures used for science;
– four ATs are in operation on several fixed configurations;
– four Delay Lines are in operation for UTs and two delay lines for ATs;
– mIDI has been offered since April 2004 on UTs and October 2005 on ATs, and AMBER on UTs since October 2005 and on ATs since April 2007.
– 38+40 nights of VLTI science operations in P76 (Oct 05–Mar 06), showing that the portion of night dedicated to interferometry operation is not tiny;
– six operations astronomers, three fellows, several telescope and instrument operators run the VLTI;
– more than 60 refereed papers have been published so far, showing that VLTI is now operating as a mainstream facility.

Operation of the VLTI at Paranal has been completely embedded in the VLT operation and is considered an additional telescope. The shift leader is responsible on a 24 h-basis on call both for the VLT and the VLTI. Two day astronomers are dedicated for all UTs calibration and one for the VLTI daily calibration plan. One engineer per UT is responsible for the operation and one for the VLTI coordination of daytime access. During the night, there is one night astronomer per UT and one for the VLTI and same for telescope/instrument operators.

The performance of the instruments (sensitivity, stability,...) is currently limited by the actual performance of the infrastructure. In December 2004,

Fig. 4. MIDI instrument

Fig. 5. AMBER instrument

ESO reviewed the existing VLTI infrastructure to establish its performance, operability, maintainability, and capability to host PRIMA. Although scientifically successful, the VLTI infrastructure required certain subsystems to be brought to robust operation and some others to be fully commissioned. Therefore, the Paranal team accepted the infrastructure and launched the Interferometry Task Force (ITF) in April 2005 to seek the understanding necessary to make improvements to the system. To streamline the operations and allow further deployments—in particular, before PRIMA deployment and second generation instrumentation—one needed to know that VLTI can fringe track. Although it has been shown that fringe tracking on the ATs, than on the UTs, was successful, ESO needs time to implement all modifications.

3 Some Hints for Observing with the VLTI: A Guide for Future Observers

In this section, I would like to address some practical issues that arise when submitting a VLTI proposal to ESO. The VLTI science operations scheme follows and is fully integrated into the regular VLT operations scheme from the initial preparation of the proposal to the delivery of the data. In particular, the same kind and level of service and support is offered to users of VLTI instruments as to users of any other VLT instrument. For example, one can ask either for the visitor mode or the service mode to carry out observations on the VLTI instruments, AMBER and MIDI. VLTI follows the VLT data flow: proposal form, observation block preparation and execution, FITS data, and archive.

All relevant information is provided by ESO through standard documents and via the ESO web pages: call for proposals, instrument web pages, instrument user manuals and template manuals, general- and instrument-specific proposal observation preparation instructions. Everything is accessible from the ESO web pages[3]. A good reference is also the paper *Observing with the VLT Interferometer* in *The ESO Messenger* **119**, 14 [12].

There is often a misconception in the community that there is very little telescope time for interferometry. In the period P77 (1 April 2006–30 September 2006), there were 29 programs approved on MIDI (19 programs in service mode for a total of 85 h on UTs and 324 h on ATs and 10 programs in visitor mode for a total of 3.7 nights) and 15 programs approved on AMBER (12 programs in service mode for a total of 51 h on UTs and 3 programs in visitor mode for a total of 4.4 nights). A total of about 20 nights have been dedicated to interferometry with the UTs within one semester corresponding to more than 10% of the VLT time. My message therefore is that the future interferometry users should not hesitate to apply for UT time on the VLTI instruments, since it is really science quality that drives the allocation of time.

---

[3] `http://www.eso.org/observing`.

To prepare the observations, I would like to advise future users of the VLTI to read the proceedings of the two VLTI schools which were held in Les Houches in 2002 [5] and in Goutelas in 2006 [6]. They will find all the material necessary to easily prepare such observations. In order to be complete, I outline here the main tools:

- to simulate visibilities with different VLTI setups:
- `VisCalc` at ESO which is simple and uses an HTML form, returning visibility amplitudes.
- `Aspro` and `Aspro-Light` at JMMC based on Java. `Aspro-Light` is tuned every semester to match the configurations and performance offered by ESO.
- to search for calibrators:
- `CalVin` at ESO for bright known calibrators which were prepared originally for VINCI.
- `SearchCal` at JMMC for bright and fainter calibrators not necessarily known. `SearchCal` is an automated search in the catalogs to find the best suited stars for calibration.

All these tools are based on the same engine, the GILDAS software developed by IRAM for the *Plateau de Bure Interferometer*, which give additional complementary information. The two ESO calculators `VisCalc` and `CalVin` are accessible at the following address:

$$\texttt{http://www.eso.org/observing/etc,}$$

and the JMMC packages are available from the JMMC homepage:

$$\texttt{http://www.jmmc.fr}$$

The process of observing at the VLTI follows the general VLT data flow. First, the user has to prepare a proposal and this stage is called the *phase 1 proposal, preparation*. The information required at this stage should contain, in addition to the **scientific rationale** and the **immediate objectives** of the proposal, the **definitions of runs**, where each run corresponds to one instrument (MIDI or AMBER) and one and only one VLTI baseline configuration. The time required per run, the preferred month of observations, the seeing, and the sky transparency must be specified at this stage too. [The proposal includes also a **target list** with time on target, V magnitude, coordinates, the **instrument configuration** with the spectral resolution, specific modes, the **interferometric table** with the V magnitude, the magnitude at the wavelength of observation, the size, the baselines, the expected visibilities and the expected correlated fluxes, i.e., the flux times and the visibility]. One advice is to verify the ESO archive before writing such a proposal to check that the observations requested have not been performed before, since the proprietary period at ESO is only one year!

If the proposal is accepted, then the user is requested to enter the *phase 2 proposal preparation* using the tool of the same name: P2PP. At this stage, the user is invited to enter the coordinates of the target list and of the calibrator stars and to specify the local sidereal time of the observations. The instructions differ depending on the instruments, but they are well documented in the instrument manuals. The result is the delivery of *observation blocks* (OBs), which can be observed either in service mode or in visitor mode. Of course in the second case, the flexibility is higher, since there is a real-time interaction with the astronomer on duty. However in service mode, if special good conditions are requested, the chances of success are improved since the observations are not programmed for a given night but for given conditions.

After the OBs have been processed (meaning the observations have been carried out), the quality of data is checked once the data arrive in the archive in Garching. If the runs are found compliant with the request and completed, a data package is sent to the principal investigator which contains

- the raw data associated with the run, including the acquisition images from both telescopes, fringe tracking data,...
- pipeline-processed data associated with the run;
- obtained transfer function for the respective night as computed by the pipeline on all calibrator stars of the night.

The raw data of all calibration stars are public and can be requested from the ESO archive[4].

Finally the user should reduce the raw data, since the automatic data reduction is not optimal. For this, one should get the off-line data reduction packages from the instrument teams:

- MIDI: the MIA package at `http://www.mpia.de/MIDI` or the Meudon package at `http://www.jmmc.fr/data_processing_midi.htm`
- AMBER: the `amdlib` library and the `ammYorick` interface at the AMBER website: `http://amber.obs.ujf-grenoble.fr`

The users are encouraged to contact the instrument teams for any queries.

4 Conclusions

Brand new science in the field of *young stellar objects* is within reach, with the advent of sensitive optical long baseline interferometers: morphology of disks, disk/star connection, dust composition of circumstellar material, implications of these measurements for the initial conditions of planetary formation. Combining NIR and MIR measurements on these targets will help us understand the physics and chemistry involved in these regions. Spectral resolution allows winds and jets to be probed in their launching zone. Interferometry also provides means to actually measure precisely the dynamical masses of stars either in binary systems or in singles system with a disk. The reader is invited to

---

[4] `http://archive.eso.org`.

read the review [4] to get an overview of all results obtained so far as well as of the prospects for the future.

To achieve such science, the *Very Large Telescope Interferometer* is a unique opportunity open to everyone. The mid-infrared MIDI instrument and the near-infrared AMBER instrument are offered to the ESO astronomical community for regular service mode and visitor mode observations. In fact, since there are no other similar instruments in the world, they can be requested by anybody. The same kind of level of support is offered to users of the VLTI instruments as to users of any VLT instrument. The complexity of the VLTI is hidden to regular users. Only the main instrument modes and parameters need to be chosen. The observation preparation is rather simple compared to other VLT instruments. However, the VLTI user should be aware of the complexity of interferometry and the caveats for the analysis and interpretation of data.

Acknowledgments

Most of the material presented in this manuscript comes from the Goutelas VLTI school[5], in particular from the lectures of M. Schöller, M. Wittkowski and O. Chesneau.

References

1. Chesneau, O.: MIDI – Obtaining and analysing interferometric data in the mid-infrared. In: Malbet, F., Perrin, G.(eds.) Observation and Data Reduction with the Very Large Telescope Interferometer, New Astronomy Review, Elsevier Science, (2007, in press)
2. Haniff, C.: An introduction to the theory of interferometry. In: Malbet, F., Perrin, G.(eds.) Observation and Data reduction with the Very Large Telescope Interferometer, New Astronomy Review, Elsevier Science, (2007, in press)
3. Leinert, C., Graser, U., Przygodda, F., et al.: Astrophys. Suppl. **286**, 73–83 (2003)
4. Millan-Gabet, R., Malbet, F., Akeson, R., et al.: The circumstellar environments of young stars at AU scales. In: Reipurth, B., Jewitt, D., Keil, K., (eds.) Protostars and Planets V, pp. 539–554. University of Arizona Press, Tucson (2007)
5. Perrin, G., Malbet, F.: Observing with the VLT interferometer EAS Publications Series, EDP Sciences, Les Ulis, France (2003)
6. Malbet, F., Perrin, G.: Observation and Data reduction with the Very Large Telescope Interferometer, New Astronomy Reviews, Elsevier Science (2007)
7. Petrov, R., Malbet, F., Weigelt G., et al.: Astron. Astrophys. **464**, 1 (2007)

---

[5] `http://vltischool.obs.ujf-grenoble.fr.`, `http://www.vlti.org`.

8. Schöller M.: The very large telescope interferometer: current facility and prospects. In: Malbet, F., Perrin, G. (eds.) Observation and Data reduction with the Very Large Telescope Interferometer, New Astronomy Review, Elsevier Science, (2007, in press)
9. Tatulli, E., Duvert, G.: AMBER data reduction. In: Malbet, F., Perrin, G. (eds.) Observation and Data reduction with the Very Large Telescope Interferometer, New Astronomy Review, Elsevier Science,(2007, in press)
10. Wallander, A., Bauvir, B., Dimmler, M., et al.: A status update of the VLTI control system. In: Lewis, H., Raffi, G. (eds.) Advanced Software, Control, and Communication Systems for Astronomy. Proceedings of the SPIE, pp. 21–31 (2004)
11. Wittkowski, M.: MIDI and AMBER from the user's point of view. In: Malbet, F., Perrin, G. (eds.) Observation and Data Reduction with the Very Large Telescope Interferometer, New Astronomy Review, Elsevier Science, (2007, in press)
12. Wittkowski, M., Comerón, F., Glindemann, A., et al: Observing with the ESO VLT Interferometer, The ESO Messenger **119**, 15 (2005)

Presentation of AMBER/VLTI
Data Reduction

E. Tatulli[1], F. Malbet[2], and G. Duvert[2]

[1] INAF-Osservatorio Astrofisico di Arcetri, Istituto Nazionale di Astrofisica, Largo
   E. Fermi 5, I-50125 Firenze, Italy
   etatulli@arcetri.astro.it
[2] Laboratoire d'Astrophysique de Grenoble, UMR 5571 Université Joseph
   Fourier/CNRS, BP 53, F-38041 Grenoble Cedex 9, France
   Firstname.Lastname@obs.ujf-grenoble.fr

Abstract. This course describes the data reduction process of the AMBER instrument, the three-beam recombiner of the Very Large Telescope Interferometer
(VLTI). We develop its principles from a theoretical point of view, and we illustrate
the main points with examples taken from AMBER real observations. We particularly emphasize that the AMBER data reduction process is (i) a fit of the interferogram in the detector plane, (ii) using an *a priori* calibration of the instrument, where
(iii) the complex visibility of the source is estimated from a least-square determination of a linear inverse problem, and where (iv) the derived AMBER observables
are the squared visibility, the closure phase, and the spectral differential phase.

1 Introduction

The AMBER instrument is the first generation of near-infrared three-beam
recombiner of the Very Large Telescope Interferometer (VLTI), which offers a
combination of characteristics that, though found distributed in other interferometers, are all brought together in the same instrument for the first time.
These characteristics can be categorized as follows. AMBER/VLTI is

(1) a single-mode waveguided interferometer, in order to perform spatial fil-
    tering of the turbulent wavefront
(2) a multiaxial "all-in-one" recombination scheme [14], where the fringes are
    spatially coded on the detector and recorded all together in the same
    interference pattern
(3) a 2/3 beam-recombiner in the J ($1.25\,\mu$m), H ($1.65\,\mu$m), and K ($2.2\,\mu$m)
    bands, which can make use of the 8-m Unitary Telescopes (UTs) or the
    1.8-m auxiliary Telescopes (ATs) provided by the VLT and can achieve a
    maximal spatial resolution of $\theta \sim 2\,$mas
(4) a spectrograph, allowing spectral resolutions of $\mathcal{R} = 35, 1500, 10000$.

E. Tatulli et al.: *Presentation of AMBER/VLTI Data Reduction*, Lect. Notes Phys. **742**,
257–272 (2008)
DOI 10.1007/978-3-540-68032-1_12   © Springer-Verlag Berlin Heidelberg 2008

As a consequence, from a signal processing point of view, the process of image formation in the case of AMBER has to be described in 3 majors steps: *(i) spatial filtering, (ii) beam recombination, and (iii) spectral dispersion*, as summarized in Fig. 1. First, the beams from the three telescopes are filtered by single-mode fibers to convert phase fluctuations of the corrugated wavefronts into intensity fluctuations that are monitored. At this point, a pair of conjugated cylindrical mirrors compresses, by a factor of about 12, the individual beams exiting from fibers into one dimensional elongated beam to be injected into the entrance slit of the spectrograph. For each of the three beams, beamsplitters placed inside the spectrograph select part of the light and induce three different tilt angles so that each beam is imaged at different locations of the detector. These are called photometric channels and are each relative to a corresponding incoming beam. The remaining parts of the light of the three beams are overlapped on the detector image plane to form fringes. The spatial coding frequencies of the fringes f are fixed by the separation of the individual output pupils. They are $f = [1, 2, 3]d/\lambda$, where d is the output pupil diameter. Since the beams hit a spectral dispersing element (a prism glued on a mirror or one of the two gratings) in the pupil plane, the interferogram and the photometries are spectrally dispersed perpendicular to the spatial coding. The dispersed interferogram arising from the beam combination, as well as the photometric outputs, are recorded on the infrared detector. The individual image that is recorded during the detector integration time (DIT) is called a *frame*. A cube of frames obtained during the exposure time is called an *exposure*.

This lecture aims to tackle the two major questions that arise when dealing with the data reduction of the AMBER instrument: "What is the influence

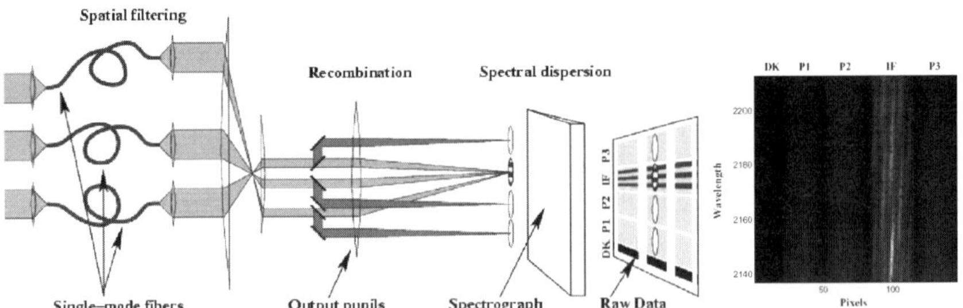

Fig. 1. (**Left panel**) Sketch of the AMBER instrument. Light enters the instrument from the left and propagates from left to right until the raw data are recorded on the detector. Further details are given in the text; (**Right panel**) AMBER reconstituted image from the raw data recorded during the 3-telescope observation of the calibrator HD135382 in February 2005, in the medium spectral resolution mode. DK corresponds to a dark region, Pk are the vertically dispersed spectra obtained from each telescope, and IF is the spectrally dispersed interferogram

of the AMBER-specific instrumental design on the recorded interferometric signal?" and conversely, "How can we use the specificities of this signal to derive an optimized signal processing of the data?" Following the previous enumeration of AMBER characteristics, these two questions can be addressed according to 3 themes that will be the framework of this coursebook:

(1) "What is the effect in single-mode waveguides on the interferometric signal and how can we make use of this in the amber data reduction process?" (Sect. 2)

(2) "What is the proper AMBER interferometric equation and what is specific about it?" (Sect. 3)

(3) + (4) "What are the AMBER observables and how can we estimate them?" (Sect. 4)

2 Spatial Filtering and Visibility

The advantage of using the practical characteristics of single-mode fibers to carry and recombine the light (as opposed to bulk optics), first proposed by [2] with his conceptual FLOAT interferometer, is now well established. Furthermore, in the light of the FLUOR experiment on the IOTA interferometer, which demonstrated the "on-sky" feasibility of such interferometers for the first time [4] showed that making use of single-mode waveguides could also increase the performance of optical interferometry, due to the remarkable properties of spatial filtering, which change the phase fluctuations of the atmospheric turbulent wavefront into intensity fluctuations. Indeed, and as schematically shown by Fig. 2, the effect of the single-mode waveguide is to only propagate in its core the part of the incoming electric-field projected on its first mode. As a result, the wavefront at the output of the fiber is perfectly plane. In the image plane (see Fig. 3), this means that the shape of the signal at the output of the waveguide is *fully deterministic* (as a matter of fact quasi-Gaussian, e.g. [7]), as opposed to multimode instruments where the short-exposed images present the famous randomly distributed speckle

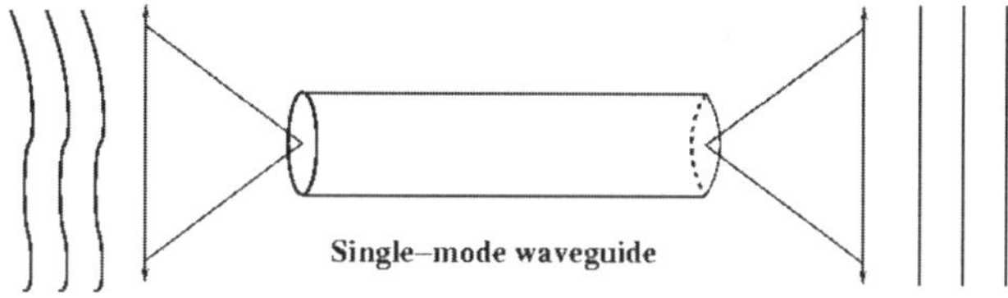

Fig. 2. Schematic pupil plane view of the effect of single-mode waveguides on the turbulent wavefront: at the entrance of the fiber, the wavefront is corrugated by the atmospheric turbulence + static aberrations of the instrument. At the output, only the projection of the electric field on the first mode of the waveguide has propagated and as a consequence, the wavefront is plane

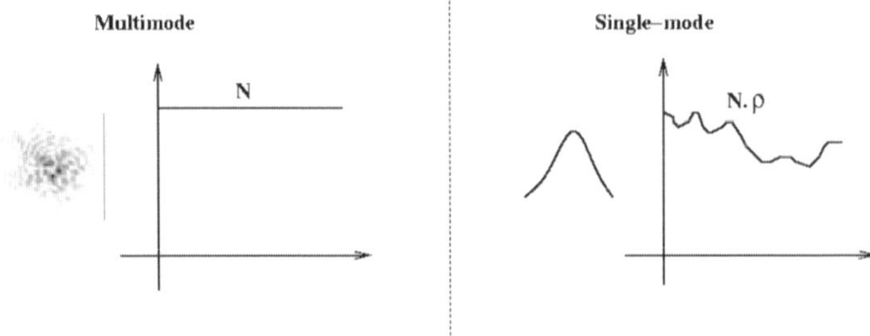

Fig. 3. Schematic image plane view of the effect of single-mode waveguides on the image. In the multimode case (**left**), the image is the well-known randomly distributed speckle pattern. But the total number of photons in the image is constant (scintillation neglected). In the single-mode case (**right**), the image is deterministic (Gaussian-like shape) but the total number of photons depends on the coupling coefficient (ρ) which varies with the turbulence

pattern. And the fact that at the fiber's output the intensity profile is deterministic, that is *calibratable* is of great significance to the AMBER data reduction process. Indeed this important information can be used as an *a priori*, there by improving the performance of signal processing. We will come back to this point in the next chapters.

The obvious trade-off in producing perfectly stable intensity profiles from single-mode waveguide is that only a fraction of the light, namely, the coupling coefficient, is enters the fiber (once again, as opposed to multimode experiments where the number of photons remains constant if we neglect the scintillation; see Fig. 3). This coupling coefficient depends not only on the source's extent [5] but also on the turbulence, more precisely the Strehl ratio [3]. This explains why single-mode interferometers also require telescopes equipped with Adaptive Optics systems, in order to optimize the fraction of flux entering the fiber. Nonetheless, this coupling coefficient property has a strong impact on the estimated visibility. Indeed, this latter will be biased both by geometric (static) and atmospheric (turbulent) effects. To cope with this situation [4] proposed to monitor the coupling coefficient fluctuations in real time due to dedicated photometric outputs and to perform instantaneous photometric calibration. And indeed, he experimentally proved that doing so, single-mode interferometry could achieve visibility measurements with precisions of 1% or lower *on compact sources*.

Later, from its theoretical work, [6] has finally shown that, in the general case of partial correction by adaptive optics and for a source with a given spatial extent, the measured instantaneous visibility could be expressed in the general case as $V_{ij} \propto \mathrm{TF}[O_\star(\alpha)L^{ij}(\alpha)]$, that is, the Fourier Transform of the object brilliance distribution *multiplied by the instantaneous interferometric antenna lobe, which is turbulent*. This easily explains why and how visibility is biased by geometric effect (lobe effect such as in radio-interferometry) and by the turbulence, and why, as one of the consequences, the field of view of single-mode interferometers is limited to one airy disk, that is, $\Theta \sim \lambda/D$, where D is the diameter of one telescope. This also tells us that strictly speaking, the modal visibility is not the source visibility. However, [15] confirmed from a analytical approach that, in the specific case of compact objects such as dealt by [4], the benefit of single-mode waveguides is substantial, not only in terms of the signal-to-noise ratio of the visibility but also of the robustness of the estimator. From now on, we will consider our observable to be the complex modal visibility.

3 The AMBER Interferometric Equation

For the sake of clarity, the notations that will be used throughout this section to derive the AMBER interferometric equation are summarized in Table 1.

3.1 Building the AMBER Interferometric Equation

The following demonstration is given, considering a generic $N_{\text{tel}} \geq 2$ telescope interferometer. In the specific case of AMBER, however, $N_{\text{tel}} = 2$ or $N_{\text{tel}} = 3$. Each line of the detector being independent of another, we can focus our attention on one single spectral channel, which is assumed to be monochromatic here. The effect of a spectral bandwidth on the interferometric equation is treated in Sect. 4.3.

Table 1. List of notations and conventions used to describe the AMBER interferometric equation and subsequent derivations

Quantity	Definition
k in index	pixel coordinate
α_k	sampling
i, j in exponent	telescope(s)number(s)
p_k^i	photometric channel
i_k	interferometric channel
N	total source photon flux
t^i	total transmission of the ith optical train
a_k^i	intensity profile for the interferometric channel
b_k^i	intensity profile for the photometric channel
f^{ij}	frequency coding
ϕ_s^{ij}	instrumental phase
C_B^{ij}, Φ_B^{ij}	polarization contrast and phase
$F^i = Nt^i$	photometric flux
$F_c^{ij} = 2N\sqrt{t^i t^j} V^{ij} \mathrm{e}^{i(\Phi^{ij} + \phi_p^{ij})}$	coherent flux

Table 2. (Left) image of the detector for one spectral channel; (Right) equations ruling photometric (when illuminated) and interferometric channels. From top to bottom: one, two, and all beams lit

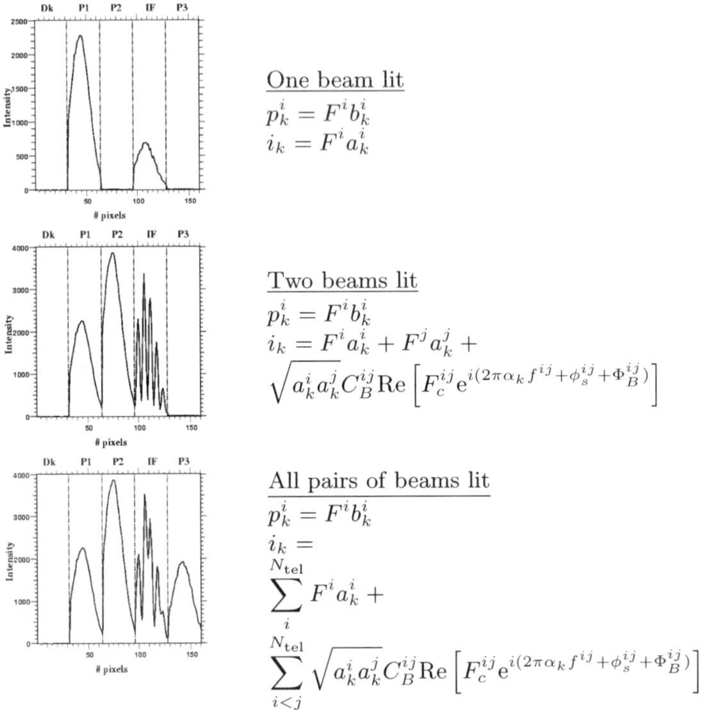

One beam lit

$$p_k^i = F^i b_k^i$$
$$i_k = F^i a_k^i$$

Two beams lit

$$p_k^i = F^i b_k^i$$
$$i_k = F^i a_k^i + F^j a_k^j +$$
$$\sqrt{a_k^i a_k^j}\, C_B^{ij} \mathrm{Re}\left[F_c^{ij} e^{i(2\pi\alpha_k f^{ij} + \phi_s^{ij} + \Phi_B^{ij})} \right]$$

All pairs of beams lit

$$p_k^i = F^i b_k^i$$
$$i_k =$$
$$\sum_i^{N_{\mathrm{tel}}} F^i a_k^i +$$
$$\sum_{i<j}^{N_{\mathrm{tel}}} \sqrt{a_k^i a_k^j}\, C_B^{ij} \mathrm{Re}\left[F_c^{ij} e^{i(2\pi\alpha_k f^{ij} + \phi_s^{ij} + \Phi_B^{ij})} \right]$$

Interferometric Output

- *One beam lit (Table 2, top)*: when only the ith beam is illuminated, the signal recorded in the interferometric channel is the photometric flux F^i spread on the Airy pattern a_k^i.

- *Two beams lit (Table 2, middle)*: When beams i and j are illuminated simultaneously, the coherent addition of both beams results in an interferometric component superimposed on the photometric continuum. The interferometric part, i.e., the fringes, arises from the amplitude modulation of the coherent flux F_c^{ij} at the coding frequency f^{ij}.

- *All beams lit (Table 2, bottom)*: Such an analysis can be done for each pair of beams arising from the interferometer. As a result, the interferogram recorded on the detector can be written in the general form:

$$i_k = \sum_{i}^{N_{\mathrm{tel}}} a_k^i F^i + \sum_{i<j}^{N_{\mathrm{tel}}} \sqrt{a_k^i a_k^j}\, C_B^{ij} \mathrm{Re}\left[F_C^{ij} e^{i(2\pi\alpha_k f^{ij} + \phi_s^{ij} + \Phi_B^{ij})}\right] \qquad (1)$$

Here, ϕ_s^{ij} is the instrumental phase taking possible misalignment and/or differential phase between the beams a_k^i and a_k^j into account, and C_B^{ij} and Φ_B^{ij} are, respectively, the loss of contrast and the phase shift due to polarization mismatch between the two beams (after the polarizers), such as the rotation of the single-mode fibers might induce. This equation is governing the AMBER fringe pattern, that is the interferometric channel of the fourth column of the detector (e.g. in figures of Table 2). The first sum in (1), which represents the continuum part of the interference pattern, is called the DC component from now on, and the second sum, which describes the high frequency part (that is the coded fringes), is called the AC component of the interferometric output.

Photometric Outputs
Due to the photometric channels, the number of photoevents $p^i(\alpha)$ coming from each telescope can be estimated independently with

$$p_k^i = F^i b_k^i \qquad (2)$$

The previous equation rules the photometric channels (resp. columns 2,3, and 5 of the detector, when illuminated; see Table 2).

3.2 Analyzing the AMBER Interferometric Equation

In order to analyze deeper the content of the AMBER interferometric equation and to use this information to derive the optimized processing of the data, one has to separate in (1) the astrophysical and instrumental parts. Furthermore, if we put the DC component on the left side of the equation in order to have on the right side an expression which is linear with respect to the coherent flux, it is:

$$i_k - \overbrace{\sum_{i}^{N_{\mathrm{tel}}} F^i a_k^i}^{\text{DC component}} = \overbrace{\sum_{i<j}^{N_{\mathrm{tel}}} \left[c_k^{ij} R^{ij} + d_k^{ij} I^{ij}\right]}^{\text{AC component}}, \qquad (3)$$

where

$$c_k^{ij} = C_B^{ij} \frac{\sqrt{a_k^i a_k^j}}{\sqrt{\sum_k a_k^i a_k^j}} \cos(2\pi\alpha_k f^{ij} + \phi_s^{ij} + \Phi_B^{ij}) \qquad (4)$$

$$d_k^{ij} = C_B^{ij} \frac{\sqrt{a_k^i a_k^j}}{\sqrt{\sum_k a_k^i a_k^j}} \sin(2\pi\alpha_k f^{ij} + \phi_s^{ij} + \Phi_B^{ij}) \qquad (5)$$

are, such as in amplitude modulation techniques of telecom data, called the carrying waves. They only depend on *the characteristics of the instrument*, hence are deterministic[1] and therefore calibratable. On the opposite, the quantities:

$$R^{ij} = \sqrt{\sum_k a_k^i a_k^j \mathrm{Re}\left[F_c^{ij}\right]}, \quad I^{ij} = \sqrt{\sum_k a_k^i a_k^j \mathrm{Im}\left[F_c^{ij}\right]} \tag{6}$$

are proportional to the real and imaginary part of the coherent flux. From (3), it can be seen straight away that a linear relationship between the *continuum corrected interferograms* and the *complex visibilities* (i.e., from the coherent flux) can be derived, provided

1. the DC component can be estimated: This can be achieved if we know the ratio v_k^i – which only depends on the instrument – between the measured photometric fluxes P^i and the corresponding DC components of the interferogram, that is,

$$a_k^i F^i = P^i v_k^i, \tag{7}$$

   where P^i is the estimated photometric flux, integrated over the pixels

$$P^i = F^i \sum_k b_k^i \tag{8}$$

2. the characteristics of the instrument, that is, the carrying waves c_k^{ij}, d_k^{ij} are known

As a consequence, and as will be discussed in-depth in the next section, the AMBER data reduction process consists in *modeling the interferogram* in the detector plane. This requires a previous calibration of the instrument, characterizing the function v_k^i, and the carrying waves c_k^{ij}, d_k^{ij}.

4 The AMBER Data Reduction

Hence, the AMBER data reduction requires 5 successive steps:

1. Cosmetic *(flat-field, sky...)* which converts the infrared CCD in photons counting, a usual procedure that will not be described in further detail here
2. Calibration of the instrument:
   fraction of flux v_k^i and carrying waves c_k^{ij}, d_k^{ij}
3. Estimation of the photometric F^i and coherent fluxes F_c^{ij}
4. Estimation of the observables
5. Biases correction.

4.1 Calibration of the Instrument

The calibration procedure is performed due to an internal source located in
the *Calibration and Alignment Unit* (CAU) of AMBER [12]. It involves acquiring a sequence of high signal-to-noise ratio calibration files, whose successive configurations are summarized in Table 3 and explained below. Since
the calibration is done in the laboratory, the desired level of accuracy for the
measurements is insured by choosing the appropriate integration time. As an
example, typical integration times in "average accuracy" mode are (for the
full calibration process) $\tau = 17\mathrm{s},\ 30\mathrm{s},\ 800\mathrm{s}$ for, respectively, low, medium,
and high spectral resolution modes in the K band and 100 times higher for
the "high accuracy" calibration mode. The sequence of calibration files has
been chosen to accommodate both two and three-telescope operations. For a
two-telescope operation, only the first 4 steps are needed.

1. v_k^i *estimation: steps 1 and 2 (and 5 when in 3-telescope mode)*
   For each telescope beam, an image is recorded with only this shutter
   opened. The fraction of flux measured between the interferometric channel
   and the illuminated photometric channel leads to an accurate estimation
   of the v_k^i functions (see Fig. 4, left).
   characterization of c_k^{ij} and d_k^{ij} : steps 3/4, 6/7, and 8/9

   In order to compute the carrying waves, one needs to have two independent (in terms of algebra) measurements of the interferogram, since there
   are two unknowns (per baseline) to compute. The principle is the following: two shutters are opened simultaneously and for each pair of beams,
   then the interferogram is recorded on the detector. [Such an interferogram
   corrected for its DC component and calibrated by the photometry, yields
   the c_k^{ij} carrying wave.] To obtain its quadratic counterpart, the previous
   procedure is repeated by introducing a known phase shift close to 90 degree γ_0, using piezoelectric mirrors at the entrance of beams 2 and 3.

Table 3. Acquisition sequence of calibration files

Step	Sh 1	Sh 2	Sh 3	Phaseγ_0	Step	Sh 1	Sh 2	Sh 3	Phaseγ_0
1	O	X	X	NO	5	X	X	O	NO
2	X	O	X	NO	6	O	X	O	NO
3	O	O	X	NO	7	O	X	O	YES
4	O	O	X	YES	8	X	O	O	NO
					9	X	O	O	YES

Sh = Shutter; O = Open; X = Closed.

---

[1] we recall here that due to the use of single-mode fibers, the intensity profiles a_k^i
and b_k^i are fixed and depend only on the configuration of the instrument.

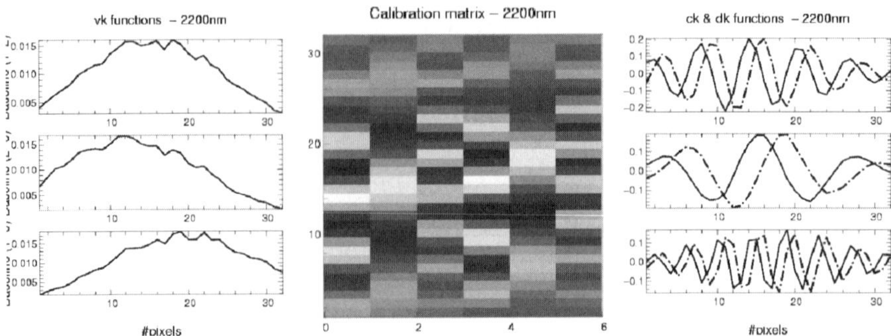

Fig. 4. Outputs of the calibration procedures. Examples have been chosen for a given wavelength $\lambda = 2.2\,\mu$m. **Left**: the v_k^i functions. **Middle**: the matrix containing the carrying waves; the first three columns are the c_k^{ij} functions for each baseline, and the last three columns are the respective d_k^{ij} functions. One can see that for each baseline, c_k^{ij} and d_k^{ij} are in quadrature. **Right**: another representation of the carrying waves. From top to bottom, both sinusoidal functions correspond to columns 1-4, 2-5, and 3-6 of the calibration matrix

Computing the d_k^{ij} function from c_k^{ij} and γ_0 is straightforward (see Fig. 4, middle and right).

4.2 Estimation of Photometric and Coherent fluxes

Photometric Fluxes – DC Subtraction

As discussed previously, the estimation of the photometric fluxes is straightforward, given the v_k^i functions (see (7)). Hence, the DC corrected interferogram, namely, m_k, can be computed (Fig. 5, left and middle):

$$m_k = i_k - \sum_{i=1}^{N_{\text{tel}}} P^i v_k^i \tag{9}$$

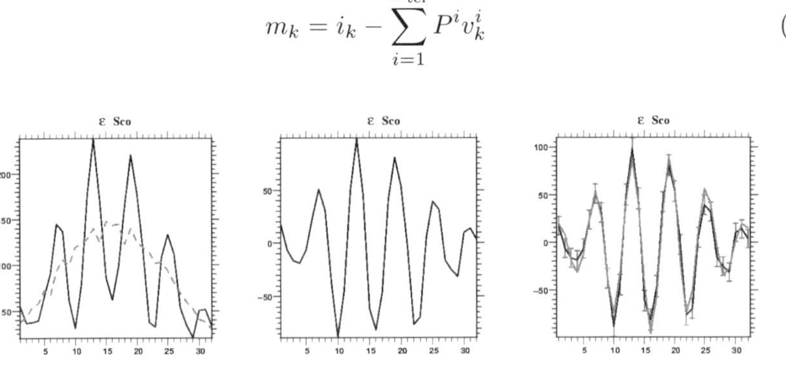

Fig. 5. Steps of the fringe fitting part of the AMBER data reduction. (**Left**) recorded interferogram i_k (DC component over-plotted); (**Middle**) DC component-corrected interferogram m_k, now centered around 0. (**Right**): fit of the m_k by the carrying waves

Coherent Flux – fringe Fitting:
Equation (9) can then be rewritten

$$m_k = \sum_{i<j}^{N_{\text{tel}}} c_k^{ij} R^{ij} - d_k^{ij} I^{ij} = \underbrace{\left[c_k^{(i,j)}, d_k^{(i,j)} \right]}_{\text{V2PM}} \begin{bmatrix} R_{ij} \\ I_{ij} \end{bmatrix}, \tag{10}$$

which defines a system of N_{pix} linear equations with $2N_b = N_{\text{tel}}(N_{\text{tel}} - 1)$ unknowns (i.e., twice the number of baselines). The matrix that contains the $c_k^{(i,j)}$, $d_k^{(i,j)}$ functions is called the V2PM and stands for *Visibility to Pixel Matrix*. Thus, estimating the real and imaginary parts of the coherent flux requires to solve the inverse problem defined by the matrix-type (10). Assuming Gaussian statistics for the noise of the measurements m_k, (10) is inverted by performing a least square fit of the m_k from the carrying waves, adjusting the R_{ij} and I_{ij} free parameters as shown in Fig. 5, right. Finally, the estimated complex coherent flux C_{ij} comes directly from

$$C^{ij} = R^{ij} + iI^{ij} = \sqrt{\sum_k a_k^i a_k^j} F_c^{ij} \tag{11}$$

4.3 Estimation of the AMBER Observables

The estimated complex coherent flux C_{ij} is directly linked to the complex (modal) visibility of the object through the following equation:

$$C^{ij} \propto F_c^{ij} = 2N\sqrt{t^i t^j} V^{ij} e^{i(\Phi^{ij} + \phi_p^{ij})} \tag{12}$$

Hence, from the previous equation one can estimate

1. the (squared) modulus of the source's visibility $|V^{ij}|$. This quantity gives information about the spatial extent of the source, with respect to the chosen baseline
2. the phase: this latter quantity cannot be computed directly because of the atmospheric differential piston ϕ_p^{ij}, which adds a random turbulent phase that cannot be dis-entangled from the source. Fortunately, there are different ways are available to obtain a partial phase information
   - Due to the use of 3 telescopes simultaneously, one can compute the so-called *closure phase*, which is independent of the atmosphere. The closure phase gives information about the geometry of the source, that is, the potential asymmetries (e.g. [9])
   - Due to the spectral dispersion offered by the AMBER instrument, the *differential phase* can be estimated as well, confirming that photocenter displacement is a function of the wavelength [1]

The Closure Phase
By definition, the closure phase is the phase of the so-called bispectrum

B^{123}. The bispectrum results in the ensemble average of the coherent flux triple product and is then estimated as

$$\widetilde{B}^{123} = \left\langle C^{12} C^{23} C^{13*} \right\rangle \tag{13}$$

The closure phase then is straightforward:

$$\widetilde{\phi_B}^{123} = \mathrm{atan} \left[\frac{\mathrm{Im}(\widetilde{B}^{123})}{\mathrm{Re}(\widetilde{B}^{123})} \right] = \varPhi^{12} + \varPhi^{23} - \varPhi^{13} \tag{14}$$

The closure phase has the advantage of being independent of the atmosphere (e.g. [13]), since the atmospheric piston is a differential quantity, which sum cancels out when using a closure relation, as is done in the three-telescope case. An example of closure phase and closure phase error bars is given in Fig. 6. So far the closure-phase internal error bars (i.e., that does not include systematics errors) are computed statistically by taking the root mean square of all the individual frames, then dividing by the square root of the number of frames.

The Differential Phase

The differential phase is the phase of the so-called cross spectrum W_{12}. For each baseline, the latter is estimated from the complex coherent flux taken at two different wavelengths λ_1 and λ_2:

$$\widetilde{W_{12}^{ij}} = \left\langle C_{\lambda_1}^{ij} C_{\lambda_2}^{ij\,*} \right\rangle \tag{15}$$

And the differential phase is

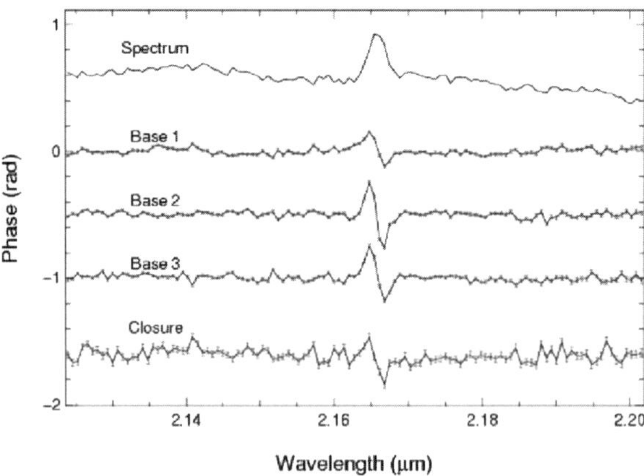

Fig. 6. Example of differential phases (after removing atmospheric piston) and closure phase computation on an observed object with a rotating feature in the Brγ emission line in α Arae, [8]

$$\widetilde{\Delta\phi_{12}^{ij}} = \text{atan}\left[\frac{\text{Im}\left(\widetilde{W_{12}^{ij}}\right)}{\text{Re}\left(\widetilde{W_{12}^{ij}}\right)}\right] \tag{16}$$

In first order approximation, the differential phase of Eq. (16) is a linear function that takes the generic form $\Delta\phi_{12} = \phi_1 + 2\pi\left(\sigma_2 - \sigma_1\right)\delta$, where $\sigma = 1/\lambda$ is the wavenumber. Its slope $\delta = \delta_p + \delta_o$ depends on the sum of atmospheric piston δ_p, which varies frame by frame, and of the linear component of the object differential phase δ_o. It can be estimated by performing a linear fit of the differential phase, as illustrated in Fig. 7. In order to distinguish between the atmospheric piston δ_p and the linear component of the differential phase δ_o, the fitting can be performed by only using spectral channels corresponding to the continuum of the source (i.e., outside spectral features), where the object differential phase is assumed to be zero. An example of a differential phase, where the atmospheric piston has been removed using the fitting technique described previously is shown in Fig. 6. Currently, like the closure phase, the internal error bars are computed statistically assuming that the differential phases are statistically independent frame to frame.

The squared visibility

By definition, the squared visibility is the ratio between the squared coherent flux and the geometrical product of the photometric fluxes. Hence, (6) and (7) state that the expression of the squared visibility writes

$$|V^{ij}|^2 = \frac{|F_c^{ij}|^2}{4F^iF^j} = \frac{R^{ij^2} + I^{ij^2}}{4P^iP^j\sum_k v_k^i v_k^j} \tag{17}$$

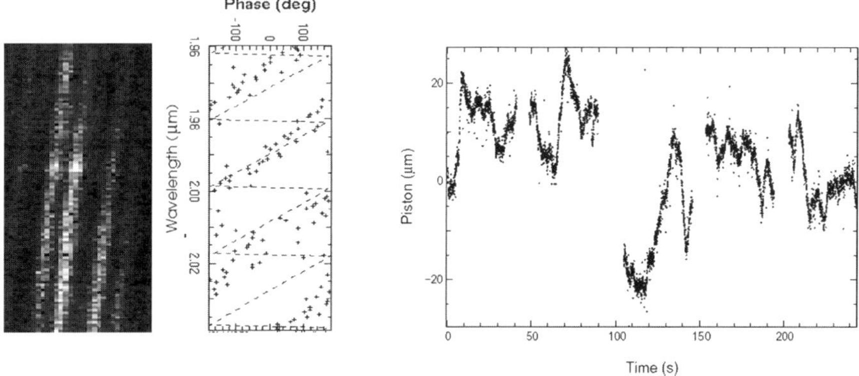

Fig. 7. Piston estimation from the fringe pattern. From left to right is (i) the raw fringe pattern, the corresponding phase; (ii) the estimated linear component of the phase from the least square fit; and (iii) a piston time-sequence over 250 seconds. Note that the piston rms is around $15\,\mu$m, which agrees with the average atmospheric conditions recorded in Paranal

However, one has also to take into account the different biases that are introduced in the estimation of the squared visibilities. The sources of these biases are

- the visibility V_c^{ij} of the internal source (CAU), which mediates in the calibration process but not during the observation. This bias is a fixed value and is easily calibratable, e.g. by observing a reference source.
- the quadratic estimation of the coherent flux. Indeed, taking the ensemble average of the squared modulus of the coherent flux introduces an additive bias (Bias $\left\{ R^{ij^2} + I^{ij^2} \right\}$) due to the zero-mean photon and detector noises [11]. This bias is the quadratic sum of the errors of the measurements $\sigma^2(m_k)$ projected on the real and imaginary axis of the coherent flux. More precisely, if ζ_k^{ij} and ξ_k^{ij} are the coefficients of the generalized inverse of the V2PM matrix, matrix, R^{ik} and I^{ij} verify the respective following equations:

$$R^{ij} = \sum_{k=1}^{N_{\text{pix}}} \zeta_k^{ij} m_k, \quad I^{ij} = \sum_{k=1}^{N_{\text{pix}}} \xi_k^{ij} m_k \tag{18}$$

Hence, the quadratic bias can then be estimated and subtracted, using

$$\text{Bias}\{R^{ij^2} + I^{ij^2}\} = \sum_k \left[(\zeta_k^{ij})^2 + (\xi_k^{ij})^2 \right] \sigma^2(m_k) \tag{19}$$

with

$$\sigma^2(m_k) = \overline{i_k} + \sigma^2 + \sum_{i=1}^{N_{\text{tel}}} \left[\overline{P_i} + N_{pix}\sigma^2 \right] (v_k^i)^2 \tag{20}$$

- non-zero optical path difference (OPD): since the coherence length $\mathcal{L}_c$ is finite, a non-zero OPD induces a loss of spectral coherence that translates into a multiplicative attenuation (ρ_p) of the visibility, which can be computed frame by frame, and corrected using the following formula:

$$\rho_p = \left| \text{sinc} \left(\pi \frac{\delta_p + \delta_o}{\mathcal{L}_c} \right) \right| \tag{21}$$

δ_p and δ_o being estimated from differential phase measurements, as discussed previously. Note that this effect is usually negligible in medium and high spectral resolution modes, providing typical excursion of the atmospheric piston at Paranal.

- fringe motion during the integration time leads to fringe blurring, hence contrast loss (ρ_{jit}), which depends on the features of the turbulent atmosphere, mainly its coherence time. This effect is calibrated by observing a reference star, assuming that the parameters of the turbulence remains the same.

In summary, the estimated visibility verifies the following equation:

$$\underbrace{\frac{\widetilde{|V^{ij}|^2}}{V_c^{ij^2}}}_{CAU\ visibility} = \frac{\left\langle R^{ij^2} + I^{ij^2}\right\rangle \overbrace{-\operatorname{Bias}\left\{R^{ij^2} + I^{ij^2}\right\}}^{\text{quadratic bias}}}{4\left\langle P^i P^j\right\rangle \sum_k v_k^i v_k^j} \underbrace{<\rho_p^2>}_{\text{piston bias}} \overbrace{<\rho_{jit}^2>}^{\text{jitter bias}}$$

$$(22)$$

The associated error of the visibility is computed from the semi-empirical formula, using a second-order development of the estimator, following [10]:

$$\frac{\sigma^2(\widetilde{|V^{ij}|^2})}{\widetilde{|V^{ij}|^2}^2} = \frac{1}{M}\left[\frac{\left\langle |C^{ij}|^4\right\rangle_M - \left\langle |C^{ij}|^2\right\rangle_M^2}{\left\langle |C^{ij}|^2\right\rangle_M^2} + \frac{\left\langle P^{i^2} P^{j^2}\right\rangle_M - \left\langle P^i P^j\right\rangle_M^2}{\left\langle P^i P^j\right\rangle_M^2}\right],$$

$$(23)$$

where M is the number of frames used to compute the visibility.

5 Conclusion

In this paper, we have described from a theoretical point of view, the reduction steps involed in VLTI/AMBER interferometric observations. The main specificities of the algorithm to be borne in mind are the following: the AMBER signal processing is (i) a fit of the interferogram in the detector plane, (ii) using an *a priori* calibration of the instrument, where (iii) the complex visibility of the source is estimated from a least-square determination of a linear inverse problem, and where (iv) the derived AMBER observables are the squared visibility, the closure phase, and the spectral differential phase.

References

1. Chelli, A., Petrov, R.G.: Model fitting and error analysis for differential interferometry. I. General formalism. A&AS **109**, 389–399 (1995)
2. Connes, P., Shaklan, S., Roddier, F.: A Fiber-Linked Groundbased Array. In: Goad, J.W. (ed.) Interferometric Imaging in Astronomy, pp. 165–+ (1987)
3. Coudé du Foresto, V., Faucherre, M., Hubin, N., Gitton, P.: Using single-mode fibers to monitor fast Strehl ratio fluctuations. Application to a 3.6 m telescope corrected by adaptive optics. A&AS **145**, 305–310 (2000)
4. Coudé Du Foresto, V., Ridgway, S., Mariotti, J.M.: Deriving object visibilities from interferograms obtained with a fiber stellar interferometer. A&AS **121**, 379–392 (1997)
5. Dyer, S.D., Christensen, D.A.: Pupil-size effects in fiber optic stellar interferometry. Opt. Soc. of Am. J. A **16**, 2275–2280 (1999)
6. Mège, P., Malbet, F., Chelli, A.: Stellar Interferometry with optical waveguides. In: Societe Francaise d'Astronomie et d'Astrophysique (ed.) SF2A-2001: Semaine de l'Astrophysique Francaise, pp. 581–+ (2001)

7. Mège, P., Malbet, F., Chelli, A.: Interferometry with single mode waveguide. In: Traub, W.A. (ed.) Interferometry for Optical Astronomy II. Edited by Wesley A. Traub. Proceedings of the SPIE, Vol. 4838, pp. 329-337 (2003)., pp. 329–337 (2003)

8. Meilland, A., The AMBER consortium: First direct detection of a Keplerian rotating disk around the Be star α Arae using the VLTI/AMBER instrument. A&Ap p. 4848 (2007)

9. Monnier, J.D.: An Introduction to Closure Phases. In: Lawson, P.R. (ed.) Principles of Long Baseline Stellar Interferometry, pp. 203–+ (2000)

10. Papoulis, A.: Probability, Random Variables and Stochastic Processes. New York: McGraw-Hill, 1984, 2nd ed. (1984)

11. Perrin, G.: Subtracting the photon noise bias from single-mode optical interferometer visibilities. A&Ap**398**, 385–390 (2003). DOI 10.1051/0004-6361:20021667

12. Petrov, R.G., the AMBER consortium: AMBER, the near infrared spectro-interferometric three telescopes VLTI instrument. A&Ap 6496 (2007)

13. Roddier, F.: Triple correlation as a phase closure technique. Opt. Commun. **60**, 145–148 (1986). DOI 10.1016/0030-4018(86)90168-9

14. Tatulli, E., LeBouquin, J.B.: Comparison of Fourier and model-based estimators in single-mode multi-axial interferometry. MNRAS**368**, 1159–1168 (2006). DOI10.1111/j.1365-2966.2006.10203.x

15. Tatulli, E., Mège, P., Chelli, A.: Single-mode versus multimode interferometry: A performance study. A&A **418**, 1179–1186 (2004). DOI 10.1051/0004-6361:20034090

High Angular Resolution Observations of Disks

L. Testi

INAF–Osservatorio Astrofisicso di Arcetri
lt@arcetri.astro.it

Abstract. In this lecture, I will review the properties of protoplanetary disks as derived from high angular resolution observations. I discuss how the combination of several different high angular resolution techniques allow us to probe different regions of the disk and to derive the properties of the dust when combined with sophisticated disk models. The picture that emerges is that the dust in circumstellar disks surrounding pre-main sequence stars is in many cases significantly evolved compared to the dust in molecular clouds and the interstellar medium. It is, however, still difficult to derive a consistent picture and timeline for dust evolution in disks, as the observations are still limited to small samples of objects.

L. Testi: *High Angular Resolution Observations of Disks*, Lect. Notes Phys. **742**, 273–285 (2008)
DOI 10.1007/978-3-540-68032-1_13

1 Introduction

Circumstellar disks are expected to form as a natural consequence of the star-formation process, as high angular momentum material in the parent molecular core cannot fall directly onto the central star but accumulates on a flattened structure perpendicular to the average original angular momentum of the prestellar core [13, 28]. Indeed, observationally, disks are found to be present around most young stellar objects. Disk-like structures are found around protostars of every mass, when sufficient angular resolution and sensitivity observations are available, from the lowest mass brown dwarf systems [24, 32] to massive protostars [4, 7].

Disks play an essential role in the formation of stars as material loses angular momentum in the disk and is accreted onto the central forming star. When the central star is assembled, and the main accretion phase, if finished, is within the circumstellar disk, then planets are expected to be assembled. In fact, around many young pre-main sequence stars protoplanetary disks are detected. The presence of disks around TTauri-like stars (TTS) and Herbig Ae/Be stars (HAeBe) was originally inferred from the presence of infrared excess in these systems and then eventually demonstrated through direct imaging at millimeter wavelengths [20, 27].

In this lecture, I will primarily concentrate on high angular resolution observations of protoplanetary disks surrounding pre-main sequence stars. I will give examples of observations obtained using many different techniques that have been discussed in this class, in particular adaptive optics-assisted (see lectures by Esposito), infrared interferometry (see lectures by Malbet and Tatulli) and millimeter interferometry (see lecture by Beuther) observations. I will discuss how different observations allow us to probe the dust component of these disks, especially regarding the properties and evolution of the dust toward the first phases of the planetary formation process. Gas, and in particular molecular gas, is the main constituent of these disks, and several high angular resolution observations are essential for our understanding of the kinematical, physical and chemical properties of the gaseous component of disks. I do not have time to discuss this aspect here and refer to the recent review in [11] for more details.

2 Structure of Circumstellar Disks

Some basic constraints on the structure of circumstellar disks can be derived by modeling their spectral energy distribution (SED). As a first approximation, the SED of a circumstellar disk can be modeled as the sum of the contribution of annuli, each emitting as black body at a local temperature $T_\mathrm{d}(r)$, where r is the distance from the central star. Under these assumptions the SED of a circumstellar disk can be written as

$$F_\nu = \frac{\cos\theta}{D^2} \int_{r_i}^{r_o} B_\nu(T_{\mathrm d})(1 - e^{-\tau_\nu})2\pi r \mathrm{d}r \; , \tag{1}$$

where B_ν is the Planck function, and τ_ν is proportional to the dust opacity at the frequency ν and the dust surface density distribution. The local temperature at each radius is determined by the balance between cooling due to the emitted radiation and the heating. The two main heating sources are the direct radiation from the central star and the viscous dissipation. Both these processes predict temperature distributions $T_{\mathrm d} \sim r^{-q}$ with $q = 0.75$ (see [22] for a detailed discussion).

These simple parametric models have been widely used to infer global properties of disks around young stars. The general result (e.g. [3]) has been that such models account well for overall shape of the observed SEDs in T Tauri and HAeBe systems, but the temperature profile $T_{\mathrm d}(r)$ has to fall off much more slowly than predicted by the simplistic models described above. To fit the observed data, the derived value of q is close to 0.5.

The most successful solution to this inconsistency is the class of disk models that include a *flaring* outer disk [18] and an optically thin disk atmosphere [8]. These models predict that the disk-opening angle or the ratio between the scale height and the radius increases toward the outer disk. The grazing angle at which the stellar radiation impinges on the disk changes with radius, allowing for an increase of the heating of the outer regions of the disk. The optically thin (to the disk radiation) atmosphere absorbs the stellar radiation and is warmer than the disk (optically thick) interior at the same radius.

These disk models have been extremely successful in explaining a number of observational properties of disks that range from the overall shape of the SED to the scattered light images of disks. These are, thus, the reference models that are used as benchmark for the observations. The disk structure predicted by these models has important implications on which regions of the disk are probed by different observational techniques.

Scattered light emission in the visible and near-infrared are a sensitive probe of the small dust grains population in the upper layers of the disk atmosphere, while emission in the mid-infrared features of the silicates probe the emission of dust grains in the atmosphere. The disk midplane, which contains the bulk of the disk mass, can only be probed directly at millimeter and longer wavelengths, where the emission becomes optically thin. In the following section, I will discuss how the use of different high angular resolution observations allow to constrain the structure and physical properties in different regions of the disk.

2.1 The Disk Atmosphere

The properties of dust grains and macro-molecules in the disk atmosphere can be probed indirectly by observing the scattered light emission from the central star or directly observing the emission features in the mid-infrared.

By comparison with disk models, both observables also allow to constrain the geometrical structure of the disk atmosphere, such as the disk flaring.

In this lecture, I will concentrate on the dust emission diagnostics and refer to [33] for a recent review on the properties of disks derived from scattered light images at various wavelengths.

2.2 Diamonds and PAH

Very small dust particles and macro-molecules emit a rich spectrum of features in the mid-infrared, most of which have not been univocally identified so far. It is generally believed that the unidentified features at 3.3, 7.7–7.9, 8.6, 11.3, and 12.7 µm are associated with transiently heated large Policyclic Aromatic Hydrocarbons (PAH). These have been widely detected in HAeBe systems [1]. The exact region of the system in which these are located has been debated for some time, until diffraction-limited 10–12 µm observations with large telescopes [6] and adaptive optics-assisted L-band spectroscopy have resolved the emission as predicted by flared disk models [14, 15]. In one system, HD97048 (see Fig. 1), an additional feature was detected and resolved at

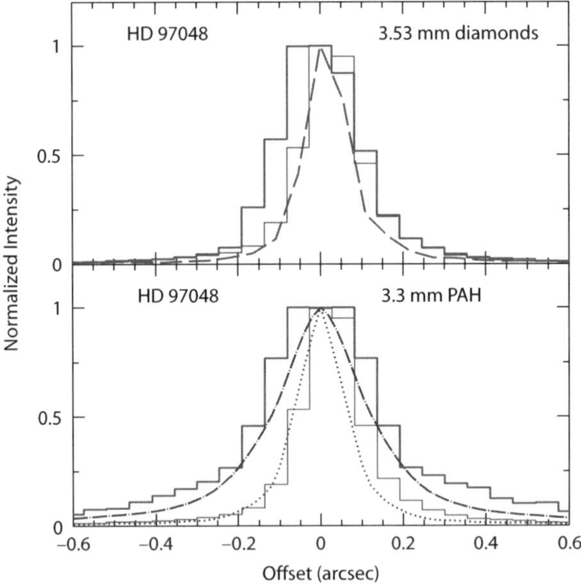

Fig. 1. NAOS/CONICA VLT observations of the 3.6 µm "diamonds" and 3.3 µm PAH features and the adjacent continuum in the HD97048 intermediate mass system (adapted from [14, 15]). The upper panel shows the intensity profile of the continuum subtracted diamond feature as a thick line histogram, the adjacent continuum profile as a thin histogram, and the profile of an unresolved star as dashed line; the diamond emission is clearly resolved. In the bottom panel the PAH and continuum profiles are compared with disk model computations for the PAH line (*dot-dashed*) and the continuum (*dotted*)

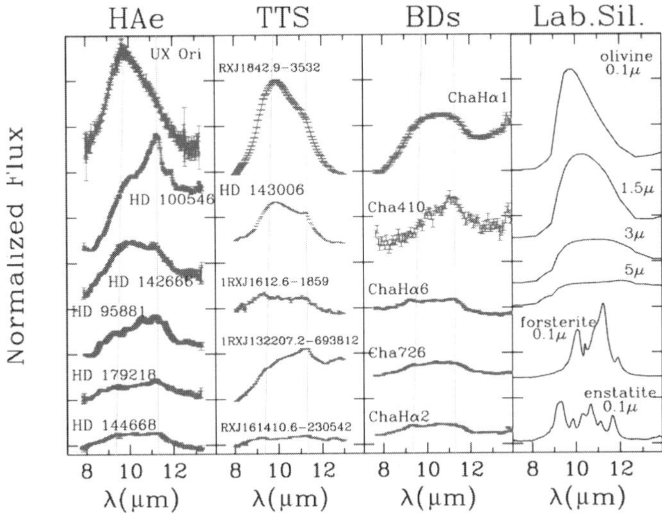

Fig. 2. Compilation of observations of silicates profiles in HAebe, TTS, and BD systems and in the laboratory (adapted from [25] and references therein)

3.6 μm, which is suggested to be associated with C–C stretch in diamond-like carbon grains.

The presence and abundance of these macro-molecules in the disk atmospheres has a strong impact on the gas heating and chemistry in the disk as they contribute to a significant fraction of the gas heating via the photoelectric effect. They may also affect the formation rate of molecular hydrogen on the grain surfaces. These grains need to be taken into account in most accurate disk models; however, one of the most serious limitations, in doing this is that observations of these grains cannot probe the population in the disk interiors, and models have to rely on assumptions on the abundance throughout the disk.

2.3 Silicates

Astronomical silicates have one of the most prominent emission features at 10 μm; one of the successes of the flared disks models with atmosphere is the natural explanation for the emission observed in this feature in a large variety of circumstellar disks [8]. In recent years with high quality mid-infrared spectra becoming available first with ISO and more recently with Spitzer, it has been possible to attempt to understand the diversity of the profiles observed in various regions.

As reviewed in [25], the modeling of silicate profiles in disks, as well as comparison to laboratory measurements, can give indications on the degree of "crystallinity" and on the size of the emitting particles. The observed systems show a range of properties with grains similar to those present in the diffuse

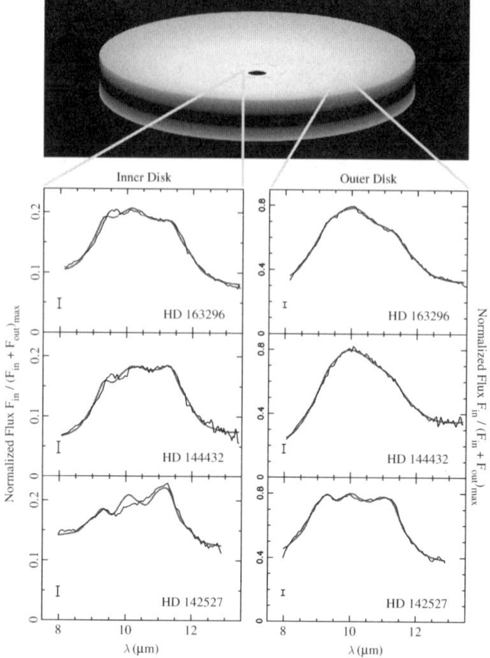

Fig. 3. VLTI/MIDI observations of the silicate profile in the three HAeBe systems HD163296, HD144432, and HD142527. In the top panel a flared disk is sketched, in the bottom panels the MIDI observations of the inner disks are compared to the emission from the outer disk derived by subtracting the interferometric spectrum from single telescope spectra (adapted from [5])

interstellar medium to grains that have undergone a significant processing, both in terms of crystallization and growth. It is still difficult to properly understand the zoo of properties observed, and in particular the expectation that dust processing evolves with time, i.e., with the age of the system, is still not evident from the current observations.

High angular resolution observations with the VLTI (see contribution from Malbet, this volume) allows one to investigate the properties of the silicate profile as a function of the distance from the central star. In Fig. 3, we show the results of [5] who demonstrated that the dust in the inner regions of disks is more processed than in the outer region. This is consistent with the expectations that the evolution of dust is faster closer to the star [12].

3 The Inner Disk

In the previous section, we have seen how high angular resolution observations allow us to probe the dust in the upper layers of the disks. The dust in the disk atmosphere is in many cases very evolved and large (up to a few microns); crystalline grains are found, especially in the inner regions of the disk atmospheres. The atmosphere of the disks, however, contains only a tiny fraction of the total disk mass, and the planet formation process is thought to occur on the disk midplane.

The disk midplane can only be probed at long wavelengths, where the disk is optically thin, as we will discuss in Sect. 4, or at the very inner regions of the system, where the radiation from the central star photoevaporates the dust grains. The properties (size, geometry) of the inner edge of the disk are shaped by the properties of the dust grains and how they interact with the direct stellar radiation. The region of the disk where this process takes place is very close to the central star. For a typical HAe system, this region is less than one Astronomical Unit from the stellar photosphere. At the distance of the nearest star-forming regions ($\sim 100 - 140$ pc), this corresponds to an angular size of up to 10 milliarcsec. With the current generation of large telescopes,

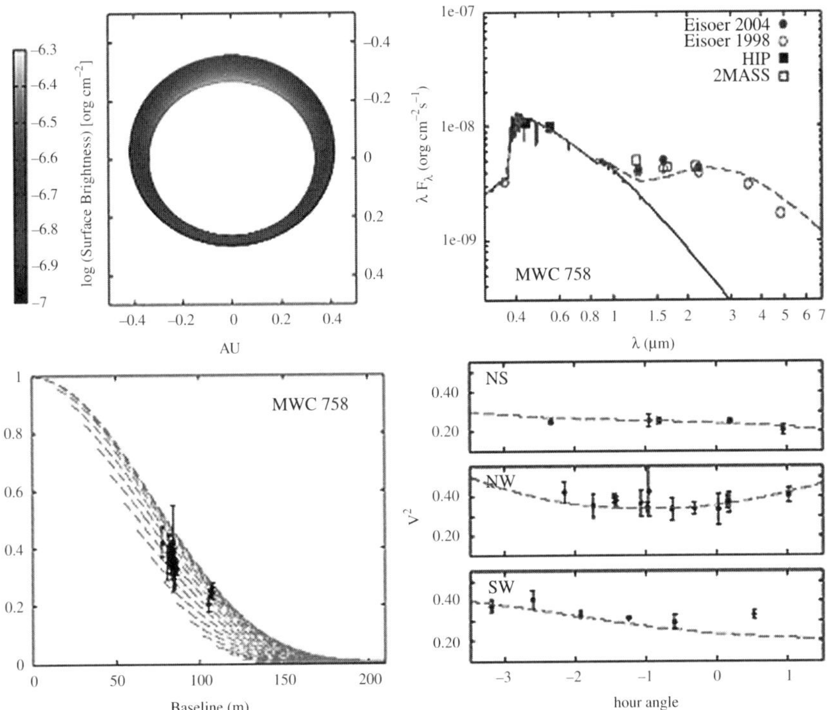

Fig. 4. The inner edge of the MWC758 disk. The top-left panel shows the model image of the inner edge of the disk (following the models of [16]); the top-right panel shows the fit to the observed SED; the bottom panels show the fits to the visibility observations: visibility as a function of baseline length (**left**) and as a function of hour angle for the three PTI baselines (left). The figure has been adapted from [17]

these sizes are beyond reach even with adaptive optics systems and need to be explored using near-infrared interferometers.

With the first near-infrared interferometers coming on line and producing scientific data, the properties of the inner regions of the disk have been the subject of intense recent modeling efforts. Observations of the SEDs of HAeBe systems and early near-infrared interferometric observations have highlighted the possibility of the presence of an inner "puffed-up" edge [10, 21]. More recently [16] have produced a self consistent model of the inner regions of the disk. This model naturally explains the size and shape of the inner edge of the disk in terms of the stellar photospheric parameters and the properties of the dust grains. These models were used by [17] to interpret the visibilities observed with near-infrared interferometers of several disks around young stellar objects (see Fig. 4 for example).

The combination of these advanced observational and modeling techniques have allowed us to constrain the properties of dust grains on the disk midplane, in the inner regions of a small sample of HAe systems. [17] demonstrated that in almost all systems investigated, the observations are consistent with the presence of grains much larger than the interstellar grains. The analysis of the near infrared data allow us to set a minimum size of the order of ~ 1 µm for the grains that dominate the population of dust in the inner disk.

The availability of the AMBER instrument at the VLTI is expected to allow for a substantial improvement in the study of the inner disk in a large sample of intermediate mass pre-main sequence systems and possibly a number of T Tauri systems. Due to its spectroscopic capabilities, AMBER will also allow a step forward in our understanding of the gaseous component on the inner disk and the relationship between disk and jet. Some initial experiments with AMBER [19, 29] have confirmed that in HAeBe stars the Brγ emission is not associated with the magnetosphere of the central star or inner gaseous disk but with the base of the wind/jet. Moreover, the hydrogen recombination line observations have shown that the wind/jet is launched from a region of the disk similar or slightly more extended than the dusty inner disk rim (see Fig. 5 and [29]).

4 The Disk Midplane

While the study of the inner disk allows us to probe the proprties of the dust grains on the midplane close to the central star, most of the disk mass which is on the midplane at large radii can only be probed at millimeter and longer wavelengths. At these wavelengths the disk becomes optically thin to its own radiation, and we can observe the emission from the bulk of the solid material.

Assuming thermal emission from an isothermal, optically thin ensemble of dust, the observed flux at millimeter wavelength can be written as

$$F_\nu = \frac{1}{D^2}\, \mathrm{B}_\nu(T_\mathrm{d})\, k_\nu M_\mathrm{d}\,, \tag{2}$$

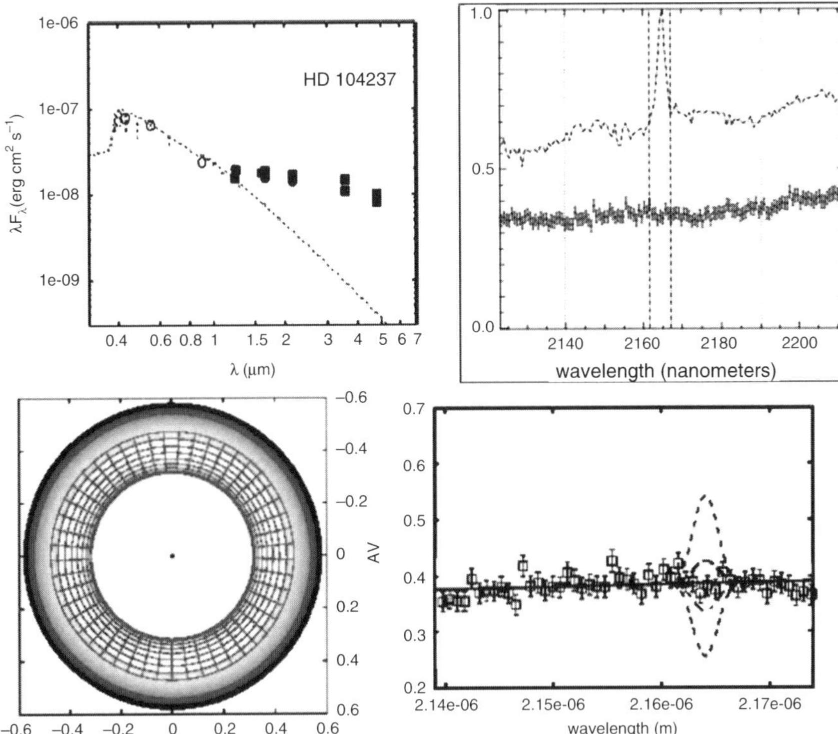

Fig. 5. AMBER/VLTI high angular resolution observations of the inner disk in the HD104237 HAe system. The top-left panel shows that the spectral energy distribution of the system, the near-infrared excess exceeding the photospheric emission (*dotted*) is due to the disk inner rim; the top-left panel shows the AMBER/VLTI total power spectrum (*dashed*) and interferometric differential visibilities (*red points with errorbars*), which show *no variation* across the prominent Brγ line; in the bottom panels visibilities predicted by different models of the Brγ-emitting regions are compared to the observed differential visibility, the models consistent with the observations predict an emitting region essentially coincident with the disk inner dusty rim (*see sketch on the bottom-left panel*). The figure has been adapted from [29]

where $T_{\rm d}$ and $M_{\rm d}$ are the dust temperature and total mass, D is the distance to the observer, $B_\nu(T_d)$ the Planck function at the appropriate frequency and temperature, and k_ν is the dust opacity per unit mass. At millimeter wavelengths the dust opacity as a function of frequency can be approximated with a power law $k\nu \sim \nu^\beta$, and the Planck function can be well approximated by the Rayleigh-Jeans function, hence

$$F_\nu \sim T_{\rm d}\, M_{\rm d}\, \nu^\alpha \tag{3}$$

with $\alpha = 2 + \beta$. This implies that the shape of the spectral energy distribution at millimeter wavelengths can be used to derive the dust opacity power law index β, while the total flux measured at a given wavelength is proportional to the product of temperature and mass.

The value of β depends on the type of dust grains in the ensemble: composition, shape, size, and combination of these. The mixture of grains that fits the properties of the interstellar medium correspond to a value of β close to two ($\alpha \sim 4$). If, however, the dust grains become much larger that the wavelength at which the fluxes are measured, then the dust opacity becomes gray (as only the geometrical cross section of the grains is relevant) and α value approaches 2 ($\beta = 0$). Even if the exact value of beta depends on a variety of dust properties that are hard to constrain (see Fig. 6 for some examples of β computations for different properties of the grain population), the general result that a low value of β is only consistent with the presence of large grains is a solid one (se also [9]).

Obviously this is a powerful probe for the presence of very large grains in circumstellar disks as discussed in [2]. The results of the first millimeter and submillimeter survey for grain growth in circumstellar disks is reported in this

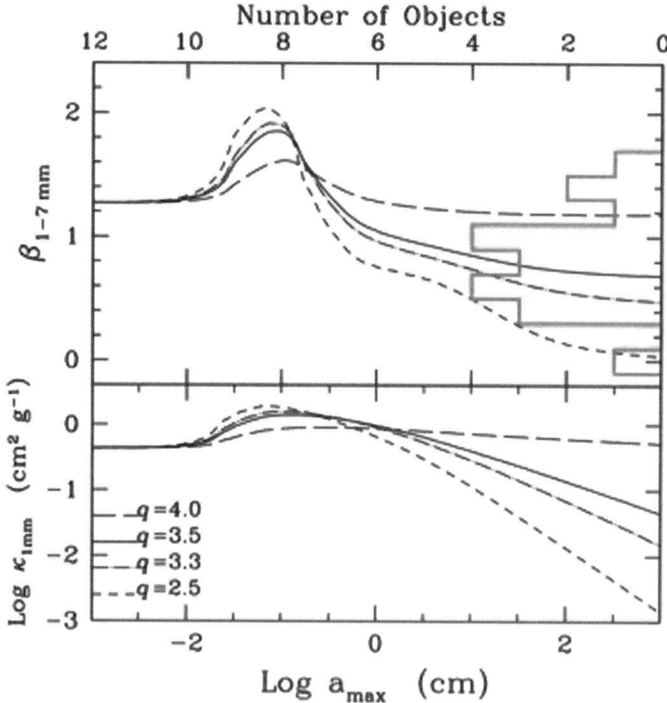

Fig. 6. Dust opacities per unit mass (**bottom panel**) and power law exponent (**top panel**) as a function of the maximum size for the dust grains and for various dust size distributions. The histogram on the right side of the top panel illustrates the values of the index beta measured in a sample of protoplanetary disks around T Tauri and HAe stars. Adapted from [25]

paper. In practice, as already discussed in [2], the situation is more complex, as disks are not isothermal ensembles of optically thin dust.

Even if the assumption of an average temperature, as discussed in [23], is not a poor approximation of (1), to give a rough estimate of the disk mass and to obtain an accurate determination of the value of β, it is necessary to use more sophisticated disk models that take into account the temperature profile and the presence of an optically thick inner region of the disk. As discussed in [30], low values of α approaching 2 may be an indication of low values of β, hence grain growth, or may be the consequence of unexpectedly high optical depth disks (see Fig. 7).

To resolve these ambiguities, fit proper disk models, and derive accurate values of β, it is necessary to resolve the disk emission at millimeter wavelengths and to use this additional constraint to solve the ambiguities. Tho achieve this, it is necessary to obtain angular resolutions of the order of $\sim$1 arcsec or beter, corresponding to linear resolutions of the order of 100 AU in the nearest star forming regions. These angular resolution at millimeter wavelengths can only be achieved by large radio interferometers (see also the lecture by Beuther). An example of such a study is the one on the CQ Tau system by [31] (see also Fig. 8). The combination of the millimeter spectral index from 1 to 7 mm and high angular resolution VLA observations at this wavelength allow to obtain an accurate measurement of β and to derive the presence of very large (centimeter size) grains in the disk midplane.

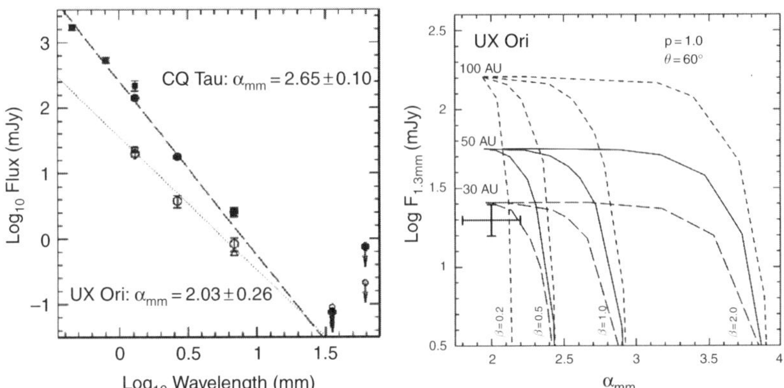

Fig. 7. (**Left panel**) millimeter-centimeter wave spectral energy distribution of the UX Ori and CQ Tau systems; (**right panel**) disk model families for the millimeter integrated flux and spectral index, for a given value of the dust opacity index β and disk radius, disk model predictions move along the lines depending on the total disk mass, the higher the mass the higher the mm flux until the disk becomes optically thick and the flux saturates. Low resolution observations that do not constrain the disk size are consistent with either large, mostly optically thin disks with low values of β or small, optically thick disks with any value of β. Adapted from [30]

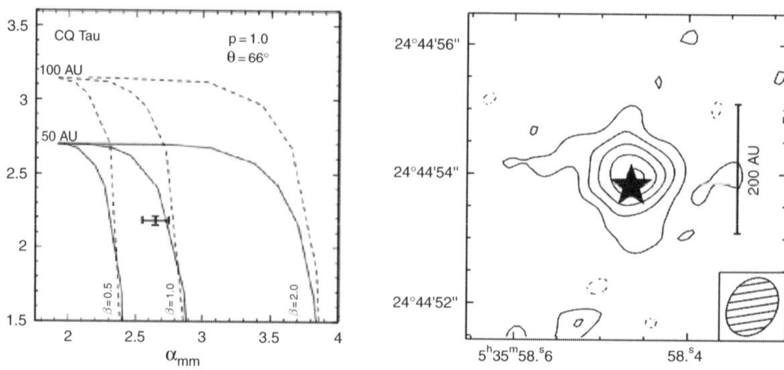

Fig. 8. (Left panel) family of models as in Fig. 7 for CQ Tau; (right panel) spatially resolved 7 mm continuum VLA map of the disk surrounding CQ Tau, the resolved image allows us to constrain the minimum disk radius and to restrict the families of models consistent with all the data. The result is that in the CQ Tau system the *average* value of β is well constrained to be ~ 0.6. Adapted from [30] and [31].

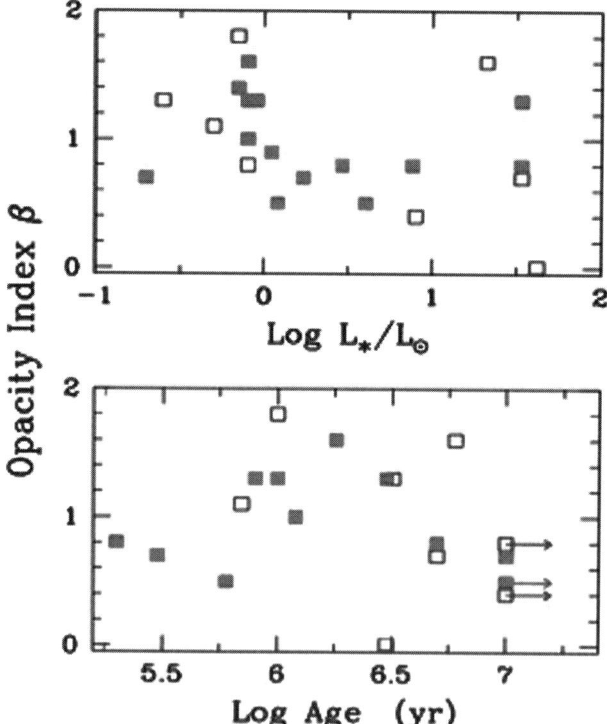

Fig. 9. Measured values of β as a function of luminosity (**top panel**) and age (**bottom panel**) of the central star. Adapted from [25].

There is now a growing number of systems for which accurate measurements of β are available (see [25] for a recent compilation). The sample studied so far is significantly biased as the goal of most of the searches was to identify systems with large grains (see e.g. [26]). So far all searches for correlations of the dust properties in the midplane with other properties of the system have not given convincing results. As an example, in Fig. 9, the run of β as a function of luminosity and age of the central star is shown, and no correlation is found contrary to the expectation that grain growth proceeds with the age of the system toward the planetary formation phase.

There are several explanations for this failure. The most obvious is that the samples studied so far are very small and biased for the trend to emerge. Obviously this is an area that needs improvement, and surveys with current and future millimeter and radio interferometers are planned to this effect. There is, however, the more interesting alternative that the lack of (expected) correlation is due to some physical effect, for example the growth to large particles may be a fast process occurring in the earliest stages of the system (prior to the stage we are observing now) and then the next step, the growth to planetesimals and planets, may be a much more difficult process that requires substantial time to occur.

References

1. Acke, B., van den Ancker, M.E.: Astron. Astrophys. **426**, 151 (2004)
2. Beckwith, S.V.W., Sargent, A.I.: Astrophys. J. **381**, 250 (1991)
3. Beckwith, S.V.W., Sargent, A.I., Chini, R.S., Guesten, R.: Astrophys. J. **99**, 924 (1990)
4. Beltran, M.T., Cesaroni, R., Neri, R., et al.: Astron. Astrophys. **435**, 901 (2005)
5. van Boekel, R., Min, M., Leinert, Ch., et al.: Nature **432**, 479 (2004)
6. van Boekel, R., Min, M., Waters, L.B.F.M., et al.: Astron. Astrophys. **437**, 189 (2006)
7. Cesaroni, R., Felli, M., Testi, L., et al.: Astron. Astrophys. **325**, 725 (1997)
8. Chiang, E.I., Goldreich, P.: Astrophys. J. **490**, 368 (1997)
9. Draine, B.T.: Astrophys. J. **636**, 1114 (2006)
10. Dullemond, C.P., Dominik, C. Natta, A.: Astrophys. J. **560**, 957 (2001)
11. Dutrey, A., Guilloteau, S., Ho, P.T.P.: Interferometric spectroimaging of molecular gas in protoplanetary disks. In: Reipurth, B., Jewitt, D., Keil, K. (eds.) Protostars and Planets V, p. 1425. University of Arizona Press, (2007)
12. Gail, H.-P.: Astron. Astrophys. **413**, 571 (2004)
13. Galli, D., Shu, F.: Astrophys. J. **417**, 220 (1993)
14. Habart, E., Natta, A., Testi, L., Carbillet, M.: Astron. Astrophys. **449**, 1067 (2006)
15. Habart, E., Testi, L., Natta, A., Carbillet, M.: Astrophys. J. **614**, L129 (2004)
16. Isella, A., Natta, A.: Astron. Astrophys. **438**, 899 (2005)
17. Isella, A., Testi, L., Natta, A.: Astron. Astrophys. **451**, 951 (2006)
18. Kenyon, S.J., Hartmann, L.: Astrophys. J. **322**, 293 (1987)
19. Malbet, F., et al.: Astron. Astrophys. (2007, in press)

20. Mannings, V., Sargent, A.I.: Astrophys. J. **490**, 792 (2000)
21. Monnier, J.D., Millan-Gabet, R.: Astrophys. J. **579**, 694 (2002)
22. Natta, A.: Star formation. In: Casoli, F., David, F., Lequeux, J. (eds.) Infrared Space Astronomy, To-Day and To-Morrow, (EDPSciences/Springer Verlag) (2000)
23. Natta, A., Grinin, V., Mannings, V.: In: Protostars and Planets IV, Mannings, V., Boss, A., Russell, S. (eds.) p. 559, University Arizona Press, Tucson (2000)
24. Natta, A., Testi, L.: Astron. Astrophys. **554**, 1087 (2001)
25. Natta, A., Testi, L., Calvet, N. Henning, Th.,Waters, R., Wilner, D.: Dust in proto-planetary disks: Properties and evolution. In: Reipurth, B., Jewitt, D., Keil, K. (eds.) Protostars and Planets V, p. 767, University of Arizona Press (2007)
26. Natta, A., Testi, L., Neri, R., Shepherd, D.S., Wilner, D.: Astron. Astrophys. **416**, 179 (2004)
27. Sargent, A.I., Beckwith, S.V.W.: Astrophys. J. **382**, L31 (1991)
28. Shu, F.H., Adams, F.C., Lizano, S.: ARAA. **25**, 23 (1987)
29. Tatulli, E., Isella, A., et al.: Astron. Astrophys. (2007 in press)
30. Testi, L., Natta, A., Shepherd, D.S., Wilner, D.: Astrophys. J. **554**, 1087 (2001)
31. Testi, L., Natta, A., Shepherd, D.S., Wilner, D.: Astron. Astrophys. **403**, 323 (2003)
32. Testi, L., Natta, A., Oliva, E., et al.: Astrophys. J. **571**, L155 (2002)
33. Watson, A.M., Stapelfeldt, K.R., Wood, K., Ménard, F.: Multiwavelength imaging of young stellar object disks: Toward an understanding of disk structure and dust evolution. In: Reipurth, B., Jewitt, D., Keil, K. (eds.) Protostars and Planets V, p. 523, University of Arizona Press (2007)

Index

Printing: Krips bv, Meppel, The Netherlands
Binding: Stürtz, Würzburg, Germany